PROGRAMMING AND CUSTOMIZING THE HC11 MICROCONTROLLER

PROGRAMMING AND CUSTOMIZING THE HC11 MICROCONTROLLER

TOM FOX

McGraw-Hill

New York   San Francisco   Washington, D.C.   Auckland   Bogotá
Caracas   Lisbon   London   Madrid   Mexico City   Milan
Montreal   New Delhi   San Juan   Singapore
Sydney   Tokyo   Toronto

McGraw-Hill

A Division of The McGraw·Hill Companies

2 3 4 5 6 7 8 9 0   AGM/AGM   9 0 9 8 7 6 5 4 3 2 1 0 9

ISBN 0-07-134406-3 (set)
    0-07-134407-1 (hc)
    0-07-13405-5 (cd)

The sponsoring editor for this book was Scott Grillo. The editing supervisor was Caroline Levine, and the production supervisor was Sherri Souffrance. It was set in Times New Roman per the TAB4 Design by Paul Scozzari and Joanne Morbit of McGraw-Hill's Professional Book Group composition unit, Hightstown, New Jersey.

Printed and bound by Quebecor/Martinsburg.

This book is printed on recycled, acid-free paper containing a minimum of 50% recycled, de-inked fiber.

This book is dedicated to Susan,
the author's beloved wife,
who went to Heaven in July of 1999

CONTENTS

ACKNOWLEDGMENTS

This is where the publisher allows the author's humility to make itself known. True humility, however, is just another word for true intelligence. It is obvious to a truly perceptive, thinking person that what he or she succeeds at can only partially be attributed to that person alone. The old adage, "A book cannot be written by a person in a vacuum," is so obvious here as to not need repeating, although I just did. For instance, what would a book on the HC11 microcontroller be about if there were no HC11?

Because of all this, the first people I wish to thank are those at Motorola who provided the inspiration as well as the sweat and, yes, probably tears, that were needed to bring about the genesis of the HC11. Of course, since this book makes extensive use of the BUFFALO monitor, the author wishes to thank especially Tony Fourcroy, who spent many, many hours working on the monitor's source code and then spent even more time revising and revising it until it was as near perfection as a program can get.

In addition to the fine people at Motorola, I wish to thank Maxim Integrated Products and National Semiconductor for their help in obtaining and using the data sheets for their great products used throughout the book and CD-ROM.

Art Salsberg deserves a special thanks for his help and encouragement and especially for his kind and compassionate criticism through the years. He can be regarded as the "executive producer" of the personal computer since he was the editor of *Popular Electronics* when the Altair 8800 computer was first introduced to the world on the cover of the January 1975 issue. (By the way, a small book could written on exactly how the Altair 8800 made it to the cover of that particular issue.)

Similarly, I want to thank Al Burawa and Joe Desposito for their help and understanding when faced with those goofs they uncovered in my manuscripts and diagrams.

Often small kindnesses are ignored. I would like to mention one that pertains to this book and thank Marilyn Edwards for extending those workshop article deadlines so I could devote more time to this book and my other endeavors.

I have worked with numerous editors in my life. In general, I would rate them from good to great. Scott Grillo would have to be rated "great." He has been enormously helpful in many ways, and he has usually answered my e-mail notes more quickly than he would have received my letters if I sent them via USPS! While I wouldn't blame him for feeling "bugged" by all my notes, he has always been patient and understanding.

A special thanks to Daniel Craver for helping with the solid-state wind sensor's sensor assembly and mounting brackets.

The Heath company, producers of the famous Heathkits, must be thanked for the enjoyment and hands-on education they have provided to this author through the years.

Finally, I wish to thank those closest to me. First, my daughters, Bernadette, Rebekah, and Catherine for their encouragement, interest, finding lost stuff, and helping my wife about the house. I must thank my two hard-working sons, Mark and Matthew, who not only helped with the photography but wired and built versions of many of the experiments and projects in the book. Their discovery of many little inconsistencies in the construction details have helped you, the reader, more than you'll ever know!

I want to especially thank my lovely wife Susan. She has helped in all phases of the book, from the original proposal up to, but not including, these acknowledgments. She has been secretary, bookkeeper, editor, and still a great mom and wife. She has not only proof-read much of the book's material but has painstakingly protected in plastic every single page of the original completed manuscript! Although these practical matters have been of immense help, I want to thank her most for her encouragement, and especially her love, without which this book would not be possible.

Tom Fox

INTRODUCTION

This book is about microcontrollers—Motorola's 68HC11 series of microcontroller units (MCUs) to be more specific. In my mind, MCUs are the neatest thing to sit on my electronics workbench since the junction transistor. Why? There are many reasons, but perhaps the most intriguing one is that they seem "intelligent"—sometimes, in fact, even more intelligent than those who design with them! Other, perhaps more practical but less fascinating features of MCUs are that they are exceptionally versatile and that in many designs they are the cheapest way to accomplish something. Isn't it interesting? These are probably the same three reasons the computer has entered into so many of our lives! There is no coincidence here. Computers are to pencil pushers what microcontrollers are to those who cherish the odor of hot solder! By the way, while I am a person who can get titillated by the feel of a soldering gun in my hand (MCU-controlled ones preferred, of course), I am also a dedicated pencil pusher who is a doting computer devotee.

A brief look at the history of the MCU must start with the first microprocessor unit (MPU). This first MPU, Intel's 4004, was introduced in 1970. At first, sales of this revolutionary chip were slow. However, as design engineers, many of whom still had a crush on the 12AU7 dual triode vacuum tube, started to recognize the "magic" that the MPU was capable of performing, orders picked up rapidly. Intel, realizing there was a growing market for MPUs, quickly released the much improved 8008 and then the *big one*—the 8080! The 8080, of course, is the great, great-grandfather of the MPU that is likely sitting on your computer's motherboard right now—the famous Pentium or one of its namesakes.

At about the same time that Intel started turning out the 8080 in mass, Motorola introduced the 6800. The 6800 never became popular as Intel's 8080 MPU for personal computers, but it did become an industry standard because of its simplicity—both in architecture and in ease of circuit design. Unlike the 8080, only five chips were needed to form a complete working 6800 system. While the 6800 itself never was used widely in PCs, a close cousin (6809) was used in Radio Shack's Color Computer and a distant cousin (6502) was used in a computer you also may have heard of—the Apple II. An offspring of the 6800 (6808) was used in probably the most famous commercially produced nontoy robot manufactured in the twentieth century—Heath's Hero 1. This robot even made cameo appearances in such major motion pictures as *The Last Starfighter* and guest starred in the television series "The Whiz Kids." It also was a frequent guest on "Mr. Wizard's World."

Even before Motorola's first MPU entered the marketplace, designers went to work on single-chip computers, often referred to as MCUs. The first MCUs were nothing more than MPUs with a built-in clock (the first MPUs required an external clock), minimal RAM (64 bytes), tiny ROM (1 kilobyte), and perhaps a primitive programmable timer. Two early MCUs were Intel's 8048, the forerunner of the famous 8051, and the 6801, the forerunner of my beloved 68HC11.

As soon as costs came down and enough design engineers became fascinated with microcontrollers, they started appearing in an abundance of electronic devices. In fact, they seemed to sprout out of just about every gadget imaginable—even some in which a mechanical thermostat or other simple controller probably would have been a better choice.

Microcontrollers—Beware! They Are Everywhere! Everywhere!

You press the VCR's remote control power switch. Nothing happens. Nothing! The VCR appears to have played its last tape. You pull the plug and sadly take the lifeless machine to your workbench and plug it in. It works perfectly! The next day you are stopped at a stop light. The road is icy. As you start up, your car's wheels spin like crazy. Luckily, after a few seconds, you get going without any apparent problem—that is, until you notice a weird red glow lighting up the car's interior. The *brake* light is on. A vision of green slips of paper floating out of your pocket and blown by the wind comes to mind. You pull up to a garage, and the mechanic pulls off the battery cable for a minute, reconnects it, and hands you a bill for $35. Since the scary red glow is now gone, you feel a bit relieved even though the man you hand the money to is wearing the funniest smile you ever saw. You do not smile back. Instead, you start to think, "What is going on? Something strange seems to be taking place with electric gadgets these days. They just do not act the way they used to."

It's true. Electrical gadgets just aren't the same as they were 20 years ago. What's the difference? Microcontrollers. They are everywhere today—including VCRs and a car's ABS (antilock brake system). While I just mentioned two potential problems that can occur with microcontrollers—especially when computer operating properly (COP) systems are either nonexistent or poorly designed—the happy fact is that these "incidents" seem to be happening less frequently today than when microcontrollers were first used. Things are improving. Trust me. And you can improve things even more once you learn about the intricacies of microcontrollers and start designing with them yourselves.

What Are Microcontrollers?

Think of a laptop computer. Then picture it being shrunk to a size of $^{3}/_{4}$ in square by $^{1}/_{10}$ in high. This tiny wafer can now be called a *microcontroller*. Of course, no commercially available microcontroller is really just a shrunken laptop computer. There are differences. Nonetheless, both are fundamentally the same. True microcontrollers are complete computers in themselves—except for the built-in power source, of course. The fact that they have no keyboard or built-in display means beans. They have the capability to take information from the outside world and then work on that information intelligently. However, they have another capability—they can do something useful with that information. For instance, microcontrollers can input information about the oxygen level in a car's exhaust, the air and engine temperatures, the barometric and manifold pressures, vehicle speed, etc. and determine in milliseconds how much gasoline and air to inject into the engine's cylinder to minimize pollution and maximize gas mileage.

What Are Embedded Controllers?

Sometimes, the term *microcontroller unit* (MCU) is used synonymously with *embedded controller*. However, not all embedded controller applications use MCUs, and not all MCUs are used as embedded controllers.

Embedded controllers are MCUs or MPUs that are *embedded* into a product, and the user of the product may not even know there is a computer inside. Two products already mentioned, VCRs and automobiles, use embedded controllers. A few other examples include microwave and some deluxe toaster ovens, TVs, cameras, fish locators, toys, and even some computer mice.

Now that we know what we are talking about when referring to microcontrollers, let's get more specific. Abstract talk breeds confusion—concrete examples are the grandfather of true education. This will be my motto throughout this book.

Preferences, Preferences

Everyone has his or her own preferences. For instance, some people like pizza loaded with pepperoni and olives, some like only cheese, and some, somewhere, I am sure, love a pizza smothered with anchovies. Preferences, preferences! When it comes to writing letters, legal briefs, proposals, or even books, some people still use a typewriter or even a pen, but most people today prefer to hit keys on a computer keyboard and let a word-processing program do much of the work for them.

There are many types of microcontrollers around. Here, too, there are preferences. Which one is best? Well, this is something like asking the question, "Is a cheese pizza a better choice than a pepperoni pizza?"

My preference for microcontrollers, you probably already guessed, is the HC11 series of MCUs. Why? To be honest, my preference probably has something to do with the fact that the MPU I cut my first baby tooth on was its grandfather—the famous 6800. After learning about the 6800, I left it behind when I got enough confidence to design my own projects and quickly migrated to the 6802 and 6808 MPUs.

When the 6801/6803 MCUs appeared, I jumped into using that series in several new designs—as did General Motors (GM) engineers. Then, as the story goes, these same GM engineers sat down with some people from Motorola and gave them a wish list of features they would like in a microcontroller—my beloved HC11 was supposed to be the result of this meeting of brilliant minds. It just so happens that the software written for the 6800/02/08 and 6801/03 series of chips is upward compatible with the HC11! Enough said? Not really! Since I am a bit prejudiced here, my comments may not be as valid as those of others. This said, it is believed by many in the industry that the 6800's architecture is one of the most straightforward, and thus it is probably the easiest MPU/MCU series to understand. The HC11 retains the same basic architecture as the 680X at its heart, although its many additional features can complicate matters if you look at the whole thingamajig at once. Since I firmly believe in the KISS ("Keep it simple, stupid") philosophy, I will open up the HC11's innards one step at a time. In fact, I might go a bit overboard on this philosophy, because I will start this book with a simple project that was inspired by computer hobbyists of a generation ago.

1

WHAT IS A COMPUTER?

A microcontroller unit (MCU) is essentially a "computer on a chip." But what is a computer? During the "electronic hobbyist's golden decade" (1955–1965), some of the most fascinating projects described in electronic hobbyist magazines and books were simple electromagnetic or electronic computers. Unlike the computer projects described in magazines in the mid-1970s and later, these early computers often consisted merely of a bank of relays or four to eight bistable flip-flops in series with an old phone dial as an input. For output, they used either no. 48 bulbs for the transistorized or relay version or NE-2 bulbs for the tube version. About the only thing this "computer" could do was translate decimal numbers to binary numbers and add them. For instance, if you dialed "3," the first from the right, and the second light would go on—all others would remain off. This indicated that "3" in decimal was "11" in binary. If you dialed "5," lights 1 and 2 would go out, and light 4 would go on—this indicated "8" in binary. Of course, since a dial doesn't turn that fast, you could "see" the action taking place, with each light going on for an instant and then off as the "computer" counted upward in binary. This visible action meant that the process was slow, but it also added to the "computer's" mysterious charisma.

If you hooked up eight of these flip-flops in series (which, in those days, verged on a supercomputer), you were able to record a number as large as 11111111 in binary, which is 255 in decimal. Now, doesn't something look familiar here? Eight flip-flops in series—along with 8 indicator lights in a row. What we have is an 8-bit register—very similar in concept to the 8-bit registers in 8-bit microcontrollers. *Quick question:* What is the size of most registers in the HC11 and other 8-bit microcontrollers?

To get the concept of an 8-bit register etched in your mind, let's describe how to wire up a twenty-first-century version of the mid-twentieth-century hobbyist computer. Here we will use flip-flops in integrated chips (ICs) instead of making them from junction transistors, electron tubes, or even electromagnetic relays. Before I show the schematic and how to breadboard it, let's look at solderless breadboards and power supplies.

Solderless Breadboards

Solderless breadboards (also referred to as *experimenter's breadboards*) could be described as "high-tech Legos"—and are almost as easy to use. See Figs. 1-1, 1-2, and 1-3. If you use one of these, circuits can be tested quickly and easily. These breadboards come in many sizes and shapes and are produced by several companies. Radio Shack, for one, sells several sizes and types of these boards.

Solderless breadboards have plug-in points that are 0.1 in apart and accept DIP ICs, diodes, $1/4$-W resistors, most capacitors, low-power transistors, light-emitting diodes (LEDs), 22-gauge wire, and many other parts that use 22- or 24-gauge leads. For more information, refer to the instructions that usually come with the boards. These boards are simple to use.

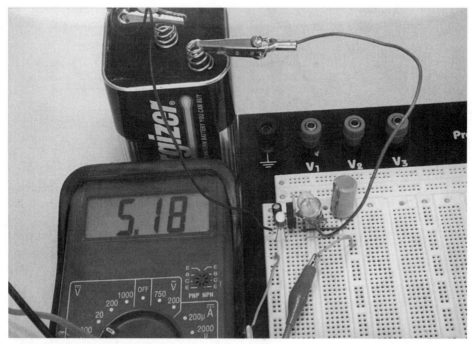

FIGURE 1-1 (*A*) Close-up of the breadboarded 5-V power supply whose schematic is given in Fig. 1-4. Also shown is a digital VOM indicating the supply's output voltage.

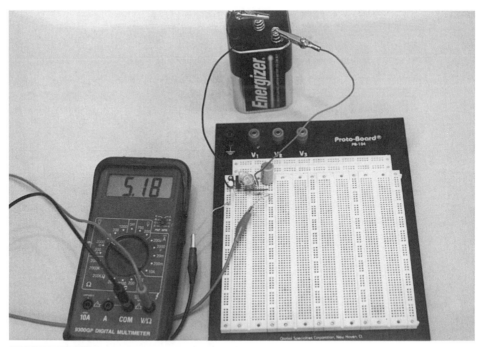

FIGURE 1-1 *(Continued)* (*B*) Distant view.

FIGURE 1-2 (*A*) Close-up of a breadboarded circa 1965 hobbyist computer.

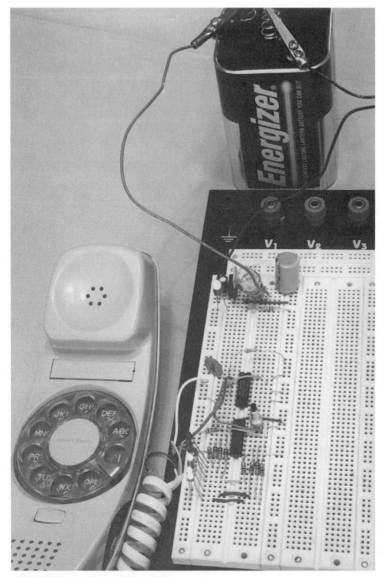

FIGURE 1-2   (*Continued*) (*B*) Distant view.

Power Sources

While you can power the first experiment (circa 1965 hobbyist computer) with two D-cells in series (3 V) or a 6-V battery, like most experiments and projects in this book, it is best powered by a regulated 5-V power supply capable of supplying 100 mA.

Keep in mind that this first experiment is a "get acquainted with the book's technique" project. While the circuit's operation is informative and helps ingrain the all-important concept of an 8-bit (1-byte) number into your noggin, even more important, it enables you to gain experience with a solderless breadboard. What is most important, however, it gives me an excellent opportunity to discuss a piece of equipment you should own (not *must*, just *should*)—a regulated 0- to 12-V power source capable of producing 250 mA. See Fig. 1-3. Ideally, this power supply should have the capability of monitoring the current drawn and the voltage produced. An alternative to this 0- to 12-V regulated supply is a +5-V regulated supply. If you have trouble finding either power supply at a reasonable price, you may be interested in this next section.

AN INEXPENSIVE +5-V REGULATED POWER SUPPLY WITH CURRENT MONITOR

This simple power supply can be used to power most projects and experiments in this book. Figure 1-4 shows a schematic of a +5-V power source capable of supplying at least 100 mA continually and more than twice this for short periods of time. You can use either a solderless breadboard or a printed circuit board for this power supply. Figure 1-1 provides two views of this circuit breadboard. If you breadboard the power supply, take care when inserting the LM2931T leads in the board—they are slightly on the large size.

If you get serious and make a printed circuit (PC) board for this power supply, follow Fig. 1-5 for the solder-side foil pattern. Figure 1-6 provides the component mounting guide so that you know which part goes where. For tips on making PC boards, see Appendix E.

If you do use a PC board, it is wise to lay down the LM2931T and use a screw and nut to fasten it to the board. The reason for this is that the foil on the solder side aids in dissipating heat. If you are using the optional PR2 flashlight bulb in the circuit, first solder 1-in lengths of no. 22 wire to its two connections before inserting it in either the breadboard or the PC board.

Notice that this power supply uses an LM2931T low-dropout regulator IC. Like the HC11, this IC was designed originally for automotive applications and has a whole bunch of safeguards built in. According to the data sheet, this chip is almost impossible to destroy, even if you don't know what you are doing! For instance, the data sheet claims that you can connect the power supply backward or connect it to 25 V or even temporarily to 60 V and not only will blue smoke not be seen or an acrid burnt odor fill your sinuses, but the chip should still regulate nicely. This tiny piece of high-tech sand is supposed to act like nothing ever hit it! The manufacturers claim that you can even stick it in a circuit backward, and as long as you take it out before it starts smoking too much, no damage is done. Awesome! While I haven't been this rough on the chip, I have used it extensively without a problem. Did I mention that it also has short-circuit and thermal-overload protection and that its dropout voltage is typically less than 0.5 V?

Notice the optional PR2 flashlight bulb in the circuit. Its purpose is as a current monitor. If you omit the PR2, make sure you connect a jumper wire in its place. Under normal low-current operation, the PR2 should *not glow.* It takes about 100 mA to light the filament just a little. If you see the filament lit, shut the power down and check for trouble. There is no doubt that this bulb is quite useful—but it isn't as good as a milliammeter! Did I mention that it is cheap, though?

One caution when using the LM2931T in your design: Make sure you connect at least a 100-μF capacitor across its output. This unusually large capacitor is required for stability.

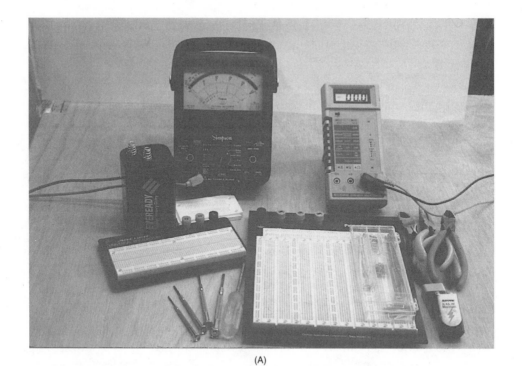

(A)

(B)

FIGURE 1-3 (*A, B*) Necessary tools and equipment such as breadboards, long-nose pliers, soldering iron, VOM, batteries, etc.

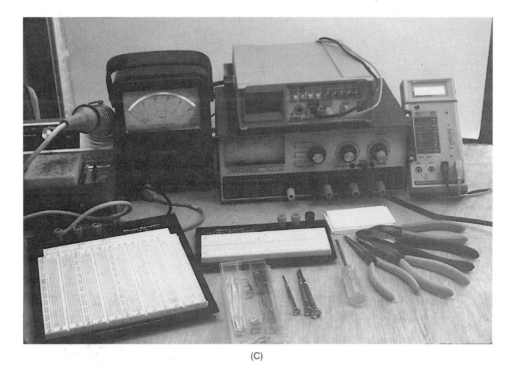

(C)

FIGURE 1-3 *(Continued)* (*C*) **Regulated power supply and precision digital voltmeter.**

The actual energy for this power supply can come from a multitude of sources: (1) a 6-V lantern battery, (2) an old 12-V car battery that on a good day can turn over a lukewarm Yugo, or (3) a 7- to 13-V "wall wart" ac-to-dc adapter that is cheap and usually convenient. "Wall warts" often come with small electronic gadgets such as radios and cassette recorders. You also can purchase one from Radio Shack or just about any electronics supply store.

Note that the power supply circuit in Fig. 1-4 is only rated at 100 mA (continuous use). However, this is good enough for almost everything in this book—the exception may be the HC11-based solid-state wind-direction indicator described in Chaps. 11 and 12. However, all projects (Mag-11, MagPro-11 and MagTroll-11) include a built-in voltage regulator, so there is no need for a regulated supply here.

The following project will give you a hands-on feel for solderless breadboards. It may even etch into your brain "the essence of digital computers."

An Updated Version of a Circa 1965 Hobbyist Computer

The schematic in Fig. 1-7 shows an 8-bit "computer." The only thing this computer can do is translate the decimal numbers on a telephone dial into binary and add them. The

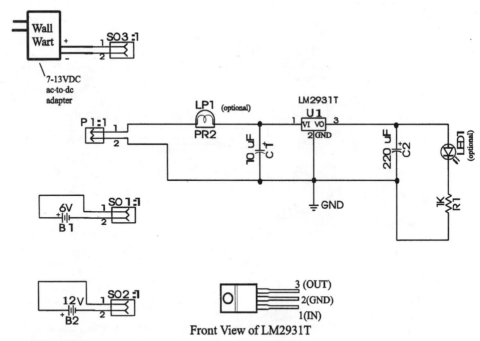

FIGURE 1-4 Schematic of inexpensive +5-V regulated power supply with optional PR2 flashlight bulb current monitor.

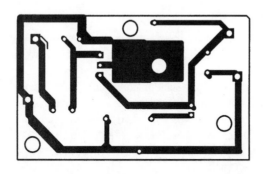

FIGURE 1-5 Solder-side foil pattern for +5-V power supply whose schematic is given in Fig. 1-4.

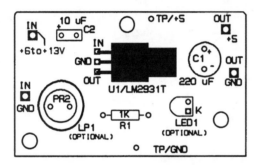

FIGURE 1-6 Component mounting guide for +5-V power supply whose schematic is given in Fig. 1-4.

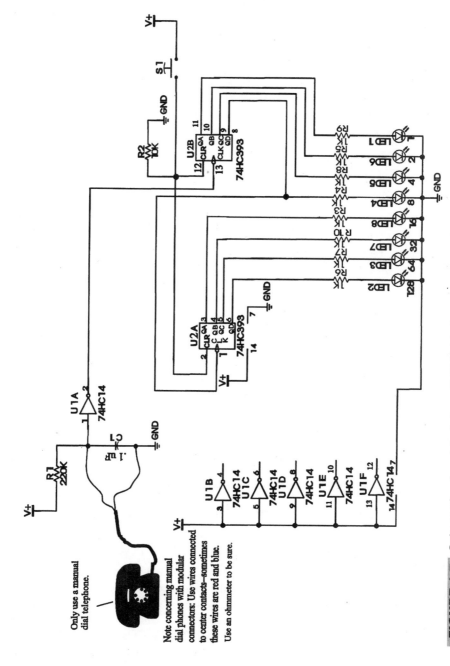

FIGURE 1-7 Schematic of updated circa 1965 hobbyist computer.

figure shows this "computer" wired up on a solderless breadboard. The input here is from an old mechanical dial telephone. If you want to wire this circuit but can't get your hands on an old telephone, you can substitute a simple SPST push-button switch. *Warning:* Push-button switches aren't as much fun as a dial telephone! This circuit is fascinating—and eye-opening.

WIRING IT UP

If you are going to actually try wiring up this circuit (highly recommended), you will need the following parts:

- An old mechanical dial telephone (an SPST push-button switch can be substituted)
- A 74HC393 dual 4-bit binary counter IC
- A 74HC14 Schmitt trigger hex inverter
- Nine 1-KΩ, $\frac{1}{4}$-W resistors
- A 10-k resistor
- A 220-k resistor
- A 0.1-μF capacitor
- Eight LEDs (low-power-type preferred)
- An SPST push-button switch (optional—it's used to clear the counter)
- A solderless breadboard
- Good selection of no. 22 hookup wire
- A +2- to +6-V power source

Using the schematic in Fig. 1-7 as a guide, wire up the circuit on a solderless breadboard. Also see Fig. 1-2. Note that the telephone is connected to the breadboard with two wires. There are several different ways to connect the telephone. The hookup shown in the figure uses a 1982 "trimline" AT&T mechanical dial telephone that has a modular connector. Connections to the circuit are made from the wires attached to the center connections. Use an ohmmeter or continuity tester to determine which two connections to use. (The ohmmeter should show a resistance change from about 400 Ω to infinity the number of times indicated on the dial.) A few minutes of tinkering with an ohmmeter should do the job here.

TRYING THE CIRCA 1965 HOBBYIST COMPUTER OUT

Now connect up power to the circuit. Some LEDs may be lit. If so, press S1, the reset switch. All LEDs should go off. Now try dialing "1." LED1 (value = 1) will light. Now try dialing "4." You will see lights quickly flash, and when they stop, LED1 and LED3 (value = 4) will be lit. This means that the computer's display (LED1 through LED8) indicates binary 101, which is 5 in decimal. Now dial "9." LED2 (value = 2), LED3 (value = 4), and LED4 (value = 8) will be lit, and all others will be off. This indicates binary 1110 or decimal 14. But now let's write 1110 as 00001110. This way of writing, with the leading zeros, makes sense because we have eight LEDs. When the LED's are lit, this indicates a 1 in our binary number, and when they are unlit, this indicates a 0. (By the way, for more on binary and hexadecimal numbers, see Appendix A). Writing binary numbers with eight digits and including leading nonsignificant 0's is typical (and logical) when dealing with 8-bit computers.

As you continue to input information (dialing) into this "computer," you will notice something interesting happen when the decimal number exceeds 255 (11111111 in binary). Guess what? The 256th pulse (100000000 in binary) turns off all the LEDs! What happened simply is that we have run out of flip-flops. This condition also occurs in 8-bit registers in MPUs and MCUs and is sometimes called *overcount*. Such MPUs and MCUs don't have a problem with this because most have a special carry register that indicates that this condition has occurred. You also can add a carry register to the circa 1965 computer by adding a type of flip-flop that has only set and reset inputs. See Fig. 1-8. Notice that once this carry register is set, the only way to clear it is to press the reset switch. As you can tell, this ninth flip-flop doesn't add a true "ninth bit" to the "computer."

This breadboarded 8-bit hobbyist computer differs in many aspects from 8-bit MPUs and MCUs of today. However, it does have one thing in common—it has an 8-bit register similar to the 8-bit registers found in 8-bit MPUs and MCUs. The concept of 8-bit registers is vital, and anyone interested in MCUs should have the image of an 8-bit register permanently etched into their brain cells (ideally into their DNA, if possible). If you don't wire the circuit in Fig. 1-7, at least look at Fig. 1-2 and imagine its operation—this should help you understand how 8-bit MCUs really and truly operate.

What You Absolutely Need to Follow in this Book

This is a hands-on book, and to really get something worthwhile out of it, you will need the following equipment:

- PC-compatible computer with a CD-ROM drive.
- Volt-ohm-milliammeter (VOM)—either digital or analog or both power supply (already discussed)
- Solderless breadboard (already discussed)
- Basic tools (long-nose pliers, soldering iron, etc.) see Fig. 1-3.

Other Helpful Stuff

Other stuff that would be helpful includes

- Suitable printer (preferably LaserJets-compatible)
- Logic probe
- EPROM programmer (discussed in Appendix B)
- Oscilloscope (used very little here—logic probe is usually more useful)
- A big junk box full of all kinds of resistors, capacitors, transistors, wires, etc. See Fig. 1-9.

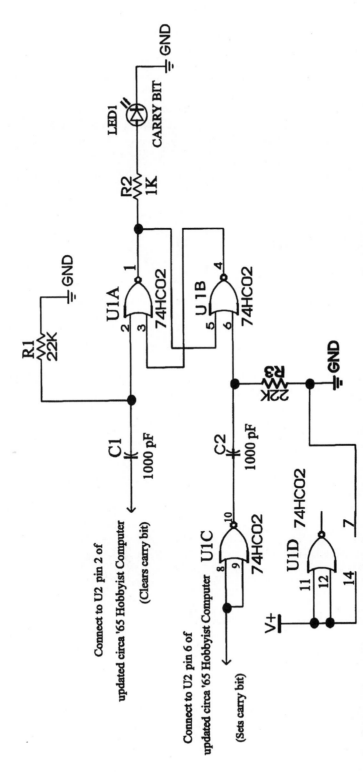

FIGURE 1-8. Schematic of add-on carry register circuit for updated circa 1965 hobbyist computer.

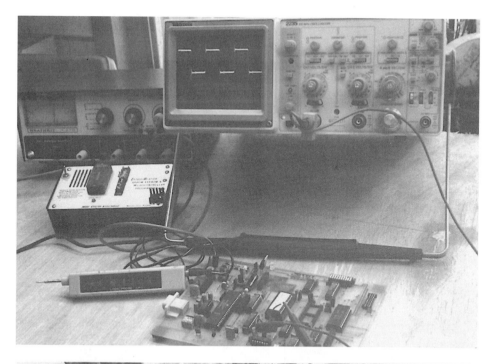

FIGURE 1-9   Helpful equipment such as a logic probe, oscilloscope, variable voltage-regulated power supply, EPROM programmer, etc.

The Architecture of a Simple Computer

The simplest device that can honestly be called a "computer" has a central processing unit (CPU), a memory, and an input-output (I/O) system. If you were to compare a computer with a person, the CPU would be the "brain" of the machine, the memory would be the "memory," and the I/O system would be the "senses/muscles."

The simplest CPU consists of a control unit that decodes and monitors the execution of instructions, an arithmetic logic unit (ALU) that does simple math and manipulates data, and at least one register, often called an *accumulator,* that functions as a temporary storage location that is controlled by the ALU when it is performing operations, such as adding two numbers. If you envision a computer as a person with the CPU's ALU and control unit as the "brain," the accumulator can be looked on as a "scratch pad" used by the person.

Using these criteria, the 8-bit circa 1965 hobbyist computer shown in Fig. 1-7 is almost, but not quite, a true computer. This basic hobbyist computer has a crude CPU because it has an ALU that can add numbers, a control unit that knows what to do with the pulses entering, and a register (the 8 flip-flops). It also has an I/O system, with the telephone dial being the input and the LEDs the output. Nonetheless, it does lack a memory—unless you fudge things a bit and think of the flip-flops as doing double duty as accumulator and memory.

Another Potentially Much More Sophisticated Computer, Although Even Simpler and…Deadly, Deadly Slow

Look at the schematic in Fig. 1-10. This simple circuit seems almost trivial. However, it serves as a complete computer as long as an intelligent person manipulates the dip-switches. In this case, the person functions as the CPU's ALU and control unit as well as the "computer's" memory. The dip-switch functions as the accumulator and the input system, while the LEDs act as the output system.

For example, say you want to "load" the accumulator with "23," which is binary 00010111. Simply put on switches 1, 2, 3, and 5 and leave off the other switches. Now say you want to subtract 5 from the accumulator. Since $23 - 5 = 18$, or 00010010 in binary, you turn off switches 1 and 3. If you make a list of instructions on paper of what you want to do, the result is an extremely slow, but sophisticated computer. For instance, a short list of instructions might read as follows:

1. Clear accumulator (turn off all switches).
2. Add 1 to accumulator.
3. Go back to instruction 2.

What will show on the LEDs if these instructions are followed? They will continue counting up in binary as fast as you can figure out the next number and flick the appropriate dip-

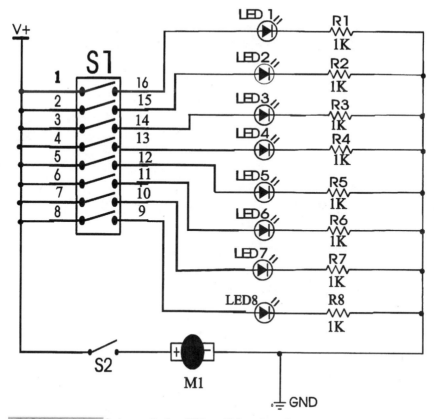

FIGURE 1-10. Schematic for DIP switch computer.

switches. In fact, they will continue doing this until you get sick and tired of flicking the dip-switches. Here, you function as the ALU and the control unit, whereas the list of instructions functions as the program memory.

Notice that the schematic in Fig. 1-10 shows an additional switch connected to a motor. This motor could run a small fan. Notice here that it is simply connected to the "computer's" power source. Now let's look at a new list of instructions assuming that we start with switch 9 being "on," which keeps us cool by blowing air at us.

1. Clear accumulator (i.e., turn off switches 1 through 8).
2. Add 1 to accumulator.
3. If LED 8 is "on," turn off motor's switch.
4. If LED 8 is "off," turn on motor's switch.
5. Go back to instruction 2.

What happens? Everything proceeds as before until a count of 128 is reached, which is 10000000 in binary, and then the fan is turned off and we start to feel warm. We then continue through the cycle and turn the motor back on when LED8 is turned off. This is all rather silly, but it is demonstrative of what a functioning computer does—follows instructions to the letter.

An activity such as this, where you control the dip-switches, is sometimes referred to as "playing computer." It is quite instructive for several reasons, including a hands-on feel for what basically is taking place in MCUs. Also, you learn an important fact: An MCU does only trivial things, but it does them so fast and does so many of them that it appears "intelligent." This could be referred to as *virtual intelligence*. To be honest, the most sophisticated computer ever made doesn't have a clue as to what it is doing—only the person who programs it does. It is the computer's speed that makes it so useful, not what it "knows." MCUs are based firmly on the KISS ("Keep it simple, stupid") philosophy.

Assumptions, Assumptions, Assumptions!

In order to write a readable book, authors must make certain assumptions about readers' knowledge. For instance, authors writing in English must make at least three assumptions right off:

1. The reader isn't severely visually impaired.
2. The reader has learned to read.
3. The reader is acquainted with the English language.

In a book such as this, the author must make many more assumptions. Three of the more important assumptions are

1. The reader is acquainted with at least the basics of electricity/electronics.
2. The reader knows the basics of using a computer as well as the basics of a programming language such as BASIC.
3. The reader is somewhat acquainted with the binary number system as well as its cousin the hexadecimal number system.

For those who feel that I may have gone too far in this third assumption, I have provided a short, practical tutorial on this subject in Appendix A.

WHAT IS MEANT BY A "1" OR A "0" IN AN HC11 CIRCUIT?

On and off. High and low. Lit and unlit. Set and clear. +5 V and 0 V. These are all examples of what I mean by "1" and "0."

A software programmer who has had no education or experience with electronics probably never thinks in terms of +5 V representing "1" and 0 V representing "0." However, if you want to actually "do something" with HC11 circuits and not just write programs for them, you should—no, you *must*—realize that 1 and 0 represent voltage levels. While there are now HC11 varieties that use a 3-V supply instead of a 5-V supply, in this book I only talk about the original HC11s that use a 5-V supply.

Well, what does a typical 5-V HC11 consider a "1," and what does it think is a "0"? Well, an HC11 uses HC (high-speed CMOS) technology. With VDD (that's the positive supply) at +5 V and VSS at 0 V (that's ground), and assuming that all load currents are less than 10 µA, Motorola guarantees the following:

1. An input voltage over 3.5 V but less than 5.3 V is always interpreted as a "1."
2. An input voltage less than 1 V but more than −0.3 V is always interpreted as a "0."
3. When the HC11 sets an output pin "high" (i.e., to "1"), its voltage is at least +4.9 V.
4. When the HC11 sets an output pin "low" (i.e., to "0"), its voltage is at or below +0.1 V.

These are the guaranteed limits. Reality is much easier to deal with. As long as the HC11's outputs are unloaded, you can assume a +5-V output voltage when its output is "high" and 0 V when it's low. In fact, actual measurements of the outputs of the 74HC393 IC, which uses basically the same technology as the HC11, revealed the following: With a supply voltage of 5.096 V connected to VDD and 0.0000 V connected to VSS, the measured output voltage when "high" was 5.095 V and when "low" was +0.0003 V.

For input voltages, anything below 2 V is usually interpreted as "low" (i.e., a "0"), and anything above 3 V is interpreted as "high" (i.e., a "1"). Between 2 and 3 V is a mysterious land where just about anything can happen—especially overheating!

If you can get your hands on a logic probe, its use is highly recommended. Try out the circa 1965 hobbyist computer again, and this time use the logic probe to check the levels at all the 74HC393's pins. An oscilloscope also can be used, but since the 1965 hobbyist computer doesn't have a clock, a logic probe is more interesting to use with this experiment.

A Brief Look at Naming Conventions Used in This Book

In this book I will indicate hexadecimal numbers with a dollar sign ($) prefix (e.g., $1000, $F800, $FFFF, and $4E) and decimal numbers with no prefix (e.g., 50, 65, 535, and 253). Binary numbers, when used in programs, will have a percent (%) prefix (e.g., %10101111 and %11111111.) Sometimes, however, when I just discuss binary numbers (and sometimes even hexadecimal numbers), I will leave the prefix out as long as the context of the discussion doesn't leave room for misunderstanding. For instance, in the phrase "the binary number 00001000, which is 08 in hexadecimal," both binary and hexadecimal numbers are indicated clearly in the sentence. I won't discuss octal numbers, so there is no need for a prefix. Octal numbers remind me of using cassette tape players to record programs—both were good ideas, but there were better ones. Now that we achieved a hands-on feel for a computer, the next step is to look at a real-life series of MCUs—the 68HC11 series. The 68HC11 series has more than 30 members, and the number is continually rising. However, for reasons of clarity, most of this book will look at only three members in detail—the 48-pin standard DIP model (MC68HC11A1P), the 52-pin standard PLCC model (MC68HC11A1FN), and the 52-pin 2-kB EEPROM model (MC68HC811E2FN). The last chapter of this book will take a brief look at other HC11 members as well as related chips such as the HC12 and HC16.

The next chapter gets right to it—an HC11 circuit that you can breadboard and test yourself.

2

INTRODUCING THE HC11 AND BREADBOARDING

A Simple HC11 Computer Circuit

So far I have attempted to keep things interesting and understandable by looking at real-life, concrete, down-to-earth, physical stuff and avoiding the abstract. I will continue doing this for a while, but soon, very soon, I will have to touch on the abstract more than you (and definitely me) will like. Take a deep breath, however, because you have a short reprieve before this scary time arrives.

A Quick Look at the MC68HC11A1P

The 68HC11A1P comes in a 48-pin DIP plastic package. This type of package makes this chip ideal for use in solderless breadboards, which were described in Chap. 1. Instead of looking at the architecture of the chip right now, which would probably just confuse you at this time and take up valuable space, let's go right ahead and wire the chip into a solderless breadboard circuit. This hands-on method provides a good way to learn about the chip's actual use. Once this is done, we will look at only the makeup of the HC11 that is needed to perform our monumental task—turning an LED on and off! Once you fully understand how to design an MCU circuit to turn an LED on and off exactly the way you want to, you will be well on your way to understanding how to use the HC11 to do just about any task you want, whether this task is simply to turn on a light when it gets dark or power a robot to bring you your afternoon tea. As mentioned in the Introduction, one of the HC11's ancestors, the 6808, was used in one of the most popular personal robots yet—Heathkit's

Hero 1. This robot even became a well-known TV and movie star. The HC11 has even more potential in a robot than the 6808 because it takes less power and can do so much more.

A schematic for a breadboarded circuit that flashes an LED is shown in Fig. 2-1. Breadboard this circuit! For some reason, your hands often import information to your mind in ways that aren't fully understood. This said, I concede your freedom to do as you wish. I also concede that it isn't absolutely essential that you actually wire up this circuit. If you believe, in the depths of your heart, that the LED will flash exactly as described, you will have accomplished about 73.5 percent of what you would have learned by actually experiencing it first hand. Because of this, let's get right into it.

Tips on Breadboarding This Experimenter Circuit

In order to wire this circuit, you will need the following:

- One MC68HC11A1P
- One MAX690ACPA
- One MC74HC373AN
- One 74HC00
- One 27C256 EPROM programmed with TEST6811 or equivalent firmware
- 13 4.7KΩ, $\frac{1}{4}$-W resistors
- One 10 MΩ, $\frac{1}{4}$-W resistors
- One 1-kΩ, $\frac{1}{4}$-W resistors
- One low-power LED
- Two 0.1-μF capacitors
- One 0.01-μF capacitor
- One 10-μF capacitor
- Two 15-pF capacitors
- One 4-MHz crystal
- Solderless breadboard large enough to handle the parts.
- Solid no. 22 insulated copper hookup wire—several colors preferred.
- Power supply capable of supplying at least 100 mA at +5 V—The supply in Fig. 1-4 will work here.

A kit of these parts, except for the breadboard, wire, and power supply, is available from Magicland. See Appendix C.

When wiring this circuit, do a relatively neat job, and keep the wires as short as possible.

In order for this circuit to operate properly, you will need a 27C256 EPROM (erasable programmable memory) that is programmed with appropriate software/firmware/sandware. [*Software* is the program that controls the functioning of the hardware. *Firmware* is a software program that has been programmed into a ROM (read-only memory), PROM (programmable memory), EPROM, etc. *Sandware* is simply my nickname for firmware in an EPROM or EEPROM (electrically erasable programmable memory). *Sandware* usually can be used interchangeably with *firmware* but is more descriptive.]

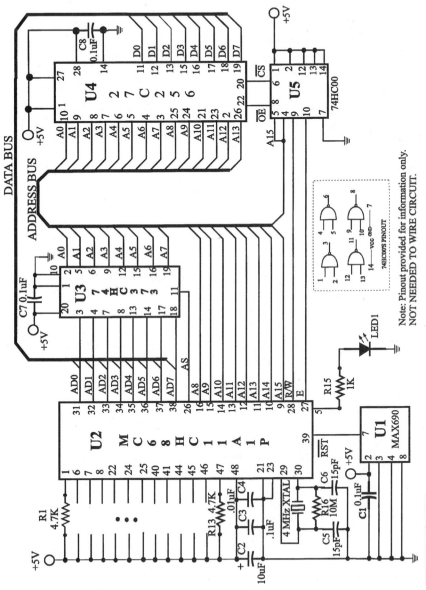

FIGURE 2-1 Schematic of breadboarded HC11 computer circuit.

For those who are unacquainted with EPROMs, ROMs, and PROMs, just think of them as a computer file that contains the MCU instructions and that has had a magical spell put on it that has transformed it from an intangible file to a silicon chip. The main difference between EPROMs and the other two "magic files on a chip" is that it is possible to erase EPROMs using ultraviolet light—ROMs and PROMs can never be erased.

One program that will flash the LED neatly is the TEST6811 program, whose MOT S format listing is given in Fig. 2-2, as well as the enclosed CD-ROM. If you have access to an EPROM programmer, you can use this listing as is to program the EPROM. With today's EPROM programmers, this job is a snap. If you don't have access to a programmer, an EPROM with the TEST6811 program already burnt in is available separately from Magicland. See Appendix B for notes about inexpensive EPROM programmers and programming them, and Appendix C for sources for preprogrammed EPROMs and other parts. Shortly I will describe how this program twinkles the LED. But for now we are on a mission—a mission to get an HC11 to blink an LED!

Checking for Blue Smoke!

If everything is connected correctly, an instant after power is applied, the LED should light. About a second later it should go off for a second and then light again and continue doing this until power is removed. If your primary aim is education and you have access to a good oscilloscope, consider yourself *fortunate* if the circuit *does not function correctly* the first time you turn it on! One learns how something works 10 times faster when that something breaks down or fails and one *successfully* fixes it than by simply studying about it. In fact, by simple definition, learning requires mistakes to be made. If you took a calculus course in college and never made an error, the course was a *huge* waste of time and money—all it showed was that you knew all that stuff already. Working with MCUs is no different. [*Quick question for calculus gurus:* What is d/dt(infinity)?]

If the LED doesn't flash as described, troubleshoot the circuit with an oscilloscope or a logic probe. *Hint:* First check the reset signal at pin 39 to make sure it is "high" (over 4 V), and then check to see the clock is working by monitoring the "E" (enable) signal at pin 27 of U2. A nice, clean 1-MHz squarewave should be seen.

Do not hesitate to try this circuit out because you feel you will have to purchase expensive parts just to experiment with. All ICs, except the 74HC00, used in this breadboard circuit are also used in the educational, yet fascinating and practical, single-board computer—Mag-11—which will be described in Chap. 4.

```
S105FFFEF80005
S123F8008E00FF8608B71000BDF8137F1000BDF81320F018CEFFFF18091808180926F83936
S9030000FC
```

FIGURE 2-2 The MOT S19 object code (.S19 file) for the TEST6811 program used in the breadboarded computer circuit.

After checking out your circuit with the TEST6811 firmware, you may want to experiment a bit with your own software/firmware/sandware. Of course, this may be really getting ahead of the game, since you may have to read a few more chapters before you can do this effectively!

What the Circuit Is All About

While going over circuit details here, refer to the schematic in Fig. 2-1 and the 68HC11A1P pin description in Table 2-1.

The clock circuit consists of XTAL, R16, C5, and C6. With a 4-MHz crystal, the frequency of the clock is 1 MHz (4 MHz divided by 4). The 68HC11A1P can operate at over 2 MHz, but I restrict it here to 1 MHz to lessen possible problems.

TABLE 2-1 MC68HC11A1P PIN DESCRIPTION

PIN NO.	OFFICIAL LABEL	BRIEF DESCRIPTION
1	PA7/PAI/OC1	Bit 7 of port A—General-purpose input or output/pulse accumulator input/output compare 1 output pin
2	PA6/OC2/OC1	Bit 6 of port A—General-purpose output/output compare 2
3	PA5/OC3/OC1	Bit 5 of port A—General-purpose output/output compare 3
4	PA4/OC4/OC1	Bit 4 of port A—General-purpose output/output compare 4
5	PA3/OC5	Bit 3 of port A—General-purpose output/output compare 5
6	PA2/IC1	Bit 2 of port A—General-purpose input/input capture 1
7	PA1/IC2	Bit 1 of port A—General-purpose input/input capture 2
8	PA0/IC3	Bit 0 of port A—General-purpose input/input capture 3
9	PB7/A15	Address line A15
10	PB6/A14	Address line A14
11	PB5/A13	Address line A13
12	PB4/A12	Address line A12
13	PB3/A11	Address line A11
14	PB2/A10	Address line A10
15	PB1/A9	Address line A9
16	PB0/A8	Address line A8
17	PE0/AN0	Bit 0 of port E—General-purpose input/A-D channel 1

TABLE 2-1   MC68HC11A1P PIN DESCRIPTION (*Continued*)

PIN NO.	OFFICIAL LABEL	BRIEF DESCRIPTION
18	PE1/AN1	Bit 1 of port E—General-purpose input/A-D channel 2
19	PE2/AN2	Bit 2 of port E—General-purpose input/A-D channel 3
20	PE3/AN3	Bit 3 of port E—General-purpose input/A-D channel 4 (*Note:* This is the pin we will use in our breadboarded circuit.)
21	VRL	Low-voltage reference for A/D converter—typically connected to circuit ground
22	VRH	High-voltage reference for A/D converter—maximum voltage (VDD+0.1 V)
23	VSS	Ground connection
24	MODB/VSTB	This dual-purpose pin sets the operating mode. (When high, the mode is set for single-chip or expanded multiplex and grounding it sets it to the special test or bootstrap mode). When high, it is used as the input for RAM standby power.)
25	MODA/LIR	Another dual-purpose pin that sets the operating mode. A high sets the operating mode for expanded multiplexed or special test. Grounding this pin selects the single-chip or special bootstrap mode. In single-chip or special bootstrap modes the address and data buses are disconnected from the external pins (they become invisible). The LIR signal aids in program debugging.
26	STRA/AS	AS—Address strobe signal
27	E	Enable signal (clock)
28	STRB/ R/W	R/W signal (data bus direction indicator) R/W "High"=read/data coming in R/W "Low"=write/data going out
29	EXTAL	Connect to crystal (also external clock input)
30	XTAL	Connect to crystal
31	PC0/AD0	Multiplexed A0/D0
32	PC1/AD1	Multiplexed A1/D1
33	PC2/AD2	Multiplexed A2/D2
34	PC3/AD3	Multiplexed A3/D3
35	PC4/AD4	Multiplexed A4/D4
36	PC5/AD5	Multiplexed A5/D5
37	PC6/AD6	Multiplexed A6/D6

TABLE 2-1   MC68HC11A1P PIN DESCRIPTION (*Continued*)

PIN NO.	OFFICIAL LABEL	BRIEF DESCRIPTION
38	PC7/AD7	Multiplexed A7/D7
39	RESET	Reset Signal
40	XIRQ	Nonmaskable interrupt
41	IRQ	Interrupt request
42	PD0/RxD	Bit 0 of port D—General-purpose I/O, also receives data input for SCI
43	PD1/TxD	Bit 1 of port D—General-purpose I/O, also transmits data output for SCI
44	PD2/MISO	Bit 2 of port D—General-purpose I/O, also master-in-slave-out for SPI
45	PD3/MOSI	Bit 3 of port D—General-purpose I/O, also master-out-slave-in for SPI
46	PD4/SCK	Bit 4 of port D—General-purpose I/O, also serial clock signal for SPI
47	PD5/SS	Bit 5 of port D—General-purpose I/O, also slave select input for the SPI
48	VDD	Positive power input (connect to +5 V)

Note about the cross-assembler used here: All HC11 assembly-language source codes in this book are compatible with the AS11 cross-assembler that is on the CD-ROM included with this book. For a few more details on this cross-assembler, see Chap. 3. Chapter 8 is especially devoted to HC11 assembly language as viewed through the eyes of the AS11 cross-assembler.

U1, a MAX690 MPU supervisory IC, is an "insurance" chip that keeps the reset input low whenever VDD is below the minimum of about 4.5 V. This occurs when the power is turned on and off and can cause corruption of several important EEPROM registers or even the 512-byte EEPROM itself.

U3 is an octal transparent D latch that is used to separate (demultiplex) the HC11's data bus from the lower-byte address bus. After this is done, you can use the data buses and address buses just like you can in the original 6800.

Note: The HC11 here is put into the normal expanded multiplex mode in this circuit by holding pins 24 (MODA) and 25 (MODB) high by connecting them, through 4.7-kΩ resistors, to +5 V. In this mode you can think of the HC11 as a deluxe version of the 6800. More on operating the HC11 in different modes later.

The low-power LED is connected, through a 1-kΩ resistor, to pin 5 of the HC11. Pin 5 is connected to bit 3 of port A (hexadecimal address 1000), which is configured here as a general-purpose output. Thus, if the firmware in the EPROM instructs the HC11 to write a binary number such as 00001000 or 11111111 to port A, LED1 will light. More on this later.

Capacitors C1 through C4, C7, and C8 are bypass capacitors whose purpose is to make the circuit more reliable. It is always wise to use more bypass capacitors than seem necessary. It is almost impossible to use standard design procedures to accurately determine the minimum number of capacitors needed. Generally, find the recommended number of capacitors and then double it. Next, look for a supplier who will provide good-quality capacitors cheaply. Summing up, use lots and lots of bypass capacitors—in the long run, this is the cheapest way to go.

Notice that pins 1, 6, 7, 8, 40, 41, 42, 43, 44, 45, 46, and 47 of the MCU are terminated in resistors connected to +5 V. Under certain circumstances, these pins can function as inputs, so they shouldn't be left floating if not used.

How Does the HC11 Know What To Do?

While the HC11 is the "brain" of the circuit, the program in the EPROM (U4) tells the "brain" what to do. Simple address decoding for U4 is accomplished with two NAND gates, with one used as a simple inverter. The inverter's input is connected to address line A15, and its output is connected to U4's chip select input (pin 20). Whenever A15 is "high," U4 is selected. However, no data get to the data bus until U4's output enable line also goes "low"—this is a job for the other NAND gate. This other NAND gate has one of its inputs connected to the HC11's enable (basically a clock) line and the other input to the R/W line. When both these lines are "high" and A15 is "high," U4 puts its data on the data bus (i.e., U4 is being "read") and is then able to issue its built-in instructions to the HC11's CPU.

Tip:

> Neglecting to include the enable signal in the address decoding is one of the most frequent design omissions by neophyte HC11 designers.

You might ask the following: "How does the HC11 know where the first instruction of the program in the EPROM is located? It needs this information so that its CPU knows where to start." This is an excellent question because it is so important. If the CPU doesn't get the first instruction in the right order, everything is meaningless, and you wind up with nonsense and what appears to be a nonfunctioning circuit. As you may know, the 8080A and related Intel chips start looking, after RESET, at location 0 for the program's start.

While starting at 0 is intuitively pleasant, Motorola, with the HC11 and most of its 8-bit predecessors, handles where to start a bit differently. You, the software designer and not the chip designer, decide which memory address you want to locate the first instruction. This is called the *vector method*. Here, the software designer puts the starting address in the last two locations of memory—$FFFE and $FFFF. [Remember, I use the dollar sign prefix ($) to indicate that the number is hexadecimal. See Appendix A for more on hexadecimal numbers.]

The following is vitally important, so pay strict attention: *At reset, the HC11, when connected to a circuit such as that in Fig. 2-1, will start executing instructions that have their most significant byte (MSB) of address stored at $FFFE and their least significant byte (LSB) at $FFFF.*

Here we store $F8 at $FFFE and 00 at $FFFF. This simply means that we plan on starting our program at $F800. Although at first this may seem unnecessarily confusing, once you get the hang of Motorola reset and interrupt vectors, it seems that nothing could be simpler!

A Brief Aside—The Wild, Wild Program Syndrome

As you probably have noticed, I have emphasized here how the CPU knows where to start executing its list of instructions. If the CPU doesn't get this right, everything else is meaningless. A problem similar to this, which I have referred to in the past as the "wild, wild program syndrome," can occur when there is, for some reason, a glitch in the MCU circuit that causes the CPU to skip an instruction or two. This can happen when a voltage spike occurs on the power supply line (sometimes caused by lightning). Such a problem can cause the troubles in the embedded controller products mentioned in the Introduction. The solution here is to get the CPU to start over again so that it starts executing the first instruction on its list. This usually can be accomplished simply by removing power for a minute and then reapplying it. Computer operating properly (COP) circuits are supposed to handle this problem automatically. Nonetheless, while COP circuits are steadily improving, they are far from foolproof. More on ways to reduce the "wild, wild program syndrome" later.

A First Quick Peek at What's Inside an HC11

Of course, to achieve a working MCU circuit, you will need a functioning program in the EPROM. Since all you want to know is that the circuit is working, all the program needs to do is instruct the 68HC11 to turn on the LED, wait about 1 second, and then turn it off—and repeat itself *ad infinitum*. This program can be extremely simple. However, before going into the program itself, it would be helpful to look at a simplified programming model of the HC11, as shown in Fig. 2-3. Unlike the block diagram of the HC11 given in Chap. 3, this one is designed to be gentle so as not to scare you. Note that only four registers of the HC11 are shown in Fig. 2-3. These are the only registers used by our program. (This isn't strictly true, since the X index register is used by the delay subroutine and isn't shown. However, this simplification is justified assuming that you can "imagine" that the JSR DELAY instruction is a "virtual" HC11 instruction that delays it for about 1 second. Of course, this JSR DELAY instruction really isn't part of the HC11 instruction set, but for the moment, just sit back and fantasize that it is.)

One way of picturing how the CPU functions is by thinking of the accumulator in our model as a "magical" register that contains an 8-bit binary number. This amazing register can do just about anything to this 8-bit number the program in the EPROM wants it to do! In contrast to the accumulator, the program counter (PC) is a relatively dumb 16-bit register that simply keeps track of the locations of the instructions of the program. The PC register can be looked on as the HC11's "bookmarker." While the PC register is vital, for now you can forget that it even exists. While you're *reading* a book, a bookmark is usually forgotten about and not needed. It is only when you start reading again, after an interruption, that it becomes important.

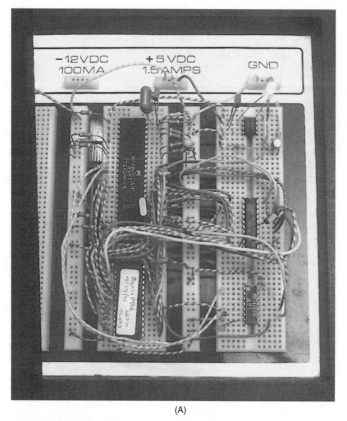

(A)

FIGURE 2-3 (A) Circuit in Fig. 2.1 breadboard.

The stack pointer (SP) register is another register that is necessary whenever a subroutine is used or an interrupt (other than a power-on reset) occurs. If you ignore the SP, you might have a mess.

Tip:

Forgetting to initialize the stack pointer is probably the most frequent error made by neophyte HC11 machine and assembler language programmers.

Port A—One of the HC11's Access Lanes to the Real World

The most interesting register in Fig. 2-4, at least to me, is port A. This register is connected between the CPU and the rest of us. This is the register with which we can do something really interesting. (This register is a great example of why MCUs are so much more fun to play with than their more abstract cousin the MPU.) This register is located at $1000

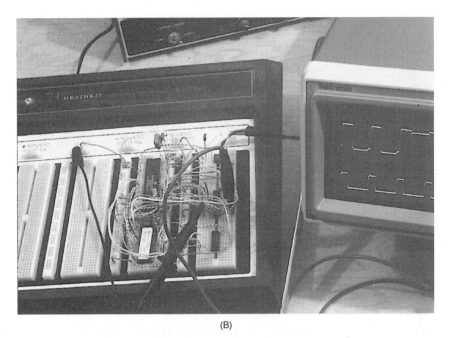

(B)

FIGURE 2-3 (B) Oscilloscope screen. (Bottom trace on screen is enable line; trace on top is an address line.)

on the address map. While port A is a multipurpose register, all we are concerned with at the moment is bit 3. (Remember, bit 3 is the fourth significant binary bit—the one fourth from the right—not the third. Bit 3, you know, is the one that has the "value" of 8. The reason for this apparent numbering confusion is that the bit on the extreme right *of a binary number is called bit 0, not bit 1.*) Thus, if you instruct the accumulator to store the binary number 00001000, which is 08 in hexadecimal as well as decimal, at $1000 (port A), then pin 5 in this schematic goes "high" because it is connected to port A's bit 3. Now let's get right to the program to see how we can do something really exciting—flash an LED!

A Step-by-Step Look at What Makes the LED Blink

Tip:

If the following is a bit hard to understand, this simply means that you haven't had experience with Motorola's 6800 series of MPUs.

Before I discuss the source code in Fig. 2-5, you should be aware of the fact that load and store instructions don't actually *move* data as you would think from their names. Rather, load and store instructions *copy* data. Logically speaking, LOAD could be replaced with

"COPY memory data to accumulator" and STORE with "COPY accumulator's data to memory."

Referring to the assembly language source code in Fig. 2-5, the operation starts after the RESET line goes "high." The source code starts with the lines

```
*FIRST WE STORE THE STARTING ADDRESS AT $FFFE-$FFFF
*WHICH IS THE ADDRESS OF THE RESET VECTOR
          ORG    $FFFE
          FDB    $F800
```

These two lines of code (the lines that start with an asterisk are ignored not only by the assembler and HC11 but also by us) form a sort of virtual program whose sole purpose is to place the data $F800 in the reset vector located at $FFFE and $FFFF—these data ($F800) constitute the address where the program starts.

With housekeeping chores now done, we can start our actual program with the ORG $F800 statement. This statement simply tells the assembler program (not the CPU) that we want to start our program at $F800. Remember, the assembly language statements ORG $FFFE *and* FDB $F800 *already produced machine language code that told the CPU that we are starting our program at $F800.* The ORG $F800 statement simply tells the assembler the same thing.

The first actual instruction directed at the HC11 is the LDS #$00FF statement. This statement causes the stack pointer to be loaded with the top address of internal RAM, which is $00FF. (The # prefix here means that we want to load the stack pointer *immediately with*

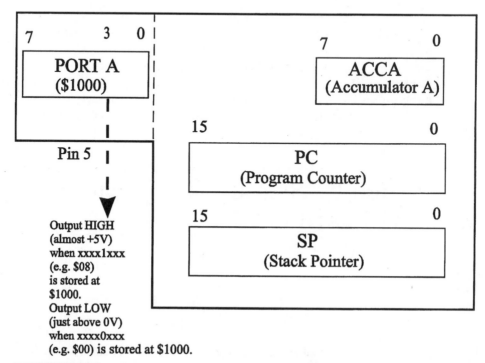

FIGURE 2-4 Simplified programming model of HC11

```
*TESTHC11.ASM program short program for turning
*on and off the LED controlled by bit 3 (00001000)
*of port A which is located at address $1000
*FIRST WE STORE THE STARTING ADDRESS AT $FFFE-$FFFF
*WHICH IS THE ADDRESS OF THE RESET VECTOR
          ORG         $FFFE
          FDB         $F800
*NEXT WE TELL THE ASSEMBLER THAT WE WANT
*TO START OUR PROGRAM AT $F800
          ORG         $F800
*THEN WE LOAD STACK POINTER WITH ADDRESS OF
*THE TOP OF INTERNAL RAM
          LDS         #$00ff
*NEXT LOAD ACCUMULATOR A WITH $08
DOAGIN    LDAA        #$08
*STORING ACCUM A AT $1000 LIGHTS LED
          STAA        $1000
*WAIT HERE ABOUT 1 SECOND
          JSR         DELAY
*CLEAR $1000 — SHUTS OFF LED
          CLR         $1000
*WAIT HERE ABOUT A SECOND
          JSR         DELAY
*GOTO DOAGIN AND START OVER
          BRA         DOAGIN
*PROGRAM END delay subroutine follows
****************
**THE FOLLOWING PROGRAM SEGMENT IS THE DELAY SUBROUTINE
**For now simply think of this subroutine
**as a built-in CPU instruction whose mnemonic is DELAY
DELAY     LDY         #65535
LOOP      DEY
          INY
          DEY
          BNE         LOOP
          RTS
*THE "END" DIRECTIVE IS NOT NEEDED BY MANY ASSEMBLERS BUT WE
*INCLUDE IT HERE ANYWAY
          END
```

FIGURE 2-5 The assembly-language source code (.asm file) for the TESTHC11 program used in the breadboarded computer circuit.

the "number" $00FF and not the "data" located at the address $00FF.)

Next, we store $08 (which is 00001000 in binary) at port A ($1000) *which turns on the LED.* This is accomplished by the two instructions: LDAA #$08 and STAA $1000. Here, LDAA #$08 means load accumulator A with the number $08, and STAA $1000 means store accumulator A at memory location $1000, which is port A.

The program next causes a delay of about 1 second by jumping to the delay subroutine called DELAY (JSR DELAY). Next, *we shut off the LED* by clearing port A (CLR $1000). Again, we jump to the delay subroutine. Finally, the statement BRA DOAGIN causes the CPU to start back at the beginning, and the cycle repeats itself. Notice that the mnemonic *BRA* stands for "branch" and is similar to a GOTO in BASIC. The assembly language listing is shown in Fig. 2-5.

Note:

The character "*" in the first column of the listing indicates that comments follow. These comments are ignored by the assembler and are only added to help the person reading the source code.

Hopefully, you will have breadboarded the circuit in Fig. 2-1 and have gotten it to flash the LED as indicated. Also hopefully, you will have a good understanding of what's taking place. If these hopes have been fulfilled, then you are well under way on your quest for knowledge concerning the HC11.

Before going to the next chapter, take another look at the functioning circuit hooked up on your breadboard. If you have access to an oscilloscope, check out signals coming from the HC11. With the probe's ground connection attached to a circuit ground, carefully touch the end of the probe to pin 27 of U2. This is the enable line, which should be a nice 1-MHz squarewave. Keep the shape of the enable signal in mind. This knowledge will help you troubleshoot problem HC11 circuits. Now take a look at the HC11's address lines (pins 9 through 16) and the multiplexed data/address bus (pins 26 through 31). Also examine pins 3 through 10 (address lines A0 through A7) of the EPROM (U4). While some of the high-order address lines (like A14) may appear dead, most of the other lines will seem to be alive with complex, "intelligent looking" waveforms. Also keep these waveforms in mind because they are how things should appear when an MCU circuit is functioning properly. These memories of waveforms will help in troubleshooting problems that surely will crop up when you start designing with MCUs. By the way, you also may want to check out these same signals with a good-quality logic probe.

The next chapter provides a much more detailed look at the HC11.

3

GETTING TO KNOW THE HC11

In this chapter, things start to look more complicated. Figure 3-1 shows the block diagram for the MC68HC11Axxx version of the HC11. Scary? I think so. However, Fig. 3-2 shows the full-blown programming model for the HC11. It really doesn't look too bad, does it? Since there are few applications that will make full use of the HC11's capabilities, the programming model (Fig. 3-2) is the one to concentrate on. The hairy looking block diagram shown in Fig. 3-1 will almost always be simplified when working with the HC11—sometimes dramatically so, as was done with the first working HC11 circuit described in the last chapter. This chapter concentrates on the programming model and only slowly delves into the HC11's peripheral registers and capabilities as we start to use them. As you may realize, we have already started on our journey into the intricacies of the HC11 by looking at port A in the last chapter.

The Registers

If you are acquainted with the 6800 or its siblings, the 6802 and 6808, the programming model of the HC11 should look somewhat familiar. The primary differences are that in the 6800/02/08 series there is only one index register and accumulators A and B aren't inherently cascadable. Following is a brief description of the registers.

Accumulators A and B   These are 8-bit general-purpose registers that get most of the work done. As Fig. 3-2 indicates, accumulators A and B can be cascaded (put together) to

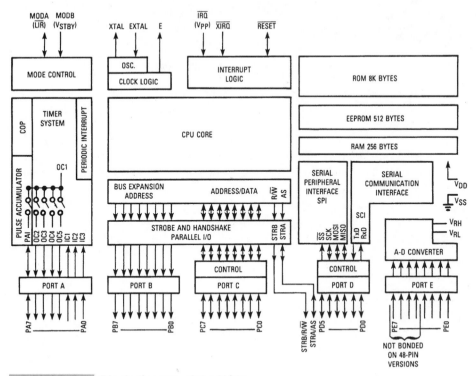

FIGURE 3-1 Block diagram of the HC11.

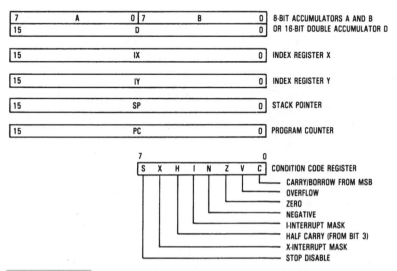

FIGURE 3-2 Full-blown programming model of HC11.

form a single 16-bit register called the *accumulator D*. As you will see in Chap. 8, there are several instructions that manipulate the 16-bit D register directly. Since the HC11 can handle 16 bits of data at a time, it can be viewed as an $^8/_{16}$-bit hybrid MCU. If its data bus were 16 bits wide, it would be a true 16-bit MCU, like its big sister the HC16.

Index registers X (IX) and Y (IY)   These two 16-bit registers are used primarily in addressing operations that are descriptively, but unimaginatively, referred to as *indexed addressing*.

SP (stack pointer)   The HC11's 16-bit stack has a last-in-first-out (LIFO) structure. The stack pointer points initially to the location above the top element of the stack. By convention, the stack "grows" from high addresses (e.g., $00FF) to low addresses (e.g., $00FE, $00FD,…, $00F0). Watch out! This growth toward lower addresses is counterintuitive. It would seem logical that the more you piled on the stack, the larger would be the address. In other words, it would seem intuitively obvious that if you set your SP at, say, $F0 and you pushed 5 bytes of data, the SP would point to $F5! This just isn't so! It actually would point to $EB! Intuition often may be helpful, but it isn't foolproof.

For those who like to compare rather esoteric concepts with real-life situations, the HC11's stack can be pictured as a pile of playing cards with "bytes of data" on the cards instead of aces, queens, etc. Say you put (push) a card with $3F down, then a card with $11 on top of it, and then another card with $0A on top of that. Now take (pull) the cards, one at a time, off the stack. What will be the first card? The second and third? That is basically how the HC11's stack works.

Practically speaking, often you can start a program by loading the SP with the top of built-in RAM, which is located at $FF, and only use RAM addresses below, say, $C0 for other uses. Once this is done, you can put the SP in the back of your mind. However, if you are using a monitor program, such as BUFFALO, you often don't need to set up the SP.

Tip:

If you feel that you do need to set up the SP, even though you are using the BUFFALO monitor, load it with $2F.

CCR (condition-code register)   This is an extremely important 8-bit register that is used directly by the programmer less often than you would think. It is used clandestinely, however, by the frequently used "smart" instructions (i.e., the "branch if so and so…" instructions). More on this register later.

PC (program counter)   This is an absolutely vital 16-bit register. It is the register that keeps everything in order. The address of the next instruction to be executed by the CPU is kept here. The PC register is then increased by the number of bytes used in the machine-language instruction. For instance, if the PC contains 100 and the machine-language instruction that starts at 100 is 3 bytes long, then the PC will increase by 3 or be at 103 to point to the start of the next instruction. (Note that the preceding statement isn't always true. For instance, it isn't true with instructions that purposely change the PC register, such as the JUMP or BRANCH instructions.)

When the CPU is reset, the PC is loaded from a specific pair of memory locations called the *reset vector*. As you may recall, the HC11's normal reset vector is located at $FFFE and $FFFF. *Quick question:* If $FFFE contains $F8 and $FFFF contains $00, what is the PC loaded with immediately after reset?

The PC reminds me of breathing—both are absolutely vital, but a constant conscious recognition of either the PC or breathing is seldom necessary. Of course, a realization of the existence and purpose of the PC is vital—especially for a fuller understanding of vectors, branches, jumps, and subroutines.

Memory Addressing

Like all 8-bit MPUs/MCUs, the HC11 has an 8-bit data bus and, like most, a 16-bit address bus. This simply means that it can address one plus 1111111111111111 in binary, which is FFFF in hexadecimal and 65536 in the "garden variety" notation referred to as decimal. This number (65536), by the way, is usually written as 64 kB (kB stands for kilobytes. 1 kB = 1024 bytes).

Note that with the HC11, data, instructions, and peripheral registers, including output ports, all use the same bank of 65536 memory locations. Also, most input-output (I/O) ports are treated the same as data locations. For instance, port A, as we saw, is located at address $1000.

Memory Locations

As has been inferred, there are 65536 (1 more than 65535, or 1 more than $FFFF) possible memory locations in a "normal" HC11 system. Another way of saying this is that the HC11 can address 64 kB of memory. Each memory location can contain 1 byte (8 bits) of information (data, part of an instruction, etc.). Traditionally, these memory locations have been thought of as "post office boxes" and their address as "P.O. box numbers" and the data in the locations as "mail." However, one rarely gets a P.O. box number confused with the mail that is in the box. Probably the reason here is that you are dealing with totally different items. In reality, these memory locations are better thought of as "e-mail addresses." With e-mail, there is a possibility of confusion. For instance, say your e-mail address is sharpe@AOL.com and you receive an e-mail note that reads simply, "sharpe@AOL.com contact me at HC11exp@AOL.com as soon as possible." Let's face it! It is possible, if you have a new e-mail address, that you will get yours confused with the sender's and send yourself an e-mail response.

A similar confusion can happen when addressing computer memory. For instance, say you want to store the number located at $FE at $FF and the number at $FE is $FD. Things can get even more confusing if you are dealing with what is referred to as *immediate addressing,* where you store actual data (not the address of the data) at a memory location.

To be honest, I believe that I have overstated the possible confusion here with regard to memory addressing. It is all relatively simple. However, keep alert at all times so that you don't make as many goofs as I have in the past—and I have made more idiotic mistakes than I care to admit!

The HC11 has six different ways to address memory. These ways are referred to as *addressing modes:*

1. *Immediate mode*   This simply means that we "immediately" load or somehow manipulate (e.g., add or subtract) a number in a register with the number specified in the instruction.

   Example   Say we want to add the decimal number 33 to accumulator A. Here, we would make use of immediate addressing. In assembly language, we would write ADDA #33. Here, the assembler understands the "#" to mean "immediately" and that we want to add the number 33 to the accumulator and *not the data at memory location 33.*

2. *Extended mode*   This is probably the most commonly used addressing mode. Here, the data we are dealing with are stored at a memory location indicated by the 2-byte number following the opcode.

   Example   Say we want to add the data stored at location 0033 to accumulator A. Here, we can use the extended addressing mode. In assembly language, we would write ADDA 0033. Notice the omission of the "#" prefix to the number following the opcode. The assembler "understands" that this omission means that we want to use extended addressing.

   One other point: If we had added the prefix "$" to the number, the assembler would understand that the number is hexadecimal and not decimal. ($33 = 51 in decimal) *Watch the prefixes!* They can be confusing and get you into trouble—bigtime.

3. *Direct mode*   This is a simplified variation of the extended mode. With this mode, a single-byte number follows the opcode. Obviously, this mode can only be used when addressing the first 256 bytes in memory, so its use is rather limited. However, *always use this mode if you can.*

   Example   As in the last example, say we want to add the data stored at location 0033 to accumulator A. However, this time we want to use direct addressing, so we only include the number 33 after the opcode. In assembly language, we would simply write ADDA 33. Since a single 2-byte number is added after the opcode, the assembler automatically knows we want to use the direct mode of addressing.

   Quick questions   Does the following instruction use extended or direct addressing: ADDA 125? How about ADDA $ADDA?

   Answers   The former uses direct and later extended addressing. Remember, $ADDA is really a number—aren't I real tricky!

4. *Indexed mode*   This mode is slightly more complex than the previous three, although it often is the way to go. This mode makes use of the index registers (X or Y). Here, the number stored in the index register is automatically added to the offset number in the instruction. The resulting sum is *the address* used by the CPU. Notice that one reason this mode is more complicated to use is that the base number of the address already must be stored in the appropriate index register.

   Example   As in the previous examples, say we want to add the data stored at location 0033 to accumulator A. This time, however, we will employ index addressing using the X index register. Actually, there are many ways to do this. I will show two. The first way is to first store 0033 at index register X and then use index addressing with zero offset to add the data at 0033 to accumulator A. For instance, the following two lines accomplish this.

```
LDX #0033
ADDA 0,X
```

The second way is just as simple. Here, however, we load 0000 at index register X and use 33 as the offset.

```
LDX #0000
ADDA 33,X
```

Notice that the offset in this example is identical to *the address*. The reason for this is that we stored zip (i.e., 0000) in the X index register.

Question   If we had stored 23 in the X register, to what would we have to set the offset to accomplish our goal? Well, I'll give you a hint—the offset rhymes with *men*.

5. *Relative mode*   This mode and the next are different from the preceding four addressing modes in that you have no choice here. These modes are *attached* to their instructions. For instance, the relative mode is used with all branch instructions. Nearly all branch instructions generate two machine-code bytes [the exceptions are 4-bit manipulation instructions, such as the branch if bit(s) set (BRSET) instruction], one for the opcode and one for the branch offset. When writing in machine language, you must use a branch offset. Branch offsets range from −128 to +127. You may be a bit confused here with the minus offsets. But keep this in mind: The branch offset is no ordinary number. The CPU recognizes the branch offset as a two's complement number. This simply means that if bit 7 (the bit at the extreme left) is a "1," the CPU thinks of the branch offset as a negative number and branches backward. If bit 7 is a "0," then the CPU thinks of it as a positive number and branches forward. In hexadecimal, any 2-byte number from 80 to FF is thought of as negative as long as you look at it as a two's complement number. (In reality, two's complement numbers differ from regular numbers only in the mind of the cognitive party—whether it be you or an MCUs CPU.) If you find the relative modes and branch offsets confusing, smile, better times are ahead! Read on.

Assemblers make it all so simple!   The preceding discussion concerning two's complement numbers and offsets applies to machine language. If you stick to assembly language, you can ignore the branch offset number and simply use labels and treat branch instructions like BASIC GOTO statements. *However, there is a bigtime "if" here.* Unlike GOTO statements, you can't "go" too far backward or forward. In fact, the farthest you can go is 127 memory locations ahead or 128 memory locations back. (These numbers refer, of course, to the program counter.) In hexadecimal, this is $7F memory locations ahead and $80 memory locations back. Do not fear, however! If you attempt to go too far, most assemblers will spit out an error message and won't go any further until you fix the code!

Example   This example is a short piece of code that causes a short delay. The BNE (branch if not equal to zero) causes a branch back to BACK if accumulator A isn't equal to zero yet:

```
*First we clear accumulator A and label the line DELAY
*this label (DELAY) isn't necessary, but it looks nice huh?
DELAY           CLRA
*then we add 1 to accumulator A and label the line BACK
BACK            ADDA #01
*If Accumulator A still hasn't gone from 11111111 to 00000000
*then branch (GOTO) to BACK
                BNE BACK
*otherwise continue program
```

6. *Inherent mode*   As its name indicates, this mode is inherent or presumed. Like breathing, it isn't necessary to think about this mode to use it. This mode is used by simple instructions such as NOP, INC, DEC, RTS, PUL, PSH, etc.
Examples

```
INCA
```

This increment A instruction adds one to accumulator A.

```
NOP
```

This no operation instruction does nothing, but it takes two clock cycles to do that nothing, which sometimes is important.

Machine Language

It's too bad, but machine language has a bad reputation—a reputation it does not deserve. The impression one receives from listening to "experts" on MCUs is that only those who have "only one oar in the water" ever use machine language. "It's too hard!" some say. Others complain that it takes too much time and that the source code is nearly impossible to make sense of—even by the author just a few minutes after writing the source code.

While I agree that the use of assembly language is preferred when writing anything other than short, simple programs, everyone who wants to get serious with MCUs should know the basics of machine language. In fact, knowing about machine language often simplifies things and helps in troubleshooting programs. For example, sometimes when dealing with memory addressing, things are simpler using machine language than assembly language! Why? Because machine code is straightforward and noncomplex—for crying out loud, the stupid but superfast CPU understands it. Right? For instance, the opcode "86" means that we want accumulator A loaded using immediate addressing, whereas "96" means that we want to use direct addressing. While it is possible to confuse the two types of addressing when writing or reading an assembly-language program, it is nearly impossible to make the same mistake using machine language.

Perhaps, just perhaps, my fondness for machine language has something to do with my first MPU design project—IT (IT here stands for "intelligent thermometer"). The complete description of my design for an intelligent thermometer appeared in the January and February 1983 issues of *Computers and Electronics*. This project used a 6802 MPU. The 6802 is nearly identical to the 6800 except that it has a built-in clock and 128 bytes of

RAM—the 6800 has neither. The program was contained in an antique 2708 EPROM that could hold up to 1kB. The entire program was written in machine language. A custom version of this intelligent thermometer found its way to one of Texaco's oil pipelines and apparently was used to determine the average temperature of the oil flowing through it. Who knows? Maybe it still is.

FUNDAMENTALS OF MACHINE LANGUAGE

First off, while true machine language uses only binary numbers, assemblers use binary numbers, hexadecimal numbers (a shortcut way of writing binary numbers), and decimal numbers. However, hexadecimal numbers generally are preferred for a number of reasons. Remember, 0001 in binary is 1 in hexadecimal, 0010 is 2, 1110 is E, and 1111 is F. For those who need a refresher on binary and hexadecimal numbers, refer to Appendix A.

Most opcodes consist of a single byte and are written in hexadecimal. An example is the opcode "86." This hexadecimal number instructs the CPU that you want to load accumulator A right away with the single-byte number following the opcode. (Opcodes dealing with the Y index register have 2 bytes.) The number following the opcode is called the *operand*. For instance, the line of code

```
86 10
```

loads accumulator A with the hexadecimal number 10, which is 16 in decimal. Keep in mind here that all numbers in machine code are hexadecimal. Here, "86" is the opcode and "10" the operand. The line of code

```
96 10
```

loads accumulator A with the data at hexadecimal address 10. And the line of code

```
B6 00 10
```

does the same as 96 10 only we are using extended addressing here.

EXAMPLE OF A SHORT MACHINE-LANGUAGE PROGRAM

Let's face it, I would be wasting your time if I went right ahead into the details of programming with machine language. However, you can get a general idea about machine-language programming by attempting a short machine-language program. What we will attempt to do here is simply to store the number 08 at the location $1000. If you followed the short test program in the last chapter, you will see that this test program will light the LED in the breadboarded circuit.

What is the first step here? Think. Getting started on a project is often the hardest part. However, you have to start somewhere, and so does the HC11. But it doesn't know where. You have to tell it where. This is done by placing the address where you want to start (usually the start of your program) at the reset vector addresses of $FFFE and $FFFF. Remember, with assembly language this was done with

```
ORG     $FFFE
FDB     $F800
```

The assembler did the rest.

In one sense, machine language is even simpler. You don't need to know what the assembler directives ORG or FDB mean. The following machine code accomplishes the task with the beauty of radiant simplicity.

```
FFFE          F8
FFFF          00
```

These two lines mean that we want to burn our EPROM with $F8 at $FFFE and 00 at $FFFF. Simple! Now the HC11 knows where to look for our program. So let's get started. Recall that the HC11 only understands the binary number system. However, the hexadecimal number system is used here as a shorthand way of writing binary numbers. Notice that I have left out the "$" prefix in Table 3-1 because it is understood that all numbers are in this shorthand way of writing binary.

What does this all mean? Well, the first three lines in Table 3-1 (BE, 00, FF) tell the CPU to load the stack pointer with the last RAM address ($FF). As mentioned earlier, it is good practice to start a program by loading the stack pointer so that it points to usable RAM memory. The next two lines (86, 08) tell the CPU to load accumulator A with $08. The next three lines (B7, 10, 00) turn on the LED by storing accumulator A's contents at port A, which is located at $1000. The last line, which is stored at address $F808 and has the opcode $3E, is a bit unusual. This is a wait instruction. Once the CPU receives this instruction, its machine state is saved in the stack, and the CPU, with its clock still running, simply sits and waits. What is it waiting for? Some sort of signal. This signal is called an *interrupt*. More on interrupts later.

Down-to-earth readers who don't find things interesting unless they can actually do it themselves are probably asking the question, "Fine, but how do I actually get these instructions to

TABLE 3-1 MACHINE-LANGUAGE LISTING TO TURN ON LED IN FIG. 2-1	
HEXADECIMAL MEMORY ADDRESS	HEXADECIMAL CONTENTS (DATA)
F800	BE
F801	00
F802	FF
F803	86
F804	08
F805	B7
F806	10
F807	00
F808	3E

the HC11 so that I can turn on the LED? I can see from the circuit in the last chapter that the program is in EPROM, but how do I get the right data in the right places?" Great question! (I know it's great because I just asked it!) The answer, of course, lies primarily with an EPROM programmer. Deluxe (*deluxe,* sadly, of course, is short for "lots of money") programmers often include a hex keyboard where you can enter the data at the address you want. The only thing a bit tricky here is that except when the EPROM is the same size as the addressing capability of the MCU, you have to change the address. For instance, a 27C256 EPROM can store $8000 bytes (32768 in decimal), or 32 kB. However, the addressing capability of an HC11 is $10000 bytes (65536 in decimal), or 64 kB. When you program the reset vector into the 27C256, you place $F8 at location $7FFE and 00 at location $7FFF of the EPROM. Basically, all you are doing is subtracting $8000 from the memory address. If you use a 27C512 EPROM, which can store 64 kB of data, no translation is necessary because the memory-addressing capability of the HC11 is also 64 kB. See! This stuff isn't really hard. Nonetheless, there are several disadvantages to this approach:

1. Programmers generally are expensive.
2. The manual mode is time-consuming and inherently error-prone.
3. Possible confusion can result from memory relocation, which is the duty of the operator.

However, there are several relatively inexpensive "burners" (EPROM programmers) around that make use of a computer's intelligence. Some of these burners connect to the serial interface or parallel port, while others are on a board you insert into an empty slot in a PC-compatible computer. All come with easy-to-use software. However, as far as I know, all require the object code to be put into some standard format such as the Intel hex format or Motorola's MOT S19 format. Since the HC11 is a Motorola chip, I will only use the MOT S19 format in this book. MOT S19 is an ASCII file that is understood by most burners. For details on its meaning, refer to Appendix D.

Notice from Appendix D that it isn't really all that simple to manually create a MOT S19 file. The main difficulty is the creation of an accurate checksum at the end of each line in the file. This checksum is used by the firmware or software in the burner to make sure that it is receiving error-free data. Happily, there is no need to create the MOT S19 file without help. For instance, nearly all assemblers and cross-assemblers, including the AS11 cross-assembler on the CD-ROM, automatically create a MOT S19 file that can be used by burners.

MORE HAZE THAT EPROM BURNERS CAN HELP EVAPORATE

One of the confusing things that can happen when writing in machine language (or even assembly language) and using an EPROM that contains less than 64 kB (e.g., a 27C256) is the reference to addresses in the code. For instance, let's look at a short code segment that uses the jump instruction (which has a $7E opcode) to jump over the next instruction:

```
. . .
F900      4C
F901      7E
F902      F9
```

```
F903      05
F904      4A
F905      5A
...
```

The opcode for the jump instruction is $7E. This opcode is stored at location $F901 (with respect to the CPU). Here we are telling the CPU to jump to location $F905. However, since we are installing this program in a 27C256 EPROM with only 32 kB, our modified code segment is actually

```
...
7900      4C
7901      7E
7902      F9
7903      05
7904      4A
7905      5A
...
```

If you attempt to make sense out of the preceding machine code, you probably won't because the listed memory addresses on the left do not correspond to your program's memory address references. This can be confusing. Trust me! Luckily, this conversion doesn't have to be done (by you) when using most EPROM programmers. For example, say you are attempting to program a 27C256 EPROM with the program listing in Table 3-1 (of course, you actually use the MOT S19 ASCII file to the programmer and not the table itself). Most EPROM programmers realize something is amiss and automatically (sometimes they warn the user first) convert the addresses so that everything works out all right. If you have a doubt about the programmer you are using, either check with the manufacturer or maybe the person you borrowed it from.

A Quick Look into Assembly Language (More Details in Chap. 8)

There is no question that most programmers can write medium- to large-sized programs faster with assembly language than with machine language. Because of this and the fact that it does aid in understanding programs (e.g., doesn't JMP SKIPIT make a bit more sense than 7E FA55?), assembly language will be used primarily throughout the remainder of this book.

While Chap. 8 will go into assembly language in moderate detail, let's get right at it here and revise our program that turns on the LED in our breadboarded circuit so that it is understood by a typical assembler program. The original machine language program was listed in Table 3-1.

First off, we have to get the CPU started in the right place. This was all explained before, but because of its importance, repetition here becomes the grandfather of understanding. Assemblers use a location counter to keep track of the place to put data or a program during the assembly process. The ORG directive sets the value of the location counter. In other words, the ORG directive tells the assembler in memory where to put the next byte (located immediately after the ORG directive) it generates.

For example, in our short program we want to set up the reset vector so that our program will start at $F800. This is done using the ORG directive along with another assembler directive called the FDB (form double byte) directive. The following two lines of source code accomplish our initial objective of setting up the reset vector:

```
ORG     $FFFE
FDB     $F800
```

The first line sets up the location counter so that it is at $FFFE, whereas the second line tells the assembler that we want the data byte $F8 at location $FFFE and 00 at the next location, which is $FFFF. Notice something a bit different here from our machine-language program. Here, we use the "$" prefix to indicate hexadecimal numbers. If we left the prefix out from the first line and simply wrote FFFE, what do you think would happen? Simple! The assembler would spit out an error message! Why? Because the absence of a prefix forces the assembler to first assume that the data are decimal. Once the assembler sees letters instead of numbers, however, it then assumes FFFE is a label and not a number. However, it looks over the program and doesn't see any label like FFFE, so it spits out the error message. Great! However, what if you wanted to start your program at hexadecimal 9000 but for some inexplicable reason you wrote

```
ORG     9000
```

Even though there is a mistake, no error message would result because the assembler thinks the data (9000) are decimal, translates the 9000 to $2328, and then sets the location counter to $2328 instead of $9000, as the programmer originally wanted. While a sharp HC11 guru might catch this goof before actually burning an EPROM, there are no guarantees!

Tip:

When working with assemblers, keep always alert and watch those prefixes!

The assembly-language source code for a program that is identical to the machine language program given in Table 3-1 is given in Fig. 3-3. An EPROM programmed directly with the machine-language source code in Table 3-1 would be identical to an EPROM programmed with the MOT S19 object code produced by an assembler that assembles the listing in Fig. 3-3.

The first two lines of the short program in Fig. 3-3, which set up the reset vector, were described already. The ORG $F800 instruction sets the assembler's location counter to $F800 so that it knows what address to start with when it makes up the first line of MOT S19 object code. The next line, LDS #$00FF, loads the stack pointer with $FF. This instruction isn't essential here, but it is a good habit to stick a number in the stack pointer so that the stack can save its contents to usable RAM and not to some mysterious memory location that exists solely in the mind of the neophyte programmer. The next two lines, LDAA #$08 and STAA $1000, get the real job done by storing $08 at $1000. Remember, port A

```
*File name: CH3_FG3.ASM
*Assembly language program that
*results in the machine language
*code in Table 3-1
        ORG      $FFFE
        FDB      $F800
        ORG      $F800
        LDS      #$00FF
        LDAA     #$08
        STAA     $1000
        WAI
        END
```

FIGURE 3-3 Listing of assembly-language version of machine-language program in Table 3-1.

is located at $1000 and $08 is 00001000 in binary. In other words, in the number $08, bit 3 is "1"! This is what turns on the LED that is connected to this bit at port A. The WAI instruction (whose opcode is $3E) was described earlier. Not all assemblers require the END directive to indicate the end of the source code, but I include it just to be safe.

Chapter 8 will go into moderate detail about writing source code in assembly language and using the assembler to create MOT S19 code that is employed by EEPROM programmers. In keeping with our hands-on policy, let's see what happens if we attempt to send the source code in Fig. 3-3 to the cross-assembler provided on the CD-ROM. (For now, think of cross-assemblers as being identical to assemblers.) Note that our first step here is to create an ASCII file identical to Fig. 3-3. Note that while we can add lines to provide documentation, such lines must start with an asterisk (*) or we will get an error message and the cross-assembler will spit back the file.

To create ASCII code, you use a text editor (even EDIT, which comes with DOS and WINDOWS 95/98 is okay) or simply use the word processor you generally use and save the file in ASCII format. It is good practice, but not essential, to save the file with a .asm file extension.

Once the file is created, you simply invoke the assembler as follows:

```
AS11 (file name)
```

For example, with our short test file, we would use the following line:

```
AS11 CH3_FG3.ASM
```

With this instruction line, the assembler will create the file CH3_FG3.S19, which can be used by most EPROM programmers. This file is reproduced below:

```
S105FFFEF80005
S10CF8008E00FF8608B710003EDB
S9030000FC
```

Let's look at this file just a bit closer. First, we can see that it is an ASCII file and not a binary file and thus can be read (and changed) with the use of a text editor or word processor. Details on the MOT S19 format will be given in Appendix D. For now, keep in mind that the fifth through the eighth (from the left) ASCII characters represent the hexadecimal starting address of the line of code. For instance, the first line of code starts with FFFE, just as our first ORG directive did. The next line's address starts with F800, just as our second ORG directive did. The last line is the S9 line and doesn't pertain to us here.

Chapter 8 takes a much more detailed, step-by-step look at the Motorola AS11 cross-assembler that is included on CD-ROM. It also examines a somewhat more complex and useful assembly-language program. However, first I will describe the construction and use of Mag-11, a single-board HC11 computer designed especially as a basic learning tool.

4

THE MAG-11 SINGLE-BOARD COMPUTER

With this chapter I start to get serious and introduce Mag-11—a single-board HC11 computer that uses an MC68HC11A1P MCU. As you will quickly see, the Mag-11 was designed with *you,* the reader, in mind.

In the first part of this chapter I describe how to build a Mag-11 and test it using my own M11DIAG2 firmware. This firmware starts off by checking HC11's all-important CONFIG register. After explaining more about this register than you might like, I use M11DIAG2's diagnostic capabilities. These diagnostic tests do more than test Mag-11 and the HC11; they introduce, through a hands-on approach, several of the HC11's features, including built-in RAM and an analog-to-digital (A/D) converter. M11DIAG2 firmware also introduces—sort of through the back door—two of the HC11's simplest but neatest features—the illegal opcode trap (IOT) and the clock monitor reset (CMR.) The discussion of the CMR introduces the HC11's fascinating STOP instruction. This STOP instruction is one way to preserve RAM memory if an HC11 system undergoes a power failure.

The chapter ends by looking at another one of Mag-11's built-in experiments—the binary readout thermometer. This unique thermometer introduces the HC11's programmable timer subsystem, since it uses the chip's input capture capabilities. This binary thermometer introduces the binary readout capabilities of Mag-11 as well as providing a reinforcing hands-on review of the binary and hexadecimal numbering systems. By the way, Mag-11 is also used extensively used in examples and experiments in Chapters 5 through 9.

Why Mag-11?

When I decided to start my first magazine series on the HC11 (which this book is based on), I considered centering it around one of the HC11 boards already on the market. I rejected this idea for several reasons:

1. Readers would be required to purchase a completed board before reading the series.
2. The reasonably priced HC11 boards around were good, but they weren't really designed for hobbyists who wanted to quickly hook up to an HC11 with just a few wires and inexpensive added parts.
3. Buying a completed and tested board short circuits the learning process itself. Using a manufactured board may help the reader learn about the HC11, but it really doesn't help much in learning about *designing with it.*

To emphasize the point, let us compare the purchase of a manufactured HC11 board with the purchase of a computer. If you buy and use an appliance-type computer that you just plug in and start using immediately, you can learn an enormous amount of things about *using* a computer. However, no matter how much you use the computer and learn to operate it, you still will know very little, if anything, about *designing* a computer. The same is true with the HC11. Since the aim of this book is to help you design new products, projects, gizmos, and toys using the HC11, it seems logical to have you build your own HC11 board.
4. And finally, I feel strongly that you shouldn't feel compelled to purchase a board or parts from one particular source. Because of this, the complete foil patterns for Mag-11 are provided in this book as well as on CD-ROM. Source code and object code for all firmware are also provided on the CD-ROM. Tips on making PC boards yourself are given in Appendix E.

Magicland's offer to sell PC boards, preprogrammed EPROMs, and even a complete kit is just that—an offer. There is no need to contact them—everything you need (except for actual parts, of course) is provided here either in the book or on the CD-ROM. This said, it must be admitted that the use of a professionally made PC board shortens Mag-11's assembly time by more than 90 percent and probably doubles its reliability. Nonetheless, I have proof that it isn't necessary to use a professionally made PC board for Mag-11—the original prototype of Mag-11 was a homemade board. The secret tips and hints, divulged in Appendix E, were developed through my own hands-on experience making prototypes of this and other boards.

A Quick Look at Mag-11 and Its Features

When I think about Mag-11 (Fig. 4-1), a tiny bit of electricity seems to tingle my fingertips. Perhaps this is so because I sometimes confuse gaining hands-on knowledge—and Mag-11 provides this to the fullest—with fun. However, is there really a difference?

As can be seen from the simplified schematic in Fig. 4-2, Mag-11 is a relatively complicated circuit. Mag-11's main purpose is as a learning tool. Mag-11's HC11 communicates with the outside world primarily through

- 9 LEDs, which are used as output data indicators
- 12 switches (one four-position dip-switch and one eight-position dip-switch), which are used as inputs
- 1 RS-232 interface, which is used as an input-output (I/O) device, and one thermistor, which is used to input temperature data

In addition, all pertinent MCU data, address, control, and I/O lines are accessible through four 20-pin sockets (also referred to as *headers*): J3, J4, J5, and J6. These sockets make it easy to do a lot of fun things with Mag-11 simply by sticking wires and/or leads from components into them. Many fascinating circuits can be wired with only a component or two. The leads of these components can be inserted directly into these 20-pin sockets. More complicated circuits require the addition of a solderless breadboard.

One fun thing we will do is make a voltmeter that reads out voltages in binary. Another simple but fascinating experiment will be a light meter that reads out footcandles in binary!

Other features of Mag-11 include

- Two types of firmware: M11DIAG2 firmware, which includes many built-in HC11 test routines as well as built-in binary readout thermometer, and the popular BUFFALO monitor, with a multitude of features such as a "cool" line assembler
- 1-MHz standard clock with 2-MHz capability
- 256 bytes of RAM with backup capability (located in the HC11) standard, and an additional 4 to 8 kB of RAM optional
- 512 bytes of EEPROM (located in the HC11)
- Four-channel, 8-bit A/D converter (located in the HC11)

(A)

FIGURE 4-1 (A) Mag-11 constructed with a professionally made double-sided printed circuit board plated through holes.

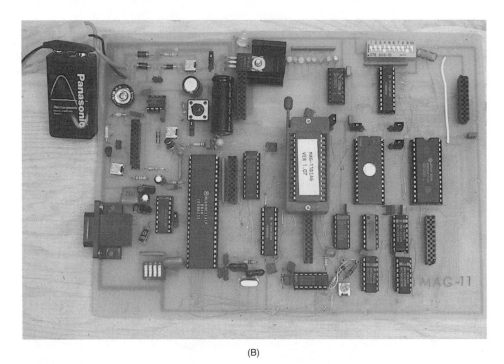

(B)

FIGURE 4-1 (B) The prototype of Mag-11 constructed with a hand-made single-sided board with many jumpers added.

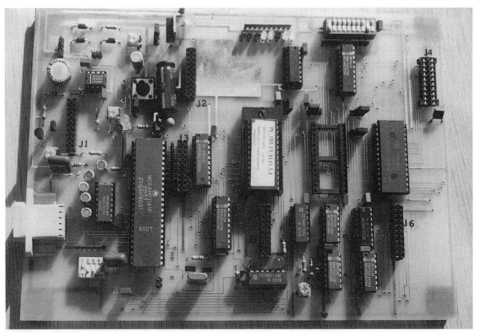

(C)

FIGURE 4-1 (C) Close-up of Mag-11.

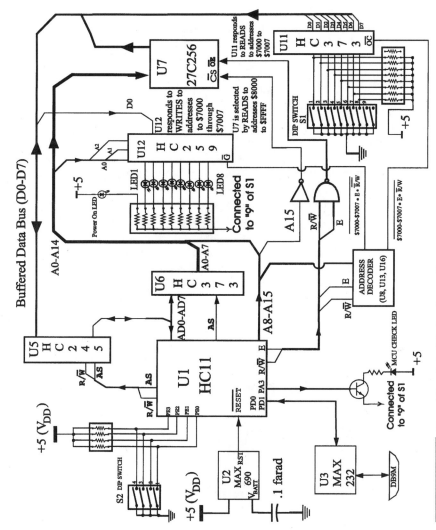

FIGURE 4-2 Mag-11's simplified schematic.

- Most of the HC11's other features, including input capture capability and several input and output ports
- A socket at U10 for an additional 27C256 EPROM (This feature allows users to write their own programs and still make use of the M11DIAG2 firmware's features.)
- Onboard supercapacitor for RAM backup
- MCU check LED
- Clock/reset fail LED indication
- Capability of operating in all four modes, including special test and special bootstrap modes
- Varied and easy-to-obtain power sources, including
  - 9-V alkaline or 7.2-V Ni-Cad battery
  - or 7 to 13 V at 100 mA unregulated dc
  - or 5 V at 70 mA regulated dc
- Connection for optional backup battery
- Expansion capability designed in

This summary of Mag-11's features may remind you of a sales pitch, and in a way, it is. Nonetheless, Mag-11 isn't the only HC11 board on the block. There are quite a few similar HC11 boards out there that are as good or better. For instance, Motorola itself sells several boards that help a designer who wants to base his or her design on the HC11. One such board is the M68HC11EVB evaluation board. At times, Motorola will sell this board at a specially low promotional price. If you can get your hands on it at a reasonable price, I highly recommend it. Whatever board(s) you get, make sure it (they) include the BUFFALO monitor. This monitor is not only extremely useful but is used incessantly throughout this book.

Let the Fun and Games Begin!— Mag-11's Construction

I will now take you step by step into this book's first major project: construction of a Mag-11, the single-board computer that takes you into the silicon depths of the HC11 microcontroller itself.

ABOUT MAG-11'S PC BOARD

Mag-11's solder-side foil pattern is shown in Fig. 4-3, while its component-side foil pattern is in Fig. 4-4. Both foil patterns for Mag-11 are also provided on the CD-ROM. You will need an HP laser-compatible printer to print the foil patterns out from the CD-ROM.

To run the foil printing program, type MG11FOIL at the DOS prompt. If you wish to use these foil patterns to make your own board, refer to Appendix E, which provides a proven method of creating your own boards. If you prefer to purchase a board or even a kit, refer to Appendix C, which gives a source for materials, including most circuit boards mentioned in the book. As mentioned in Appendix E, if you make the board yourself, it

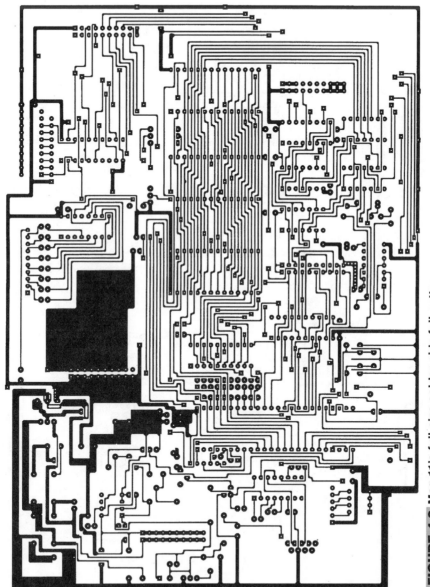

FIGURE 4-3 Mag-11's full-size solder-side foil pattern.

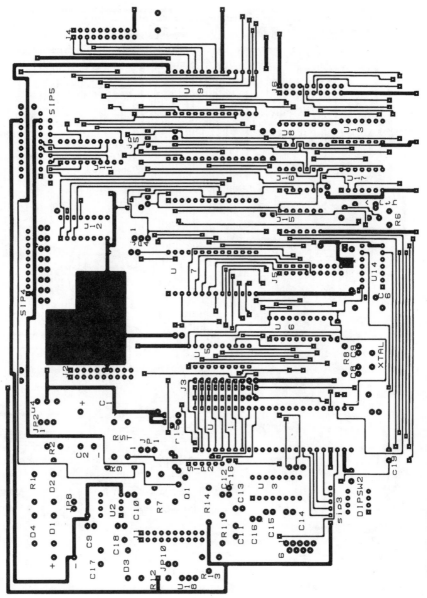

FIGURE 4-4 Mag-11's full-size component-side foil pattern.

will likely be a one-sided board, with foil on the component side being replaced by insulated jumper wires.

When constructing Mag-11, you also may want to refer to Fig. 4-1, which presents various views of a completed board. The complete schematic for the Mag-11 is not included on this book's printed pages because of its complexity. (If it were included, it wouldn't be readable!) However, the complete detailed schematic is provided on the CD-ROM.

To print Mag-11's schematic (or any other schematic referred to in this book but not printed in it) on a compatible printer, run the enclosed SCHEMATIC PRINTING program. To run this menu-driven program, type SCHEMAT at the DOS prompt. Also refer to the readme.now file on the CD-ROM for more details, including the printer drivers that are supported.

STUFFING THE BOARD

Follow the component mounting guide in Fig. 4-5 when stuffing the board with components. Also refer to Mag-11's parts list in Appendix D.

Use a heat sink for U4, which comes in a TO-220 package. It also is a good idea to use a 6-32 screw and nut to bolt down U4, with heat sink attached, to the circuit board, as shown in the figure. Use extreme care to avoid shorts. *Do not install an IC in a socket until specifically instructed to do so.*

It is not necessary to install a part on the Mag-11 that is listed as optional in the parts list. However, while most IC sockets are listed as optional, they are recommended. By the way, sockets at U7, U9, and U10 are required, *not optional.*

Note:

All LEDs used in the Mag-11 are of the high-efficiency, low-current type. While standard LEDs can be used, as long as the respective current-limiting resistor is reduced to 390 Ω, they are not recommended.

J1 through J6 are listed as 20-pin sockets in the parts list. These sockets are meant to mate with 20-pin 0.1-in matrix headers. This arrangement makes it simpler to expand the system to three layers. To use this method, all that is necessary is for the *first plug-in board* to use compatible 20-pin headers with pin lengths of 1 in or more on both sides of the header's insulator. *If the next expansion board has 20-pin sockets, it can be connected to the first expansion board without any additional wiring or connectors being necessary!* Figure 4-6 gives the connector pinout. If cables are to be used to connect the Mag-11 to expansion boards, or if it has been decided that only one layer of expansion board is necessary, J1 through J6 can be headers.

Note:

While this is the first and last time I refer to this expansion capability, you should be aware of this option.

If you examine Fig. 4-1 carefully, you will notice that the socket at U7 is of the 28-pin ZIF (zero insertion force) type. The ZIF socket facilitates the testing of Mag-11's firmware. Notice that the ZIF socket is "plugged into" a standard 28-pin IC socket. Thus

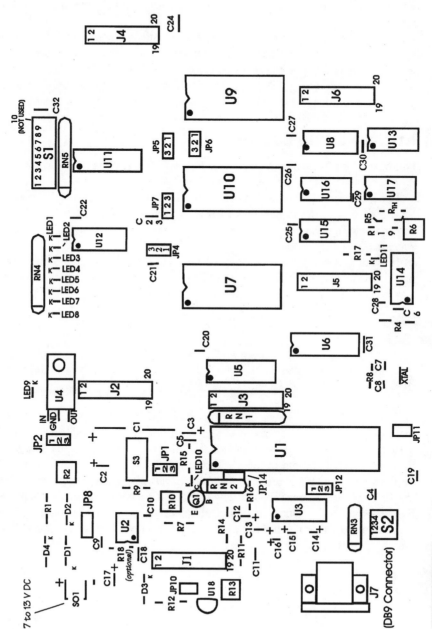

FIGURE 4-5 Mag-11's component mounting guide.

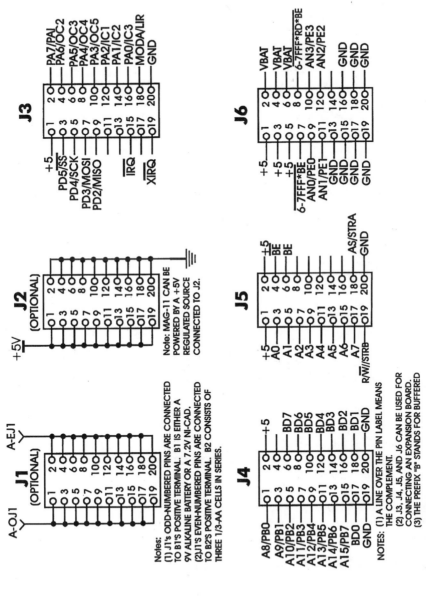

J3

+5	PA7/PA1
PD5/$\overline{SS}$	PA6/OC2
PD4/SCK	PA5/OC3
PD3/MOSI	PA4/OC4
PD2/MISO	PA3/OC5
	PA2/IC1
	PA1/IC2
$\overline{IRQ}$	PA0/IC3
	MODA/LIR
$\overline{XIRQ}$	GND

J6

+5	VBAT
+5	VBAT
+5	VBAT
6-7FFF*BE	6-7FFF*RD*BE
AN0/PE0	AN3/PE3
AN1/PE1	AN2/PE2
GND	GND
GND	GND
GND	GND
GND	GND

J2
(OPTIONAL)

Note: MAG-11 CAN BE POWERED BY A +5V REGULATED SOURCE CONNECTED TO J2.

J5

+5	+5
A0	BE
A1	BE
A2	
A3	
A4	
A5	
A6	
A7	AS/STRA
R/$\overline{W}$//STRB	GND

J1
(OPTIONAL)

A-OJ1

A-EJ1

Notes:
(1) J1's ODD-NUMBERED PINS ARE CONNECTED TO B1'S POSITIVE TERMINAL. B1 IS EITHER A 9V ALKALINE BATTERY OR A 7.2V NI-CAD.
(2) J1'S EVEN-NUMBERED PINS ARE CONNECTED TO B2'S POSITIVE TERMINAL. B2 CONSISTS OF THREE 1/3-AA CELLS IN SERIES.

J4

A8/PB0	+5
A9/PB1	BD7
A10/PB2	BD6
A11/PB3	BD5
A12/PB4	BD4
A13/PB5	BD3
A14/PB6	BD2
A15/PB7	BD1
BD0	
GND	GND

NOTES: (1) A LINE OVER THE PIN LABEL MEANS THE COMPLEMENT.
(2) J3, J4, J5, AND J6 CAN BE USED FOR CONNECTING AN EXPANSION BOARD.
(3) THE PREFIX "B" STANDS FOR BUFFERED

FIGURE 4-6 Mag-11's connector pinout.

it is apparent that this ZIF socket can be moved easily to U10 or even some totally different project without the need for an aggravating desoldering-soldering cycle.

INITIAL TESTS

Visually inspect the board for breaks or shorts in the foil. Also check for unsoldered or poorly soldered connections. Do not install ICs in their sockets yet. Connect the jumper at JP2 to positions 1 and 2; leave JP8 disconnected. Put S1(9) ON (position 9 of S1), and leave all other switches off. Connect a 9- to 12-V power source at 250 mA to SO1. (As an astute reader can tell from the schematic, there is a Schottky rectifier D4 in series with the positive line. This means that the Mag-11 will not shoot out fire and brimstone even if you connect the power the wrong way.) Ideally, to check out the new circuit, the power should come from a variable-voltage power supply with meters available to continuously monitor voltage *and current* (see Chap. 1).

Once all the initial testing has been completed, a standard power source can then be used, such as a 7- to 13-V dc "wall wart" or even a 9-V alkaline battery. The Mag-11 also can be powered directly with a +5-V regulated supply. Here, however, connect the power supply to J2 and *not* SO1 (see Fig. 4-6).

> **CAUTION:** If you connect a +5-V power source directly to Mag-11, watch out for polarity! By installing a +5-V power source directly, you are bypassing all built-in safeguards, and you can destroy most of Mag-11's components in several milliseconds.

Apply power. The power-on LED, LED9, should light—everything else should be off.

Connect a voltmeter's negative lead to a foil ground. Touch its positive lead to pin 48 of socket U1. The voltmeter should read between 4.5 and 5.5 V, a typical reading being +5.1 V. Similarly, measure the voltage at pin 28 of U7 and U10; pin 20 of U5, U6, and U11; pin 16 of U3, U8, and U12; pin 14 of U13, U14, U15, U16, and U17; and pin 2 of U2. These pins should be close to +5 V. If not, make the following tests:

1. Check the board again for continuity.
2. Touch the voltmeter's positive probe to pin 14 of U7, U9, and U10; pin 10 of U5, U6, and U11; pin 8 of U12, pins 5 and 8 of U8; pin 15 of U3; pin 7 of U13, U14, U15, U16, and U17; and pins 3 and 4 of U2. These should all be at 0 V. Anything below 0.02 V is considered 0 V here.

Remove the power source. If everything looks okay, proceed to "Initial Setup." If you experience trouble, don't fret. Instead, jump to the section "Having Trouble with Mag-11?"

INITIAL SETUP

Make sure the power is off. Install a 27128 or a 27C256 EPROM with M11DIAG2 firmware (see parts/software availability note in Appendix C) into socket U7. Be sure to use caution because EPROMs are extremely sensitive to damage from static electricity (see the note about handling static-sensitive ICs in Appendix E). Set jumpers as shown in Table 4-1.

TABLE 4-1	MAG-11'S RECOMMENDED JUMPER POSITIONS FOR INITIAL TESTS		
JUMPER	**POSITION**		
JP1	1-2		
JP2	1-2		
JP4	2-3		
JP5	1-2		
JP6	1-2		
JP7	1-2		
JP8	Leave jumper out		
JP10	Connect jumper		
JP11	Leave jumper out		
JP12	2-3		

Note: Refer to component mounting guide for locations.

Install the ICs in their respective sockets. Most ICs are sensitive to damage from static electricity. Standard precautions *must be taken* (see Appendix F for details). Place positions 1 and 9 of S1 ON [i.e., S1(1) and S1(9) ON.] All other switches should be in the OFF position.

TRYING IT OUT

First make sure that the ICs have been installed correctly in their sockets. Watch carefully the placement of pin 1. Refer to Fig. 4-5 as well as Fig. 4-1. As before, connect a 7- to 13-V dc source that can supply at least 250 mA to SO1. Watch polarity, although, as already mentioned, you won't see any blue smoke even if you connect the polarity backwards! Monitor carefully the current drawn by the circuit. If the circuit draws more than 150 mA, shut down the power immediately and check for shorts or improperly installed ICs.

As before, LED9 (the power-on LED) should light. With switch and jumper settings as indicated in the "Initial Setup" section, LED10 (the MCU check LED) should turn on for about 1 second, then off for about 1 second, and then on again *ad infinitum.* (For now, ignore LED1 through LED8.)

If LED10 does not perform as stated, observe LED11. If LED11 is lighted, or even flashes intermittently, the reset circuit may be defective, and U2 should be checked. Another possibility is a problem with the clock. In this case, use an oscilloscope to observe the enable signal at pin 27.

If LED11 does not light but LED10 is not performing as indicated (flashing on and off), remove power and go to the section "Having Trouble with Mag-11?" Otherwise, push the reset switch. If the Mag-11 is functioning properly, green LEDs LED2 and LED4 should light with all others off. These LEDs indicate that most circuits of the Mag-11 are functioning normally.

Having Trouble with Mag-11?

You can skip this section if everything is working properly.

1. If the green LED9 (power-on LED) does not light, check the power supply and the power supply's connections.
2. If red LED11 (bad reset LED) lights and *stays* lighted, there is a problem with the reset circuit. Check U2 and related circuit.
3. If LED10 (MCU check LED) flashes erratically (i.e., not at a steady rate), check the MCU's clock circuit. Here an oscilloscope will come in handy.
4. If red LED6 and/or red LED7 go on and stay on, the HC11 itself may be defective.
5. Check to make sure all ICs are inserted properly in their sockets. Watch out for bent IC pins; sometimes they are bent inward, making detection difficult.
6. If you still have no luck in getting the Mag-11 to perform up to snuff, check out the troubleshooting guide in Appendix F.

Before discussing the actual use of the Mag-11, let's look at how it works. Examining a proven design is an excellent way to get acquainted with a chip and a great starting point for your original design.

About Mag-11's Circuit

You may have heard the apparently true story of a man who called the technical support department at a large personal computer company. The man said to the soft-spoken representative:

CUSTOMER: "Listen, I just received this new computer from your company, and all I get is a blank screen."

REPRESENTATIVE: "Ah. Well sir, did you put the computer's and the monitor's power switch to the on position?"

CUSTOMER: "Sure did!"

REPRESENTATIVE: "Hmm. I assume you connected the monitor properly to the computer."

CUSTOMER: "Of course!"

REPRESENTATIVE: "Are you absolutely sure the power plug is inserted properly in the wall socket? Sometimes people don't stick it all the way in! And believe it or not, a friend of mine talked to a man complaining that his computer didn't work, and she found that the fella never plugged the computer in."

CUSTOMER (laughing): "Really? Well, I use a surge strip, and I think the strip's plug is connected properly in the wall socket, but I'll check anyway...hey Miss, it will take me a few seconds. I need to get a flashlight. All the lights in the building are out."

This dramatized and edited conversation brings up an important point: The most vital part of a working circuit is the power supply. Remember, a circuit, no matter how advanced or superior the design, will not work at all without power!

Note:

For the following discussions, refer to both the simplified schematic in Fig. 4-2 and the detailed schematic that can be printed from the CD-ROM.

Because of its importance, special effort was put into the design of the Mag-11's power-related circuit. Many options are available. Since Mag-11 uses an LM2931T low I/O DIF +5-V voltage regulator, any dc voltage, even one with a moderate ripple, can be used as long as it lies in the 7- to 13-V range and can supply 200 mA. This main power source is connected to SO1. (Note that tests indicate that the Mag-11 can be powered by a new 6-V lantern battery as well.)

Mag-11 has its own backup circuit, consisting of R1, R2, D1, D2, JP2, and JP8. All that is needed is a 7.2-V nickel-cadmium battery connected to the pins of J1. If a backup battery is installed, place jumper at JP8, and make sure the jumper at JP2 is between pins 1 and 2.

In an MCU/MPU circuit, the clock and reset circuits come next in importance. Because of its design, the HC11 requires a reset circuit that provides automatic voltage monitoring. According to the data sheet, the HC11's reset input must be pulled low whenever VCC (the +5-V power source) falls below 4.5 V. U2 (MAX690 MPU supervisory circuit) provides this critical service. Pin 7 of the MAX690 is an active low output with an internal 3-μA pull-up (probably a 1.5-MΩ resistance to +5 V). Typically, pin 7 goes low whenever VCC falls below 4.65 V. It goes high when VCC rises above 4.69 V for at least 50 ms. S3 provides manual reset.

U2's other function is to provide memory/backup battery switchover. Pin 8 is connected to the backup power source, and pin 1 is connected to the CMOS RAM that is to be backed up. Basically, U2 compares VCC with VBATT and connects pin 1 to whichever is higher. The output from U2 is connected to pin 28 of the 8-kB CMOS RAM U9, which is optional. Pin 1 is also connected, through R9, to jumper terminal block JP1. If the jumper at JP1 is in position 1-2 (normal expanded mode or normal single-chip mode), the backup voltage will power the 68HC11's internal RAM through pin 24 (MODB/VSTB pin). If the jumper is in position 2-3, then pin 24 is grounded, and the MCU will be in either special test mode or special bootstrap mode. More on the HC11's four operating modes later.

An onboard backup power source is provided by the 100,000-μF capacitor C17. Since its voltage rating is only 5.5 V, a 5.1-V zener diode D3 is connected across it for insurance purposes. The clock circuit is provided by XTAL1, C7, C8, and R8.

U18, along with associated components R12 and R13, provides an optional precision voltage reference for the HC11's internal analog-to-digital (A/D) converter. If U18 is not used, connect the jumper at JP10.

U3 is an MAX232 +5-V RS-232 driver/receiver. This RS-232 driver does not require a separate ±12-V power source to operate. It is connected to the HC11's port D PD0/RxD and PD1/TxD signals. U3's transmit line is connected to DB-9 pin 3, and its receive line is connected to pin 2. In addition, DB-9 pins 5 and 7 are connected to ground. This DB-9 connector is AT compatible. Keep in mind here that no handshaking is used in this circuit, and some systems require other pins to be active. However, eliminating this requirement through options in the communications software program is normally possible.

Notice that all LEDs, including LED9 the LED power-on indicator, are controlled by SW9 located in S1 [S1(9)]. With SW9 left open (off), the Mag-11 operates in its low-power "invisible" mode.

LED10, the MCU check LED, is connected in the collector circuit of Q1. Q1's base is connected to pin 5 of U1. Pin 5 of the 68HC11A1P is connected to bit 3 of port A (located at $1000). Thus writing to address $1000 with a xxxx1xxx (where x indicates a 1 *or* 0) will turn on LED 10. (LED10 is turned off by writing xxxx0xxx to $1000.)

Pin 6 of U1 is internally connected to bit 2 of port A. This pin functions in the Mag-11 as an edge-sensitive timer/input-capture (IC1) and is connected through C19, which along with a 10-kΩ resistance functions as a differentiator, to the output of a temperature-sensitive squarewave oscillator. This oscillator consists of three Schmitt-trigger inverters (U14) and a thermistor-resistor combination. Notice that the combination of parallel resistor R5 and series resistors R19 and R6 results in a network whose resistance varies inversely, fairly linearly, with temperature. The result is that as temperature increases, the period decreases nearly linearly. While in MPU/MCU-controlled circuits a linear response to temperature is not a requirement, it does simplify the software. After adjustment, the period at 0°F is 2503 μs (2503 cycles) with a 1-MHz clock. Each cycle corresponds to 1°F. Each *decrease* (in cycles) of the period length corresponds to an increase of 1°F. It can be seen that at a temperature of 100°F, the period length (in cycles) should be 2503 − 100, or 2403. The software converts the relative period length into a binary display of temperature in degrees Fahrenheit.

Mag-11 can be used in all four of HC11's operating modes. Of course, normal single-chip mode would be used only rarely because of its cost. In order for Mag-11 to use either its normal expanded or special test modes, it is necessary to demultiplex the low-order address lines and data lines. This is accomplished with a 74HC373 octal transparent latch at U6. Besides demultiplexing, U6 buffers the address lines A0 to A7.

Buffering for the bidirectional data lines is done by U5. U5's enable is connected to AS (address strobe line), which means the buses are isolated when AS is high (during the time address information is present on the multiplexed line). The R/W line is connected to pin 1, the direction control, of U5. This connection ensures that during read times data are flowing toward the MCU and during write times flowing away.

Note:

In most HC11 systems, there is no need for buffering. However, in Mag-11, which is meant as an HC11 experimenter's "dream board," buffering is used because one never knows what experimenters will do. Do we?

E, the enable signal, is inverted and buffered by an inverter (U15). Buffered lines in the schematic have a prefix "B" (e.g., "BE" means a buffered enable signal.)

U7, the primary EPROM, is selected whenever address line A15 is high. (A15 is inverted and connected to U7's CS pin.) When JP4's jumper is placed in position 1-2, HC11's address line A14 is connected to the EPROM. This results in U7 being located at memory addresses $8000 to $FFFF. With JP4 in position 2-3, A14 is eliminated, and U7 responds to two different addresses. For instance, addresses 1111 1111 1111 1111 and 1011 1111 1111 1111 are not distinguishable without address line A14. The result here is that U7 will respond to addresses $F800 through $FFFF *and also* $B800 through $BFFF. This rather odd decoding allows operation in the special test and special bootstrap modes because the reset vector for these modes is located at $BFFE through $BFFF. Since only 2 kB of the EPROM is used by M11DIAG2, U7 can be a 27128 EPROM as long as JP4 is set to the 2-3 position. U7's OE pin is activated when the R/W signal is high (MCU is in a read cycle) and E is high.

The optional 8-kB RAM U9 is decoded by U8. Except under unique situations, U9 will only respond to addresses $0100 through $0FFF and $1040 through $1FFF. Under normal circumstances (IRV bit 0 in the HPRIO register), it will not respond to addresses $0000 through $00FF and $1000 through $103F because these are addresses of internal RAM and internal registers. The external data bus is undriven during reads of internal addresses.

The decoding for optional EPROM/RAM U10 is fairly complicated. Basically, if jumpers are configured for a 32-kB EPROM, U8 and U13 do the primary decoding, with the result that U10 is located at $2000 through $6FFF. If jumpers are set for 32-kB RAM, decoding is simpler. Here, one can see that the CE (pin 20) of U10 is *deselected* when A15 is high (one input to the OR gate U16 is high). Also, one can see that whenever A14, A13, and A12 are all high, irrespective of A15 (address $7xxx), U10 is *deselected.* This means that U10 is selected for addresses $0000 through $6FFF (addresses $7xxx are left out). Of course, U10 will not respond to addresses 0 through $FF (internal RAM) and $1000 through $103F (internal registers).

U11, an octal transparent latch, provides the necessary control and buffering for S1. U11's OE output enable pin, which is an input, is connected to the output of a 2-in OR gate. The output of this gate is low *only* when both inputs are low. One input of this OR gate is connected to the output of an 8-in OR gate (U13). The output of U13 is low only when address lines A5 through A11 are low and A12 is high. The other input is connected to pin 7 of U8, which goes low when there is a read operation taking place on addresses $6000 through $7FFF. From this it can be seen that U11 responds to *only reads* of addresses 0111 0000 0000 0000 through 0111 0000 0001 1111 or, in hexadecimal, $7000 through $701F. In the software we will be writing, we will assume that U11 responds to $7000, although it will respond also to $7001, $7002,…, $701F. It can be seen that SW1 [S1(1)] controls bit 0 at address $7000, SW2 [S1(2)] controls bit 1 at address $7000, etc. *When the switch is closed, the corresponding bit is 0, not 1.* This fact is extraordinarily important when writing programs concerning these switches.

U12, an OCTAL bus transceiver, provides control and buffering for LED1 through LED8. The decoding scheme is similar to U11. However, *U12 only responds to writes* to addresses $7000 through $71FF, *not to reads.* Notice that only the least significant bit of buffered data line (BD0) is connected to U12. Also notice that pins A0, A1, and A2 of U12 are connected to the respective address on the address bus. Thus LED1 is located at address $7000 (as well as $7008), LED2 at address $7001, LED3 at address $7002, etc. To turn on LED1, write a $01 to address $7000; to turn on LED2, write a $01 to address $7001; etc. Of course, writing a $FF also will turn on the respective LED. *Quick question:* What do you write to the address to turn the LED off? *Hints:* If you were on a rocket ship to the red giant Arcturus, which is 36 light-years from earth, how many seconds, measured with your stopwatch, would it take to get there if the rocket traveled at the speed of light? Or, for you down-to-earth folks, what is the total number of gallons of gasoline in seven empty 20-gallon gas tanks?

For a pictorial view of Mag-11's memory map, see Fig. 4-7. Figure 4-8 gives Mag-11's jumper placement guide.

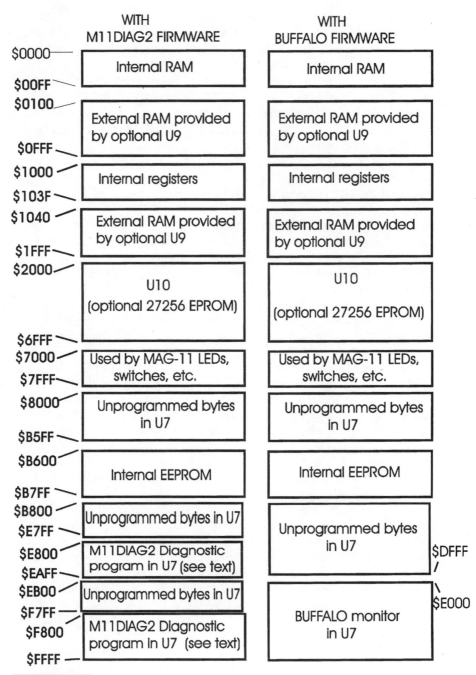

FIGURE 4-7 Mag-11's memory map.

Note: Refer to Mag-11's component mounting guide and Mag-11's schematic for the following discussion.

JP1   Place jumper between pins 1 and 2 for normal expanded multiplex or normal single-chip mode. Place jumper between pins 2 and 3 for special test mode or special bootstrap mode.

JP2   Place jumper between pins 1 and 2 if no B1 battery is installed or if B1 is a Ni-Cad or other rechargeable battery (the most common circumstance) and between pins 2 and 3 if B1 is a 9-V alkaline battery (very rare).

JP3   Not used (was used in prototype)

JP4   Place jumper between pins 1 and 2 for normal modes. Place jumper between pins 2 and 3 if U7 is a 27128 or if you want to operate the HC11 in the special test mode. Notes concerning JP4:
1. When pins 1 and 2 are jumped, U7 is completely decoded between $8000 and $FFFF.
2. When pins 2 and 3 are jumped, U7 is only partially decoded (also called redundant mapping and mirroring) and will respond to addresses $8000 to $BFFF just as it responds to addresses $C000 to $FFFF. Thus the special test mode's reset vector, $BFFE and $BFFF, will be accessed even when the program thinks the reset vector is at $FFFE and $FFFF.

JP5, JP6, JP7   Place jumper between pins 1 and 2 if U10 is a 27C256 EPROM or between pins 2 and 3 if U10 is a 32-kB RAM.

JP8   Normally do *not* install jumper here. Only install this jumper if B1 is a Ni-Cad or other rechargeable battery. If B1 is an alkaline or other primary-type battery, problems can arise if this jumper is installed. (The reason this is so is that a charging current is applied when JP8 is installed.)

JP9   Not used (was used in prototype)

JP10   Only install this jumper if U18 (LM336-5.0) is *not* used.

JP11   Install this jumper for special bootstrap or single-chip mode. Do *not* install for normal extended multiplex or special test modes.

JP12   Place jumper between pins 1 and 2 if Mag-11 is in special bootstrap mode *and* you want to jump directly to EEPROM at reset. Otherwise, place jumper between pins 2 and 3 (normal placement).

JP13   Not used (was used in prototype)

JP14   Jumper if BUFFALO monitor is used. Do *not* install if XIRQ is to be used.

FIGURE 4-8 Mag-11's jumper placement guide.

A Simplified Look at Mag-11's Built-In "Outside World" Interface

Designed into the basic Mag-11 are 9 low-power LEDs and 12 switches that allow the user to communicate with Mag-11. Since this is a simple interface, it is easy to understand. It is also versatile. The following discussion will help a potential programmer make use of this interface.

LED10   This is called the *MCU check LED* in that it is controlled directly by the MCU without the need for address decoding and thus can be used to check if the MCU is operating properly. LED10 is lighted when bit 3 of port A is set. Thus storing xxxx1xxx (e.g., LDAA #$08: STAA $1000) at address $1000 will turn on LED10. LED10 is turned off by storing xxxx0xxx (e.g., 00) at $1000.

LED1 through LED8   These LEDs are controlled by U12, a 74HC259 8-bit addressable latch. Here, only the first bit (D0) of the data bus comes into play. Each LED has a specific address:

```
LED1 <> $7000
LED2 <> $7001
LED3 <> $7002
LED4 <> $7003
LED5 <> $7004
LED6 <> $7005
LED7 <> $7006
LED8 <> $7007
```

To turn on an LED, simply store a number, with bit 0 set (i.e., a "1" in bit 0) at the respective address. For instance, storing 10110001 ($B1) at $7004 will turn on LED5. (Storing $B0 at $7004 will turn off LED5.)

DIP switch S1   The first eight switches in S1 are accessed by a read to address $7000. (Note that LED1 is only accessed by a write to this address.) Switch position 9 of S1 merely provides power to the LEDs, and position 10 is not currently used. Each switch position affects one particular bit of the data at address $7000:

```
Position 1 <> bit 0 (D0)
Position 2 <> bit 1 (D1)
Position 3 <> bit 2 (D2)
Position 4 <> bit 3 (D3)
Position 5 <> bit 4 (D4)
Position 6 <> bit 5 (D5)
Position 7 <> bit 6 (D6)
Position 8 <> bit 7 (D7)
```

Actually, the switch positions affect BDx, where the prefix "B" stands for "buffered." As you can tell from the schematic, when a switch is on, the corresponding bit is cleared (i.e., "0"). For instance, if a read to address $7000 results in the data 01101000, this means that switch positions 1, 2, 3, 5, and 8 are on, with positions 4, 6, and 7 off.

DIP switch S2 This four-position switch is connected to the first "nibble" (lowest 4 bits) of port E of the 68HC11. The second nibble (highest 4 bits) is not used in the 48-pin DIP for the simple fact there are not enough pins. Since port E is located at $100A, this switch is accessed by a simple read of port E. For example, if all positions of S2 are turned on, a read to $100A (e.g., LDAA $100A) will result in the data xxxx0000.

Tip:

To use this switch, it is best that the A/D converter not be turned on. Unless you purposely write to the option register at $1039, you do not have to be concerned.

Using M11DIAG2 Firmware

Let's look at using M11DIAG2 firmware to test out the MC68HC11A1P and get a hands-on feel for the HC11. For this discussion, refer to the component mounting guide in Fig. 4-5 and M11DIAG2's flowchart in Figs. 4-9 and 4-10. However, before we use the M11DIAG2 firmware to do interesting things, let's look at probably the most important non-CPU register in an HC11—the CONFIG register.

Let us start our description of the CONFIG register by quoting directly from Motorola's 500+ page book on the HC11, *The HC11 Reference Manual.* I will refer to this manual frequently in this book as the "HC11 bible." (This manual, as well as the *HC11 Technical Data Booklet,* can be obtained directly from Motorola. Refer to Appendix G for details on obtaining these and other HC11 data sheets, application notes, related Web sites, etc.) According to the HC11 bible, "The CONFIG register is an *unusual* control register used to enable or disable ROM, EEPROM...." Unusual? Synonyms for *unusual* include *rare, strange, fascinating,* and *mysterious.* Actually, all these synonyms fit the CONFIG register quite well.

"What's all the fuss about the CONFIG register?" you may ask. Well, the control bits in the CONFIG register are like mask-programmed options that users can program themselves. In other words, you sort of custom design the HC11 yourself.

For example, perhaps you are an insecure type who's afraid of someone stealing your unique, trail-blazing program. If so, you will no doubt want to enable the security feature by clearing bit 3 of the CONFIG register. Or maybe you are a perfectionist and, as far as humanly possible, you don't want anything to go wrong with a gadget you designed. Ever! If this fits your personality, you will likely want to enable the COP watchdog system by clearing bit 2. Or how about this! You have designed a superprogram and burned it into a 27512 EPROM that requires every usable address space possible. If this is the case, you probably will want to disable the on-chip EEPROM. Simple. Make sure bit 0 of the CONFIG register is cleared. Or maybe you are a bit easy-going, like me, and you love the built-in EEPROM and do not want the hassle that NOCOP and the security features give you. No problem here either. Just make sure bits 0, 2, and 3 are set.

So what's so unusual and even mysterious here? Well, the CONFIG register is not mask-programmed by the manufacturer, and while you can read it like any other register, you cannot write to it with a simple STORE instruction. However, as you would expect with a

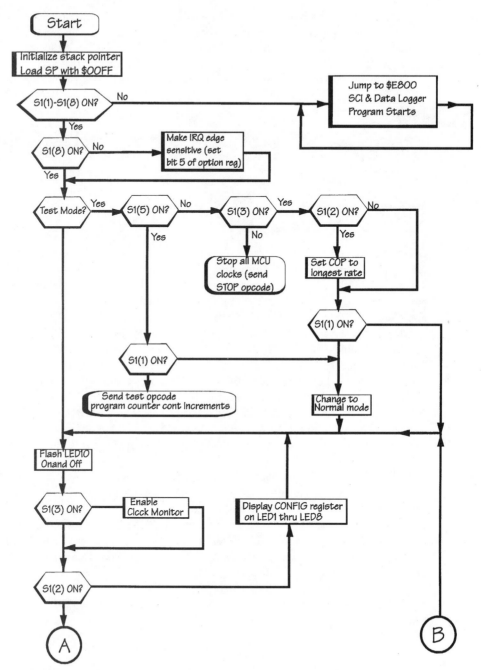

FIGURE 4-9 First page of flowchart for M11DIAG2 firmware.

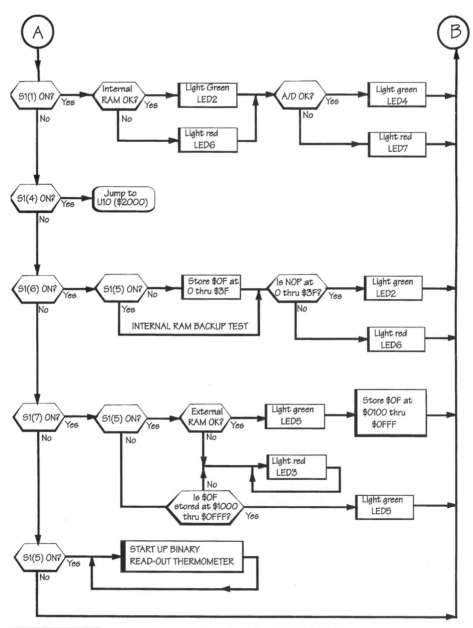

FIGURE 4-10 Second page of flowchart for M11DIAG2 firmware.

mask-programmed register, the contents of the CONFIG register will likely remain around longer than you will—unless you decide to change it, that is. The CONFIG register, you see, is EEPROM-based. To change this EEPROM byte, a special procedure is necessary that is similar to the procedure to change the HC11's 512-byte EEPROM itself. More on writing to the CONFIG register later.

BRIEF CONFIG REGISTER FACT SUMMARY

Type: EEPROM

Location: $103F

For a pictorial look of the CONFIG register, check out Fig. 4-11.

Description of bits

Bit 3: NOSEC   If this bit is cleared (i.e., "0"), the security feature is enabled. Note that this special security feature must be requested at the time a user submits a mask ROM pattern. For regular guys and gals like us, make sure this feature is disabled, i.e., bit 3 is set "1."

Bit 2: NOCOP   If this bit is cleared, the COP system is enabled. Generally, this bit should be set. There are easier ways to increase reliability. Perhaps the simplest method here is using the illegal opcode trap. This will be explained later in this chapter.

Bit 1: ROMON   If this bit is cleared, the 8-kB ROM is disabled. Unless you plan to write a program larger than 2 kB *and* you plan to produce thousands and thousands of your design, make sure that this bit is set so that the 8-kB ROM is disabled. Remember, even so-called ROM-less parts (such as the MC68HC11A1P, which is used by Mag-11) probably have a ROM inside.

Bit 0: EEON   If this bit is cleared, the 512-byte EEPROM is disabled. *Make sure this bit is set so that you can use the EEPROM.*

Note that with most members of the HC11 family, the 4 most significant bits of the CONFIG register (bits 7, 6, 5, and 4) are cleared (i.e., "0"). However, with the HC811 series, these 4 bits are used to set the location of the 2-kB EEPROM. For instance, with these 4 bits set, the EEPROM is located at $F800 through $FFFF—a very, very popular setting.

DISPLAY CONFIG REGISTER

Now that you have some idea what the CONFIG register is all about, let's have some fun and look inside the CONFIG register of the HC11 used in the Mag-11.

Register Name: CONFIG					Register Address: $103F			
BIT#	7	6	5	4	3	2	1	0
BIT NAME	—	—	—	—	NOSEC	NOCOP	ROMON	EEON
BIT'S STATUS FOR TYPICAL 68HC11A1	0	0	0	0	1	1	0	1

Note: Some HC11 derivatives (e.g., HC811) make use of bits 4 through 7.

FIGURE 4-11 The HC11's CONFIG register.

With jumpers set to the positions listed in the "Initial Setup" section, set S1(2) and S1(9) to their on positions; leave all other switches off. Not only should the MCU check LED (LED10) slowly flash on and off, but with the MC68HC11A1P as shipped, LED1, LED3, and LED4 also should go on, after a second or so. All other data-indicating LEDs should be off. This corresponds to the binary number 00001101 (0D in hexadecimal) being preprogrammed by the manufacturer into the CONFIG register. Looking back, we can see that this means that our HC11's security feature, COP watchdog feature, and ROM are all disabled. Happily, however, the EEPROM is enabled. This is the normal configuration.

TEST INTERNAL RAM AND A/D CONVERTER

Mag-11's HC11 has a 256-byte RAM and a 4-channel 8-bit A/D converter. To test these systems to make sure they are functioning properly, set S1(1) and S1(9) to their on positions; all other DIP switches should be off. If the internal RAM is functioning, the green LED2 should light; if not, red LED6 will light. If the A/D converter is operating properly, green LED4 will light; if not, red LED7 will light.

TEST INTERNAL RAM BACKUP

The preceding test checked every bit of internal RAM at every unused address. However, it did not check for memory retention when the power supply was removed. This test routine stores $0F at memory addresses from 0 to $3F and then checks to make sure locations 0 through $3F have $0F stored at them. This is a two-phase test:

Phase 1   With power on, switch S1(6) and S1(9) on, leave all other switches off. Green LED2 should light, indicating that locations 0 through $3F have $0F. If an error was detected, red LED6 will light.

Phase 2   Shut off power. Now switch S1(6) on, and leave S1(6) and S1(9) on. Wait at least 2 minutes. Turn the power on. If memory was retained, green LED2 will light. If an error or errors were detected, red LED6 will light.

SUPERCAPACITOR BACKUP—MY EXPERIENCE

I tried to determine by actual tests the length of time a $\frac{1}{10}$-F supercapacitor could supply enough energy to keep the internal RAM from losing data. Apparently, it is a long time, since it retained data for over a day. This is substantially longer than specs indicate. By the way, a momentary touch of a screwdriver at the supercapacitors terminals caused instant memory loss.

TEST EXTERNAL RAM AT U9

First, try this test with socket U9 empty. Switch S1(7) and S1(9) on, and leave all other switches off. Red LED3 should light, indicating bad or missing RAM at U9. Now turn the power off, and carefully install a 6264LP-15 8-kB RAM (or equal) in socket U9. Turn the power on. Because of the large number of bits being tested, it will take a few minutes for the test to be completed. If the IC is good, green LED5 should light; if bad, red LED3 will again light.

To check U9's backup capability, turn the power off, and then switch S1(5) on. Wait at least 2 minutes, and then switch the power on again. If backup was successful, green LED5 should light; if an error was detected, red LED3 will light.

ADDING YOUR OWN UNIQUE, TRAIL-BLAZING PROGRAM

One of the Mag-11's most delightful features is that it allows an HC11 dilettante easily to get into programming a 68HC11 in assembly or even machine language. Most of the basic, but often confusing, housekeeping chores, including loading the stack pointer and setting up critical interrupt routines, have been accomplished already by M11DIAG2 firmware. This feature leaves the pedestrian details to M11DIAG2 so that the user can get right down to writing an electrifying program.

To use this feature, you require a 27C256 EPROM that has a usable program that starts at location $2000. As explained earlier, this is accomplished in assembly language with the directive "ORG $2000." After programming the EPROM, install the EPROM at socket U10. Make sure that the power is off before installing the chip. Also make sure that the jumpers at JP5, JP6, and JP7 are all at position 1-2.

To run the program, simply switch S1(4) on and make sure that S1(1) and S1(2) are off. So that the LEDs will light, make sure that S1(9) is also on. Now press the RESET switch. In a second or so, if everything is done properly, *your program will take over Mag-11's mind!* Also, LED1 and LED8 will light. Now look at red LED6. If this red LED lights, it indicates that an illegal opcode has been detected—in other words, something is screwy—your program or U10 itself.

This brings up the real reason that I mentioned placing an EPROM at U10 so early in the book—even before I elaborated on assembly language. This discussion on the installation of an EPROM at U10 gently but concretely introduces one of the really neat, excitingly useful, yet simple features of the HC11—the *illegal opcode trap.*

THE ILLEGAL OPCODE TRAP

To start the discussion of the illegal opcode trap, let me pose a question: If you do not install an EPROM at U10 but switch S1(4) on and make sure that S1(1) and S1(2) are off, do you think the red LED6 will light? In other words, will LED6 (Mag-11's illegal opcode indicator) light if Mag-11 thinks that you, the user, have inserted an EPROM in U10 with a bug-free program embedded deep within its silicon and goes right ahead and starts to run that program despite the fact that there is no program to run? Will LED6 light or won't it? If you have access to a Mag-11 with M11DIAG2 firmware, try this experiment out and find out for yourself. If you do not try it out, the answer as to whether LED6 (the illegal opcode indicator) will or will not light is spelled out in the little puzzle: The answer is the antithesis of the German word that sounds like "nine."

Now that you know that the red LED6 (illegal opcode indicator) does light when U10 is not installed, you probably want to know why. Well, it relates to two facts:

1. The HC11 issues a nonmaskable interrupt that points to the illegal opcode vector when it is issued an illegal opcode.
2. M11DIAG2 firmware has a routine that turns on LED6 when this particular type of interrupt occurs.

"But, but," you may ask hesitatingly, "what is an illegal opcode?" It is nothing more than a combination of opcode bytes that are not defined by the HC11's CPU. An illegal opcode is to an HC11 what the word *dfganittzzich* is to you. In other words, its something that does not make sense to it. When the CPU receives this confusing instruction, it causes an interrupt—an interrupt that cannot be stopped! This interrupt causes the program to jump to the memory location stored at the illegal opcode vector that is at $FFF8 and $FFF9. In M11DIAG2 firmware, $FF is stored at $FFF8 and $94 is stored at $FFF9. This simply means that when the HC11's CPU fetches an illegal opcode (an opcode that is not defined), it will interrupt its normal program and start looking for the interrupt routine stored at the illegal opcode vector. If the HC11 is running M11DIAG2 firmware, $FF94 is stored there, so the program will jump to memory location $FF94 and start executing the interrupt routine at this address. This is somewhat similar to what happens when a system reset occurs. The primary difference is that the normal reset vectors, you will recall, are located at $FFFE and $FFFF and not $FFF8 and $FFF9.

Now that we have worked our way, in a zigzag path, to M11DIAG2's illegal opcode interrupt routine that starts at $FF94, what does this routine do? (You can look at Fig. 4-12 now if you wish.) First, it loads the stack pointer with $FF. If this sounds familiar, it is. This is the first thing that our first trivial HC11 programs did after receiving a RESET interrupt. The next thing this routine does is turn on LED6 so that we know an illegal opcode was fetched. It does this by loading accumulator A with $01 and then storing it at address $7005, which happens to be the address for LED6. The next three instructions are the "no operation" opcode; this in machine language is simply 01, with NOP being the assembly language mnemonic.

As indicated, the NOP instruction does nothing; it only takes up space and uses up two clock cycles of time. Its questionable purpose here is to provide a space for a future JUMP (JMP) or JUMP TO SUBROUTINE (JSR) instruction, which makes it simple for someone to modify this interrupt routine.

This interrupt routine ends by jumping (using the JMP instruction) to the beginning of the main program at $F800. Notice that $F800 is the same address pointed to by the big-shot interrupt vector—the RESET vector, which we looked at in detail in Chapter 2.

This illegal opcode vector brings up another point: All HC11 interrupts have a similar vector that only differs in location. For instance, the illegal opcode's vector is located at $FFF8 and $FFF9, the COP FAIL vector is at $FFFA and $FFFB, and the clock monitor vector is at $FFFC and $FFFD. These are just four of the vectors used by the HC11; there is a total of 21. Twenty-one vectors. That's a lot! Trust me!

Shown in Fig. 4-12 is a short, modified assembly-language listing segment of M11-DIAG2's firmware that shows 4 of the 21 interrupt vector assignments. This listing actually was created by the assembler and provides machine language besides the source code, which was written in assembly language. This type of listing is called, straightforwardly enough, the *assembler listing*.

Note:

The assembler listing is optional and is used primarily for discovering software errors (debugging). The primary purpose of assemblers is to provide usable object code (e.g., a file in MOT S19 format).

```
*Gives interrupt routines for Illegal Opcode trap,
*Cop Interrupt and clock monitor interrupt
*...............
0531   ff94                         ORG    $FF94
0532   ff94   8e  00   ff           LDS    #$00FF
0533   ff97   86  01                LDAA   #$01
0534   ff99   b7  70   05           STAA   LED6          ILLEGAL OPCODE
0535   ff9c   01                    NOP                  LEAVE ROOM
0536   ff9d   01                    NOP                  FOR JMP OPCODE
0537   ff9e   01                    NOP                  IF DESIRED
0538   ff9f   7e  f8   00           JMP    $F800
0539   ffa2   8e  00   ff           LDS    #$00FF
0540   ffa5   86  01                LDAA   #$01
0541   ffa7   b7  70   06           STAA   LED7          COP FAILED
0542   ffaa   01                    NOP                  LEAVE ROOM FOR
0543   ffab   01                    NOP                  JMP OR JSR OPCODE
0544   ffac   01                    NOP                  IF DESIRED
0545   ffad   7e  f8   00           JMP    $F800
0546   ffb0   8e  00   ff           LDS    #$00FF
0547   ffb3   86  01                LDAA   #$01
0548   ffb5   b7  70   07           STAA   LED8          CLOCK FAILED
0549   ffb8   01                    NOP                  LEAVE ROOM
0550   ffb9   01                    NOP                  FOR JMP OPCODE
0551   ffba   01                    NOP                  IF DESIRED
0552   ffbb   7e  f8   00           JMP    $F800
............
*VECTORS
*...........
*...........
                                    ORG    $FFF8
0574   fff8   ff  94                FDB    $FF94         ILLEGAL OPCODE
0575   fffa   ff  a2                FDB    $FFA2         COP FAIL
0576   fffc   ff  b0                FDB    $FFB0
*                                   END
```

FIGURE 4-12 Partial assembler listing of M11DIAG2's firmware.

In an assembler listing, the first number is the line number, which is assigned by the assembler. Notice that this first number, 0531, means that there were 530 lines before it.

Next in the listing is the hexadecimal address of the first byte of the opcode (e.g., ff94). Then comes the opcode or data (e.g., 8e 00 ff). To the right of this is the assembly-language program source code itself (e.g., LDS #$00FF). To the extreme right are the programmer's optional comments written in a stunted pseudo-English.

Tip:

The assembler ignores everything on the line if an asterisk *starts* that line.

It is probably obvious by now that I am deeply fond of the illegal opcode trap. It is ingeniously simple yet moderately effective. As far as I know, it is the first Motorola chip to use this convenient way of taming the "wild, wild program syndrome" mentioned earlier. While the use of the illegal opcode trap is far from being foolproof in eliminating this nasty syndrome, it often is good enough for nonmedical, noncritical consumer uses. Military, space, aviation, and medical circuits should *not* rely on it, however. *Critical applications should make use of additional safeguards such as the HC11's COP watchdog system.*

To make use of the illegal opcode trap, you must properly initialize the illegal opcode interrupt vector. Often, all you need do is store the same address at this vector that you do at the reset vector. What happens then is that if the wild, wild program syndrome slams your design, an illegal opcode will likely be fetched by the CPU, and this will cause an effective RESET because the program will start over at the beginning.

The illegal opcode trap, along with a properly initialized illegal opcode vector and a functioning COP watchdog system, can be looked on as providing the same function as a sharp human observer who keeps an eye on the HC11-designed product, and if an error is detected, he or she shuts the gadget's power off for a second and then turns it back on to get things operating properly again.

Adding Even More Reliability—
The Clock Monitor Reset

There is another simple feature you can use to make your design even more reliable—the *clock monitor reset function,* or *CMR* for short. The idea here relates to a stopped clock. If for some reason the clock stops, neither the illegal opcode trap nor the HC11's COP watchdog system can get things going. However, the CMR may be able to save the day. Notice that the CMR differs in several respects from the IOT (illegal opcode trap). First, IOT is always in effect—i.e., it does not have to be enabled—but the CMR must be. To enable the CMR, bit 3 (the CME bit) of the OPTION register located at $1039 is set. (One way, but not the best way, to do this is to store %00001000 at $1039. This is done with the two instructions: LDAA #%00001000 and STAA $1039. A better way is to use the BSET instruction and just change bit 3.) By the way, M11DIAG2 firmware enables the CMR as long as S1(3) is *off.*

CMR's other feature, which differs from the IOT, is that while the IOT's vector must be initialized—if it isn't, all hell breaks loose when an illegal opcode is detected—the CMR initializes a reset routine so that the program automatically goes to the start of the program (i.e., the same place pointed to by the reset vector).

A sharp reader may be confused here in that I earlier mentioned that there was a clock monitor interrupt vector at $FFFC and $FFFD. "If the HC11 takes care of CMR problems automatically," one might ask, "then why, oh why a CMR vector?" Well, this vector is used primarily so that the firmware can determine the cause of the reset. It often is good to know where the problem originated. For instance, M11DIAG2 firmware lights LED8 to indicate that a CMR has occurred.

Tip:

Don't ever, ever enable the CMR if the HC11's clock is slower than 200 kHz! While everything may be copacetic even if you enable the CMR with a 32.68-kHz clock, there is roughly a 50-50 chance that your system will be on futile life support!

You may want to say: "The CMR doesn't seem that useful. If the clock itself fails, the whole HC11 system is likely shot, and a simple circuit reset will not be enough to fix it. It seems," you continue, "that the CMR is just an added frill with no practical use!" Well, to be honest, when I first studied the HC11 data sheet in the late 1980s, I had this same impression. However, if you look more deeply into how an HC11 operates, you will notice that little old instruction with the descriptive mnemonic called STOP. What this instruction does is simple: It *stops* the clock. It is used primarily to save power so that the HC11's RAM contents will be saved with the power remaining in the system. This is a neat feature—but it is a potentially dangerous instruction as well. The machine code that the HC11 recognizes as the STOP instruction is $CF. What if, just what if, some random pulse, perhaps from some static charge, occurs just as the CPU is loading the load index register X instruction, $CE in machine language, and the CPU mistakenly thinks it was a $CF, which is the STOP opcode ($CF = 11001111 and $CE = 11001110)? Well, the clock would stop, which means that neither the illegal opcode trap nor the COP watchdog system would be able to save the day. The HC11, and whatever it was controlling, would just sit there. The designers of the HC11 apparently had enough forethought to see this possibility. (These designers possibly experienced this problem with the HC05 and came up with ways of avoiding it with the HC11.) First, the Motorola designers stuck in a stop disable bit in the CCR (condition code register) (bit 7). Once this bit is set, the STOP instruction is disallowed, and the STOP instruction is treated like a no-operation (NOP) instruction. Second, they came up with the concept of the CMR. If somehow the CPU comes across an erroneous STOP instruction (the exact reason is not important here), the CMR will issue a system RESET, and the HC11 will again start operating.

Now you are probably asking the question: "This means that once the CMR function is enabled, the STOP instruction cannot be used, right?" Wrong! All you have to do is immediately before issuing a STOP instruction clear bit 3 of the option register (the CME) bit. This disables the CMR so that the STOP instruction can proceed normally. After recovery from stop, make sure the CME bit is set so that the CMR will operate properly. Practically speaking, this is accomplished simply by setting the option register's CME bit (bit 3) near the beginning of the program.

More about the STOP Instruction

I hope the preceding discussion has whetted your curiosity about the STOP instruction. If so, read on. If not, you are allowed to jump to the next section.

The STOP instruction's purpose is to reduce the HC11's power consumption to an absolute minimum. The way it does this is by stopping all clocks. If you are at all acquainted with CMOS technology (and the HCMOS technology used by the HC11 is very similar), you realize that, theoretically at least, the only power consumed by CMOS circuits occurs when a changing voltage is applied. The faster the voltage changes, the more power it takes. Theoretically at least, no energy is used when a steady dc supply is applied. Practically speaking, of course, there is some dc leakage current always around. However, it is usually so small that it is often difficult to measure. Because of this, when the HC11's clock is stopped, current consumption by the HC11 is just a milliampere or two. This power reduction allows the retention of RAM memory for some time. In other words, the STOP command is an easy way to retain RAM contents using primarily software and the energy stored in the power circuit's total capacitance.

By the way, you may want to ask the following question: "You showed how to stop the HC11's clocks, but how do you get them going again?" Well, this usually occurs with a system reset or a simple power up, assuming that the STOP instruction was issued because the power failed. However, the clocks also can be started again with an IRQ or an XIRQ interrupt. The processing of the XIRQ interrupt vector depends on the state of the CCR's X bit. If X is "1," everything is normal, with the clocks starting and the CPU pointing to the XIRQ vector just like it does every time the HC11 receives an XIRQ request. If X is cleared, the clocks still start, but now the XIRQ vector is ignored and the processing simply resumes with the next opcode after the STOP opcode.

More Fun—A Built-In Experiment that Demonstrates the HC11's Input Capture System (Using Mag-11's Binary Thermometer)

To work this experiment, simply switch S1(5) and S1(9) on; all other switches should be off. Place the sensing end of an accurate thermometer next to the thermistor RTH. Adjust R6 so that LEDs LED1 through LED8 indicate, in binary, the temperature in degrees Fahrenheit. LED1 displays bit 0, LED2 displays bit 1,..., and LED8 displays bit 7 of the temperature measured in the binary system. If your familiarity with binary numbers is limited, refer again to Appendix A.

The range in decimal of this thermometer is 0 through 255°F. Below 0°F, LED1 will flash slowly. Above 255°F, LED8 will flash slowly. If calibrated between 50 and 80°F, the accuracy is roughly ±3°F between 32 and 100°F. While practical uses of this circuit are rather limited, if RTH is in intimate contact with U1, it can serve as a monitor for the MCU's temperature. U1 is rated for a maximum temperature of 185°F, which means that if LED8, LED6, LED5, and LED4 are on, the temperature of U1 is approaching its maximum rating.

Referring to Fig. 4-13, if LED7 and LED4 are on and all other data readout LEDs are off, then the thermometer would indicate a temperature of 01001000°F in binary or (64 + 8) 72°F in decimal. Another example: If LED7, LED6, and LED3 only are on, then the

bit 7	bit 6	bit 5	bit 4	bit 3	bit 2	bit 1	bit 0
LED8 0 (128)	LED7 0 (64)	LED6 0 (32)	LED5 0 (16)	LED4 0 (8)	LED3 0 (4)	LED2 0 (2)	LED1 0 (1)

Note: Decimal values of LED are shown in parentheses.

FIGURE 4-13 Mag-11's LED binary readout.

temperature would be 01100100°F in binary or (64 + 32 + 4) 100°F in decimal. What if only LED6 is on? Pretty obviously, the temperature is at the freezing point of water, or 32°F. Which LEDs would be lighted at a temperature of 80°F? (*Answer:* LED7 and LED5.)

Introducing HC11's Programmable Timer

This binary readout thermometer provides a quick and easy way to check out the HC11's sophisticated programmable timer. It also is an aid to understanding how the timer's input capture feature performs.

What does the HC11's timer do? A lot. Things like input capture, output compare capability, a real-time interrupt (RTI) function, and an 8-bit pulse accumulator. It even is the basis for the computer operating (COP) subsystem!

The binary thermometer experiment discussed here makes use of the input capture function. An experiment in Chapter 6 will detail use of the output compare functions. The pulse accumulator and COP functions will be mentioned only briefly in the same chapter, whereas the RTI will be examined in more detail.

Using the HC11's Input Capture Capabilities

Typically in the past, when it was necessary to input data such as voltage, temperature, pressure, etc. into an MPU-based circuit, the designer instinctively reached for a data book on A/D converters. With the HC11's advanced timer, it often is easier to input information using its input capture pins (pins 6, 7, and 8 of the MC68HC11A1P) than to use an external, or even an internal, A/D converter. The input capture concept already has been mentioned briefly. The built-in binary display thermometer contained in M11DIAG2 firmware uses this feature. Here I will describe the concept of input capture and how it can be used to measure temperature.

Both the input capture and output compare features make use of the HC11's timer. While the concept of this timer is almost trivial, it is a complex device. One look at its so-called simplified block diagram in Fig. 4-14 assures one of that. However, I will try to keep things simple here, just imparting enough information to use for our purposes.

In essence, the timer is a free-running 16-bit counter with a four-stage programmable prescaler. There also is an overflow function that allows software to extend the timer's range. The basic clock is the MCU's E clock. With a 4-MHz crystal, the E clock is 1 MHz. The programmable prescaler divides this E clock by 1, 4, 8, or 16 (1 is the default value chosen automatically after a system reset). From the prescaler, this clock then goes to the 16-bit free-running counter. This counter starts from a count of $0000 after reset and then counts up *continuously*. Repeat: This counter counts *continuously*. Nothing stops it (except

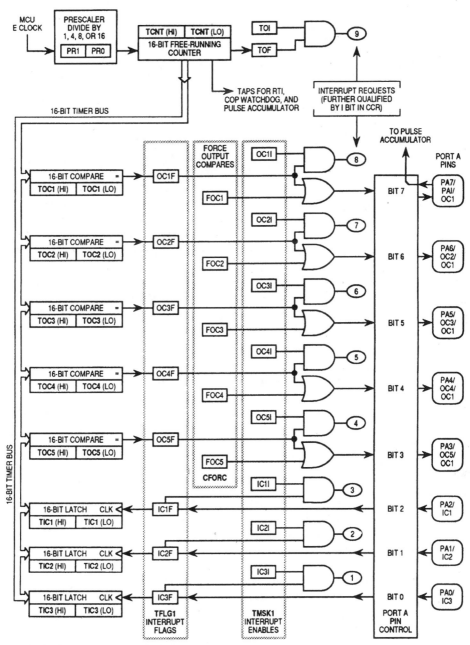

FIGURE 4-14 HC11's main timer system block diagram.

perhaps if someone steps on the board or slams down on the reset switch) while the MCU is in one of its normal modes—nothing!

This counter may be read at any time. The contents of the timer counter are in the TCNT register at $100E and $100F. It's important to read this register only with a double-byte read instruction such as LDD or LDX. If you try to read this register with a single-byte read instruction (e.g., LDAA $100F), you will most likely have errors associated with the data.

When the register reaches its maximum count of $FFFF, the counter rolls over to $0000, sets an overflow flag (bit 7 of TFLG2 at $1025), and continues counting. *Again, I repeat: During normal operation, this counter cannot be stopped. This is important to remember.*

The basic concept of the input capture is simple. All it does is use the timer to measure the length of a portion of an input waveform. The input capture function includes edge-detection logic so that the time between successive rising edges, successive falling edges, or any edge can be determined. Of course, since $f = 1 / T$, you are measuring the instantaneous frequency when you measure the period.

The MC68HC11A1P has three input-only pins that can be configured to function as input capture pins. These input pins are in port A. They are pin 8, which is the PAO/IC3 pin (bit 0 of port A or input capture 3); pin 7, which is the PA1/IC2 pin; and pin 6, which is the PA2/IC1 pin. I am going to concentrate on pin 6, which is the PA2/IC1 pin, because this pin is used by Mag-11's built-in binary thermometer. The other input capture pins perform identically.

There are three registers used by Mag-11's binary thermometer:

TIC1   This is the *input capture 1 register* ($1010 and $1011), which is used to store the value of the 16-bit free-running counter at the time an input capture occurs. This register is not affected by reset and cannot be written by software. It is not shown in pictorial form here because it can be looked on as two read-only registers located at $1010 and $1011. Normally, the TIC1 register is read with an LDD or other double-byte (16-bit) instruction.

TCTL2   This is the *timer control register 2* ($1021), which is used to enable the capture input and also determines whether capture occurs on falling or rising edges or both. Figure 4-15 gives a pictorial view of this register and indicates the significance of the various bits.

The other register used by the binary thermometer is TFLG1, or *timer interrupt flag register 1* (Fig. 4-16). It is used to indicate the occurrence of timer system events. It is used by both the input capture and output compare systems. The binary thermometer uses bit 2, the input capture 1 flag.

Before I go into its software listing, let's discuss the sensor circuit for the binary read-out thermometer. The sensing circuit itself is given in Mag-11's schematic. The circuit is a simple astable multivibrator consisting of three inverters. Because of the presence of RTH, the period decreases as the temperature rises. Here the period is approximately 2.2RC, where $C = 0.1$ µF and R is the rather complex combination of the parallel network of R5 and RTH and the serial network of R6 and R19. (Note that the *exact* frequency depends on the type of inverter and IC technology used.) R5's purpose here is to make the length of the period of oscillation more linearly related to the temperature. Although MCU circuits can handle nonlinearity quite nicely, a 3-cent resistor is the simplest solution here.

After adjustment, the multivibrator's output period at 0°F is 2503 µs, or 2503 cycles, with a 1-MHz clock. Each cycle corresponds to 1°F, and each decrease in period length (in cycles) corresponds to an increase of 1°F.

Register Name: TCTL2				Register Address: $1021				
BIT#	7	6	5	4	3	2	1	0
BIT NAME	none	none	EDG1B	EDG1A	EDG2B	EFG2A	EDG3B	EDG3A
BIT'S STATUS FOR TYPICAL 68HC11A1	0	0	0	0	0	0	0	0

Here the "1" in EDG1B and EDG1A refers to Input Capture 1. The "2" and "3" have similar meanings.

EDG1B	EDG1A	CONFIGURATION
0	0	Input capture 1 disabled
0	1	Capture on rising edges only
1	0	Capture on falling edges only
1	1	Capture on either rising or falling edge

FIGURE 4-15 The HC11's TCTL2 register.

Register Name: TFLG1						Register Address: $1023		
BIT#	7	6	5	4	3	2	1	0
BIT NAME	OC1F	OC2F	OC3F	OC4F	OC5F	IC1F	IC2F	IC3F
BIT'S STATUS RESET	0	0	0	0	0	0	0	0

Notes:

OCxF—output compare x flag

This flag bit is set each time the timer counter matches the output compare register x value. This bit is only cleared by a write of one. A write of zero has no effect.

ICxF—input capture x flag

This flag is set each time the selected active edge is detected on the ICx input line. As in the OCxF, this bit can only be cleared by a write of one.

(x here stands for 1, 2, 3, 4, or 5. For instance, for bit OC4F, x here is 4.)

FIGURE 4-16 The HC11's TFLG1 register.

Notice that if the data were used directly by the display, the result would appear non-sensical even to a "binary number nut"—as the temperature increased, the display would decrease. This problem is corrected simply by complementing the data and then subtracting 63,228 (65,535 − 2307) from the complement. Now refer to the listing in Fig. 4-17. This is only a partial listing that was taken directly from the assembly-language listing for M11DIAG2. Its only difference is that a few more comments have been added for additional clarity. These comments are preceded by an asterisk. This listing is printed here for instructional purposes only. Subroutine listings are not shown for simplicity. The subroutine names are descriptive of their function. For instance, JSR CLRLED simply turns off all LEDs, and JSR DLY500 causes a delay of about 500 ms. M11DIAG2 automatically executes this program when S1's switches 1 through 4 and 6 through 8 are off and switch 5 is on.

If this listing looks like pig-Martian to you, feel free to jump ahead to Chapter 8, which introduces HC11 assembly language. However, if you are at all familiar with the 6800

```
*..............
*..............
*CONFIGUR TCTL2 FOR FALLING EDGE CAPTURE MODE
          LDAA          #00100000B
          STAA          TCTL2
*DELAY 1 SECOND
          JSR           DLY1S
*CLEAR ANY IC1F FLAG
GOTEMP    LDAA          #$04
          STAA          TFLG1
*BE READY TO DETECT FIRST FALLING EDGE
          LDX           #TFLG1
*LOOP HERE UNTIL EDGE WAS DETECTED
          BRCLR         0,X $04 *
*WHEN FIRST EDGE DETECTED CONTINUE
*FIRST READ TIME OF FIRST EDGE
          LDD           T1C1
*SAVE TIME OF FIRST EDGE AT FIRSTE
          STD           FIRSTE
*CLEAR IC1F FLAG
          LDAA          #$04
          STAA          TFLG1
*WAIT FOR SECOND EDGE
          BRCLR         0,X $04 *
*WHEN SECOND EDGE DETECTED CONTINUE
*FIRST READ TIME OF SECOND EDGE
          LDD           T1C1
*SUBTRACT FIRST TIME FROM SECOND TIME WITH
*DIFFERENCE IN DOUBLE ACCUMULATOR. THIS
*DIFFERENCE IS PULSE LENGTH IN CYCLES OF CLOCK 'E'
          SUBD          FIRSTE
*IF TEMPERATURE IS ABOVE 255 F BRANCH TO TOOHI
          CPD           #2050
          BLO           TOOHI
*IF TEMPERATURE IS BELOW 0 F BRANCH TO TOOLOW
          CPD           #2307
          BHI           TOOLOW
*NOW COMPLEMENT DOUBLE ACCUMULATOR AND SUBTRACT
63,228
          COMA
          COMB
          SUBD          #63228
*ACCUMULATOR B NOW HAS TEMPERATURE IN DEGREES F FROM
0 TO 255 F
```

FIGURE 4-17 Partial assembly-language source code listing of thermometer with binary display that is contained in M11DIAG2's firmware.

```
*NOW WE MAKE LED1 - LED8 INDICATE TEMPERATURE IN
BINARY
     STAB        TEMPF
          LDX         #LED1
*ACTIVATE LED POINTED TO BY 'X'
NXTLED    STAB        0,X
          LSRB
*LAST LED YET?
          CPX         #LED8
*IF DONE BRANCH TO GOTEMP FOR NEXT READING
          BEQ         GOTEMP
*OTHERWISE SET UP INDEX REGISTER FOR NEXT LED
          INX
          BRA         NXTLED
*THE FOLLOWING ROUTINE SIMPLY MAKES LED8 FLASH ON AND
OFF
TOOHI     JSR         CLRLED
          LDAA        #$01
          STAA        LED8
          JSR         DLY500
          CLR         LED8
          JSR         DLY500
DOAGIN    JMP         GOTEMP
*THE FOLLOWING ROUTINE SIMPLY MAKES LED1 FLASH ON AND
OFF
TOOLOW    JSR         CLRLED
          LDAA        #$01
          STAA        LED1
          JSR         DLY500
          CLR         LED1
          JSR         DLY500
          BRA         DOAGIN
*...........................
*...........................
```

FIGURE 4-17   (*Continued*) **Partial assembly-language source code listing of thermometer with binary display that is contained in M11DIAG2's firmware.**

series of MPUs, it should make at least a little sense, except for one little instruction—BRCLR 0,X #$04 *. The BRCLR instruction is new to the HC11, and I will discuss it briefly now.

BRCLR stands for "BRanch if bit(s) CLeaR." Let's look at exactly what BRCLR 0,X #$04 * does. First, the '0,x' means that we are using index addressing with a 0 offset, which simply means that the address we are talking about is contained in the index register. (In this particular example, this is the address of the TFLG1 register, since we stored this address in the index register previously.) The #$04 is the mask, which in binary is 00000100. If the corresponding bit of data is clear, the BRCLR instruction will cause a branch. Otherwise, the program continues.

The asterisk in the instruction is interpreted by the assembler to branch to the current value of the program counter, which simply means that the program will continue looping until bit 2 of the TFLG1 register is set (which indicates that a capture has occurred). Another way to do the same thing, without using the asterisk, is to add a label to the instruction and then branch to the label. For instance:

```
LOOP    BRCLR   0,X #$04 LOOP
```

is identical to

```
        BRCLR   0,X #$04 *
```

Other Neat Things You Can Do with the Input Capture Pins

The input capture pins also can be used as simple general-purpose inputs. They can be used this way even if the input capture function is enabled. All you need to do is read port A ($1000). Here, read bit 0 of port A for logic-level data at pin 8, read bit 1 for data at pin 7, and read bit 2 for pin 6.

One possible application here is in a robot in which pin 6, 7, or 8 is connected to a front "bump" sensor that causes the MCU to issue a "backup" command to the robot's main drive motor if the robot hits something. Another possibility is in an HC11 used in a sophisticated agricultural weather-monitoring instrument in which an input capture pin could be connected to a rain-sensing circuit so that the instrument could record the start and end of a critical wetting period. However, these simple applications seem like a waste of high technology, and since they require constant polling, they also are a waste of CPU time. Nonetheless, if there is no other need for one or more of the input capture pins and the CPU has nothing better to do, such a logic-level circuit can be added economically to the device.

Another somewhat more sophisticated use for the input capture pins is as a flexible interrupt input. In order to enable masked interrupt structure, the I bit in the condition code register (CCR) must be cleared. It is automatically set after reset. Besides clearing the I bit in the CCR, the appropriate control bit in the TMSK1 ($1022) register must be set to generate a hardware interrupt request whenever the corresponding ICxF bit is set to "1." (Note that before leaving the interrupt service routine, software must clear the ICxF bit by writing to the TFLG1 register.) The TMSK1 register is shown in Fig. 4-18.

For interrupt request to take place at pin 6 (PA2/IC1), bit 2 (IC1I) of the TMSK1 register must be set. Interrupt requests for the other pins are enabled similarly. Pins 6, 7, and 8 can be configured individually as an edge-triggered interrupt with its own interrupt vector. Also, the type of edge that causes an interrupt can be specified. Refer again to Fig. 4-15

Register Name: TMSK1						Register Address: $1022	
BIT#	7	6	5	4	3	2	1	0
BIT NAME	OC1I	OC2I	OC3I	OC4I	OC5I	IC1I	IC2I	IC3I
BIT'S STATUS RESET	0	0	0	0	0	0	0	0

FIGURE 4-18 The HC11's TMSK1 register.

and register TCTL2. For an interrupt occurring at pin 8 (IC3), the interrupt vector is $FFEA and $FFEB; for pin 7 (IC2), it is $FFEC and $FFED; and for pin 6 (IC1), it is $FFEE and $FFEF. Recall that the interrupt vector is the address that is loaded into the program counter when the respective interrupt occurs.

Normally, a programmer places the starting address of the interrupt service routine at the address of the interrupt vector. For instance, say you are designing a state-of-the art, though coarse cuckoo clock that not only pops out of its bird house and cries out "cuckoo" every hour but also vocally tells you the weather outside and how you should dress. If an HC11 were used in such a gizmo, pin 8 could be connected to a simple rain sensor that immediately alerts the cuckoo clock that it has started to rain. After it detected rain, a typical obnoxious cuckoo clock might immediately report something like "Close that window there; I don't want soggy feathers! It's starting to rain, dummies!" Assuming that the software for the cuckoo clock's rain routine starts at address $D000, then one would make sure address $FFEA had $D0 stored there and address $FFEB had $00.

In the next chapter the fun continues when we connect the Mag-11 board to a PC. I hope you are patient, however. Before we do this, we must look at serial communication interfaces in general. We also will look at the HC11's own SCI.

THE SCI AND MAG-11

Which of the following statements is true?

1 Serial interfaces are really simple.
2 Serial interfaces are relatively easy to use.
3 Serial interfaces are rather complex.
4 You have to be an expert to use serial interfaces.
5 None of the above.
6 All of the above.

The answer is statement 6.

Before you accuse me of having inhaled an excess of the fumes from hot solder, hold it! The primary challenge with using serial interfaces has to do with hardware handshaking. *Hardware handshaking* is the use of dedicated handshaking circuits to control the transmission of data. Serial interfaces are basically simple—really needing only two or three wires. However, hardware handshaking quickly complicates matters, and the user has to be either lucky or an expert to get a custom system that uses hardware handshaking up and running in a jiffy.

As you may know, hardware handshaking uses the DSR and/or CTS and/or DTR and/or RQS lines to control the data flow. We will not use any of these. That should make things simpler! And apparently, as long as we keep to 9600 baud or slower, we don't need them here! Because of this, we will ignore those scary abbreviations in our discussion and will only use three lines to connect to the computer—the TxD (transmit), the RxD (receive), and the ground lines.

The Basics of Asynchronous Serial Communication

As its name indicates, serial communication is a form of communicating digitally by way of a series of pulses (i.e., *serially*, which means simply one after the other). Typically, the HC11's SCI (serial communication interface) uses 1 start bit (must be low), 8 data bits, 1 stop bit (must be high), and no parity. To keep you smiling, I will stick to being "typical" here.

The standard asynchronous standards are a bit perplexing. For instance, a *bit is high* when there is a negative pulse and *low with a positive pulse.* According to the CCITT V.28 standard, the RS232-D electrical specification (the type commonly used), a voltage more negative than -3 V at the receiver's input is interpreted as a logic 1 and a voltage more positive than $+3$ V a logic 0. Do not be too concerned here with these details, however, if you use one of the newer types of the interface chips such as Maxim's MAX232. This and similar chips take care of most of these subtle details for you. All you need is to supply the chip with $+5$ V and add a few capacitors. You connect the HC11's TxD output to the MAX232's T1IN and the HC11's RxD input to the MAX232's R1OUT! With the use of the MAX232 as mediator (maybe a better term would be *translator*), you communicate normally. For instance, to send the ASCII code for "A," which is 41 in hexadecimal or 01000001 in binary, you simply start out by sending a 0 as a start bit, then 1 0 0 0 0 0 1 0 as data (LSB first), and end with a 1 as a stop bit. Everything is nice and neat and—*very logical.*

You may be asking the question, "While MAX232 obviously is a great help here in serial communication, how exactly do you send those string of bits, and how does the receiver know what to make of them? And what about the baud rate?" Well, one way of communicating serially is by using a single bit from a general-purpose output port, such as port A, to send out the data. To receive the data, a single bit from an input port could be used—perhaps port E's AN1PE1 bit.

For instance, to send out the data, we could toggle bit 3 of port A just like we did earlier by storing $00 at port A ($1000) to clear it and storing a $08 at the same port to set the bit. To receive the data, we could set up a program loop that monitors port E's AN1PE1 bit. With the use of suitable subroutines and timing loops so that the baud rate can be set to a "typical" one, it is possible to set up a usable serial communication system.

A similar way of communicating serially has been done in the past with MPUs. However, it isn't done today with the HC11 MCU. There is no need. The HC11 has its own SCI system that does the same job more gracefully than we could with even a relatively sophisticated subroutine. The HC11's built-in SCI system will be examined in more detail shortly. For now, let's switch the focus again from our mind to our hands.

Real-Life Nuts-and-Bolts Connection from Mag-11 to a PC

Mag-11 has a serial interface. M11DIAG2 firmware will enable you to connect with a computer, via its serial interface, so that you can transfer temperature data from Mag-11 to the computer. This serial communication mode is enabled by turning all eight DIP switches of S1 to the off position. Of course, you also need a suitable cable with connectors *and* suitable software on your computer.

The CD-ROM contains three programs, M11_LITE.EXE, M11_COM2.EXE, and M11DELUX.EXE, that allow you to transfer, via a serial interface, temperature data from Mag-11 to a PC-compatible computer. M11_LITE.EXE and M11_COM2.EXE were written with Quick Basic and are relatively simple programs. M11_COM2.EXE only differs from M11_LITE.EXE in one respect: It is designed to be used with the PC's COM2 serial port, whereas M11_LITE.EXE can only be used with the COM1 port.

Both M11_LITE.EXE and M11_COM2.EXE were written explicitly as learning tools and are completely QBasic compatible. Later in this chapter I will show the source code for M11_LITE.EXE, which should help you in your quest for knowledge.

M11DELUX.EXE is the "deluxe" version of the temperature display program written in Visual Basic for DOS and has a number of features that are absent from either "lite" version, including minimum and maximum and the capability of measuring in Celsius and Kelvin. This deluxe version also shows a thermometer on the screen.

To use these programs, you will require a PC-compatible computer with at least one free serial interface. Since no handshaking is used (hardware or software), all you need connect is three wires from Mag-11's DB-9 connector to the computer's RS-232 compatible (serial interface) connector.

Pin 2 of Mag-11's DB9 connector is connected to the PC's serial interface transmit data (TD) line.

Pin 3 of Mag-11's DB9 connector is connected to the PC's receive data (RD) line.

Pin 7 of Mag-11's DB9 connector is ground and must be connected to the PC's serial interface ground (usually pin 7 with 25 pin connectors or pin 5 with 9 pin connectors.)

Also refer to the Mag-11's schematic.

Next, access Mag-11's data logger by setting all positions of DIP S1 to off, and then run the appropriate PC software from your computer. The programs are on the CD-ROM. Assuming that your CD-ROM drive is drive D, you can run M11_LITE.EXE as follows: From a DOS prompt, type, D:\chapter5\M11_LITE, and press ENTER. Follow the instructions on the screen. When you are asked to RESET Mag-11, simply press the RESET switch.

After reset, Mag-11 will send to the computer the message "You are now connected with Mag-11—firmware version M11DIAG2.x." While two baud rates are supported—4800 and 9600—only 4800 is recommended.

Notice that the program seems to run rather slowly. There is no immediate reaction to key presses when the temperature is being measured. The primary reason for this is the time the program waits for temperature readings from Mag-11.

Calibration of the thermometer is nearly identical to that of the binary thermometer that was described in Chapter 4. Place an accurate thermometer in intimate contact with Rth. Wait a few minutes for temperatures to stabilize. Adjust R6 so that the temperature on the computer screen agrees with the known accurate thermometer. Notice that if calibrated at 70°F, the accuracy of this data logger is quite good between 40 and 100°F.

A Look Inside HC11's Own SCI

Now that we have seen how the HC11's SCI operates from a hands-on, external point of view, let's look how it operates internally. *However, we will only look at the SCI's features that are required by Mag-11 to get the job done!*

Like nearly all the HC11's peripherals, the SCI is controlled by internal registers. To be specific, the registers that are needed in a program that uses the SCI are BAUD ($102B), SCCR1 ($102C), SCCR2 ($102D), SCSR ($102E), and the SCDR ($102F). We will look at each of these registers. To avoid confusion, I will be as restrained as possible—I will only discuss the bits in these registers that are used by Mag-11 firmware.

BAUD-RATE CONTROL REGISTER (BAUD)

There is definite confusion between the baud rate and the bits per second (b/ps) rate. This confusion seems to be especially significant to those who are also likely to talk, ad nauseam, on the difference between mass and weight.

Getting back to serial communication: Both *baud rate* and *bits per second* refer to the rate at which bits are transmitted. However, sometimes they are different animals. Nonetheless, in direct RS232 connections, which is what we are doing for the most part, baud rate and the bits per second rate are indistinguishable. Because of this, let's use them interchangeably. OK? I don't have a problem with it. Do you?

Located at $102B, the baud-rate control register (BAUD register for short) is the register that sets the all-important baud rate (see Fig. 5-1). For more details, consult the HC11 bible (Motorola's *HC11 Reference Manual*—literature no. M68HC11RM/AD Rev 3 or later) or an HC11 data sheet/booklet such as the *MC68HC11A8 Technical Data* booklet (literature no. MC68HC11/A8/D Rev 6 or later.) However, you really do not need these details here. For instance, with M11DIAG2 firmware, store $30 at the BAUD ($102B) register. This sets the baud rate, with a 4-MHz crystal, at 4800—twice this (9600) if you use an 8-MHz crystal. For half the baud rate (2400 with a 4-MHz crystal or 4800 with an 8-MHz crystal), store $31 at the BAUD register.

SCI CONTROL REGISTER 1 (SCCR1)

Located at $102C, this register is concerned primarily with 9-bit data characters. Since we are using 8-bit data format, this register has little importance as long as bit 4 is 0. Mag-

REGISTER'S NAME: BAUD				Register Address: $102B			
BIT#	7	6	5	4	3	2	1	0
BIT NAME	TCLR	*0*	SCP1	SCP0	SCKB	SCR2	SCR1	SCR0
BIT'S STATUS AFTER RESET	*0*	*0*	*0*	*0*	*0*	*0*	*0*	*0*

FIGURE 5-1 The BAUD register.

Register Name: SCCR2			Register Address: $102D				
BIT#	7	6	5	4	3	2	1	0
BIT NAME	TIE	TCIE	RIE	ILIE	TE	RE	RWU	SBK
BIT'S STATUS AFTER RESET	0	0	0	0	0	0	0	0

Here the "1" in EDG1B and EDG1A refers to input capture 1. The "2" and "3" have similar meanings.

EDG1B	EDG1A	CONFIGURATION
0	0	Input capture 1 disabled
0	1	Capture on rising edges only
1	0	Capture on falling edges only
1	1	Capture on either rising or falling edge

FIGURE 5-2 The SCCR2 register.

11 firmware stores "00" at this register so that you are sure that the SCI is set up for 8-bit data format.

SCI CONTROL REGISTER (SCCR2)

Located at $102D, this register is the primary one that controls the SCI (see Fig. 5-2). Since setting bits 2 and 3 of this register turns on both the SCI's transmitter and receiver, Mag-11's firmware stores $0C (00001100 in binary) at this register.

Hint:

Before setting these bits by storing data at SCCR2, you must decide whether you want your SCI-driving software to be interrupt based or whether you prefer the polling method. Since we will use the polling method, we store zero's in bits 4 through 7

SCI DATA REGISTER (SCDR)

This is the register that does the real work (see Fig. 5-3). You may be wondering how to actually send data over the SCI. Well, this is the register that does it! For instance, say you want to send a carriage return over the serial interface to the PC. The following two lines of code do just that.

```
LDAA #$0D
STAA $102F
```

Simplicity itself! Even easier than connecting to the Internet.

Now, to explain how it works. First, recall that the ASCII code for the carriage return is $0D. This is why we loaded accumulator A with $0D (LDAA #$0D). Another fact you have

REGISTER'S NAME: SCDR					Register Address: $102F			
BIT#	7	6	5	4	3	2	1	0
BIT NAME FOR TDR (READ)	R7	R6	R5	R4	R3	R2	R1	R0
BIT NAME FOR TDR (WRITE)	T7	T6	T5	T4	T3	T2	T1	T0
BIT'S STATUS AFTER RESET	U	U	U	U	U	U	U	U

Note: U means "undeterminate" and can vary from batch to batch, generally being meaningless.

FIGURE 5-3 The SCDR register.

to know is that the SCDR is located at $102F. Thus, by writing to $102F, we cause the SCI to send out a series of pulses, at the baud rate we set up, through the TxD pin, which is pin 43 of the MC68HC11A1P (pin 21 of the MC68HC11A1FN).

Now that we have seen how sending a character through the SCI is performed, let's look at receiving a character. All we have to do is read the SCDR. For instance, the following instruction transfers the character received by the SCI to accumulator A:

```
LDAA $102F
```

That's all there is to it!

You might be confused here. I explained, a few paragraphs back, that characters are transmitted by storing data at the SCDR register. Now I just mentioned that you read this same register to get information that was received by the SCI! Confused? I hope so! I was when I first looked at HC11's SCI! The solution to this mystery is trivial. While the SCDR register has only one address ($102F), it actually is two separate registers—the transmit data register (TDR) and the receive data register (RDR). As you have probably already guessed, the RDR is a read-only register, while the TDR is a write-only register.

So far it seems really simple to send one character and then maybe a few seconds later send another character. However, if you want to send thousands of characters, this just isn't practical, at least if you want to break for lunch, that is. Also, when you want to receive a character, how do you know that anything new was sent to you over the serial cable? These problems are solved by making clever use of another register called the *status register.*

SCI STATUS REGISTER (SCSR)

Located at $102E, this register is different from the preceding ones. Its primary use in Mag-11 firmware is to determine the status of the SCDR's read data register and transmit data register (see Fig. 5-4).

Register Name: SCSR			Register Address: $102E					
BIT#	7	6	5	4	3	2	1	0
BIT NAME	TDRE	TC	RDRF	IDLE	OR	NF	FE	0
BIT'S STATUS AFTER RESET	0	1	0	0	0	0	0	0

FIGURE 5-4 The SCSR register.

Bit 7 of the SCSR is called the TDRE (an acronym for "transmit data register empty") bit. When set to "1," this bit indicates that the transmit data register is empty—which means it's OK to send data.

Bit 5 of the SCSR is called the RDRF (receive data register full) bit. When set to "1," this bit indicates that the receive data register is full—which means that data was sent to you over the serial cable.

The other bits (except bit 0) also have use, but Mag-11's firmware does not use them. If Mag-11 ignores them, then you will probably be able to ignore them too and still get the HC11's SCI to do the job you want done. Again, consult the HC11 bible if you are curious about these bits or the finer details of using the SCI.

Now you should have enough information to write a short routine that will read data from the SCI and another short routine that will send a character. First, let's look at how we can read data from the SCI using the popular polling method.

Reading SCI Data

First, of course, we have to initialize the SCI by turning on the transmit and receive functions as well as setting the baud rate. This was explained previously and is repeated here in a subroutine called INITSI, which was taken from M11DIAG2 firmware:

```
*
    BAUD        EQU         $102B THE BAUD LABEL STANDS FOR $102B
    SCCR1       EQU         $102C THE SCCR1 LABEL STANDS FOR $102C
    SCCR2       EQU         $102D THE SCCR2 LABEL STANDS FOR $102D
*
    INITSI      LDAA        #$30
                STAA        BAUD    SETUP FOR 4800 BAUD WITH 4MHZ XTAL
                LDAA        #$00
                STAA        SCCR1   MAKE SURE TO USE ONLY 8 DATA BITS
                LDAA        #$0C
                STAA        SCCR2   ENABLE TRANSMIT AND RECEIVE
                RTS
    *
    *
```

As mentioned earlier, after initializing the SCI, all you have to do to read data from the SCI is to load an accumulator with the data at the SCDR located at $102F. But what if there are no pertinent data in the SCDR's receive data register (RDR)? What then? Well, you will recall, you can avoid this scenario by first checking bit 5 of the SCSR register (also referred to as the receive data register full, or RDRF, bit). The following subroutine, which is nearly identical to a subroutine in Mag-11's M11DIAG2, accomplishes this:

```
SCSR        EQU       $102E
SCDR        EQU       $102F
INSCI       LDAA      SCSR
            ANDA      #$20
            BEQ       INSCII
            LDAA      SCDR
            ANDA      #$7F
INSCII RTS
```

The first two lines of this segment of code simply tell the assembler that SCSR really means $102E and SCDR is really and truly $102F. The next pertinent line

```
INSCI       LDAA      SCSR
```

loads accumulator A with the SCI status register, whereas the following line

```
            ANDA      #$20
```

eliminates all bits of interest *except* for the RDRF bit. If this bit is "0," the next instruction

```
            BEQ       INSCII
```

causes the program to jump to the end. If the RDRF bit is set (i.e., a "1"), this means that a character has been received and can be read, which is accomplished with the instruction

```
            LDAA      SCDR
```

Notice that the line

```
            ANDA      #$7F
```

eliminates bit 7 of the data. This is the parity bit, and it is removed to eliminate possible problems. The final instruction (RTS) ends the subroutine and returns the program to the instruction immediately after the instruction that called the INSCI subroutine. (Subroutines are called with either the JSR or the BSR opcode mnemonics.)

What the INSCI subroutine did was access the SCI subsystem and return a character if one was sent. If nothing was sent, the subroutine returns with zero in accumulator A.

Well, how do we use this subroutine? The following code segment shows how:

```
*
    CHKAGAIN  JSR   INSCI
              CMPA  #00
              BEQ   CHKAGAIN
    *rest of program starts here
    *
```

All this codes segment does is call INSCI and then check if accumulator A has something in it. If it doesn't, it branches back to the start and continues looping until a character is received. If a character is received, it leaves the loop and proceeds to the remainder of the program.

What accumulator A does with its information is determined by the instructions following this code segment. One possibility, of course, is that the instructions are so designed to cause some action to be taken by Mag-11. For instance, when Mag-11 receives an ASCII "RS," which is a decimal 30, Mag-11's data logger firmware resets the minimum and maximum temperatures. After it does this, the program might simply jump back to the start, which is labeled CHKAGAIN. Now let's look at sending data through the HC11's SCI system.

Transmitting SCI Data

Once the SCI is initialized, as explained in the preceding section, you can send a character out from the HC11's TxD pin simply by storing accumulator A (which already contains the ASCII code for the character) at the SCDR register ($102F). The only problem here is that if you try to write to the SCDR at the same time the SCI is in the process of sending out the data temporarily stored in the SCDR [actually the transmit data register (TDR)—remember the SCDR register actually consists of two registers, the RDR and the TDR] and you will have problems. Well, you can just wait a bit, say, a couple of seconds, to make sure that the SCI is done. This may be acceptable if you just want to transmit a couple of characters but is stupid if you want to transmit thousands of characters. While you can use the interrupt method of determining when the TDR is empty (bit 7 of the SCCR2 needs to be set to enable this feature), it is even simpler to use the polling method and keep checking the status register's (SCSR) TDRE bit (bit 7). Once this bit is set to "1," this indicates that the TDR is empty and data can be stored there—which, of course, forces it out serially from the TxD pin.

The following program segment makes use of the subroutine OUTSCI2 (which is similar to a subroutine in Mag-11 firmware) to send a carriage return ($0D) out the TxD pin. After this, the program segment here simply sits and waits and waits.

```
*
OUTSCI2     LDAB    SCSR
            ANDB    #$80        AND ACCU B WITH %10000000
SCSR        EQU     $102E
SCDR        EQU     $102F
CR          EQU     $0D
EOT         EQU     $04
*           WAI
            LDAA    #CR
            JSR     OUTSCI2
                                SIT HERE AND WAIT FOR AN INT
            BEQ     OUTSCI2     IF TDRE BIT WAS 0 CHECK SCSR AGAIN
            STAA    SCDR        IF TDRE BIT WAS 1 TRANSMIT DATA
            RTS
```

A program segment that makes use of the subroutine OUTSCI2 to print a message such as "You are now connected with Mag-11" is given next. Notice that $04 is used to signal the end of transmission (EOT):

```
*
                LDX         #SIGNON
    NEXTCHAR    LDAA        0,X
                CMPA        #EOT
                BEQ         DONE
                JSR         OUTSCI2
                INX  ·
                BRA         NEXTCHAR
    DONE        RTS
*
    SIGNON      FCC         `YOU ARE NOW CONNECTED WITH MAG-11'
                FCB         CR
                FCB         EOT
```

The preceding code makes use of index addressing. The first line of code here loads index register X with the address of SIGNON. Then the next line (starts with NEXTCHAR) loads accumulator A, using index addressing, with the ASCII code of the first character (Y), which is stored at this address. The next line checks to see if this character is EOT (i.e., $04). Since it is a "Y" ($59) and not EOT ($04), the program continues and calls the OUTSCI2 subroutine, which outputs the "Y" from the SCI's TxD pin.

The next line, INX, adds one to the index register so that now the index register contains the address of the next character, which is "O" ($4F). Then the program branches back to the first line, and accumulator A, again using index addressing, loads the next character, which is "O" ($4F), and again the SCI sends this character out. The program keeps looping until it picks up the EOT character ($04). At that point it branches to DONE, and the program executes a return from subroutine (RTS) instruction.

A Simple Computer Program You Can Use to Connect to Mag-11's Serial Interface

This section should be of enormous interest to you hands-on, practical types. If you are of this type (and I am of this breed), the question you probably want to ask right now is this: "How can I actually connect my computer to Mag-11 so that I can communicate with it and its M11DIAG2 firmware? I want to—you know—actually see something on my screen that shows me that I am connected to the Mag-11. Something like `You are now connected....'" This has already been explained in part—at least from the viewpoint of Mag-11 firmware. What you really want to know now is how to program your computer so that *it* will communicate with Mag-11!

I will now show how to write a program that accomplishes this. However, to keep things simple, I will restrict the program to the bare minimum.

The program specifications:

1 The program will be written in QBasic/Quick Basic

2 While you will have a choice of communication ports, either COM1 or COM2, you must make your decision when the program is written—run-time choices will not be available.

3 The program will only operate at 4800 baud, so a 4-MHz crystal must be used.

4 Only Fahrenheit degrees will be displayed.

5 The program will not include any fancy stuff such as minimum or maximum temperatures or on-screen calibration.

6 The program will disable all hardware handshaking and will not use software handshaking.

Before writing this QBasic program, however, we must take an additional look at Mag-11's firmware as it pertains to the SCI system.

A Look at M11DIAG2'S Communication Program

The flowchart for M11DIAG2's communication program is shown in Fig. 5-5. (Remember, Figs. 4-9 and 4-10 showed the flowchart for M11DIAG2 as a whole.)

M11DIAG2 firmware jumps to $F800 after reset. The original version, which did not include a serial communication routine, used only memory locations between $F800 and $FFFF. However, the new version's SCI routine is located between $E800 and $F000. While the firmware starts at $F800 after reset, the firmware causes a jump to $E800 when it sees that all positions 1 through 8 of S1 are off (refer to Figs. 4-10 and 5-5).

After M11DIAG2 firmware initializes the SCI and sets it up for 4800 baud (9600 baud with an 8-MHz crystal), it sends out a "You are now connected…" message in ASCII code. The firmware then waits until it receives a pertinent character. For example, when the ACK ASCII character (06 in decimal and hexadecimal and 0000 0110 in binary) is received, the firmware measures the temperature five consecutive times and then sends the average of these measurements back down the TxD line. These data are in degrees Fahrenheit (0–255). If the temperature is 70°F decimal, 70 ($5A) will be sent to the computer. *Important:* These are raw data, *not* in ASCII.

The communication routine in M11DIAG2 responds to other characters. For instance, if it receives the RS ASCII character (decimal 30, hexadecimal 1D, binary 00011101), it resets its minimum and maximum readings it holds in its RAM. However, for simplicity purposes, we will ignore these other "features" of M11DIAG2. Just keep in mind these facts:

1 M11DIAG2 does not use any handshaking.

2 M11DIAG2 uses 4800 baud.

3 M11DIAG2 sends out a message, in ASCII, immediately after reset.

4 When M11DIAG2 receives the ACK character (06), it takes five temperature readings, averages them, and then sends out the raw (non-ASCII) data that correspond to the temperature in degrees Fahrenheit.

QBASIC PROGRAM DESIGN

I assume here that you have a "basic" knowledge of Basic. QBasic, which comes free along with the DOS 5.0 and later, has a neat feature that was absent from early versions of Basic—line numbers aren't required, and labels are supported. In my view, this was the primary problem with the "old" Basic language.

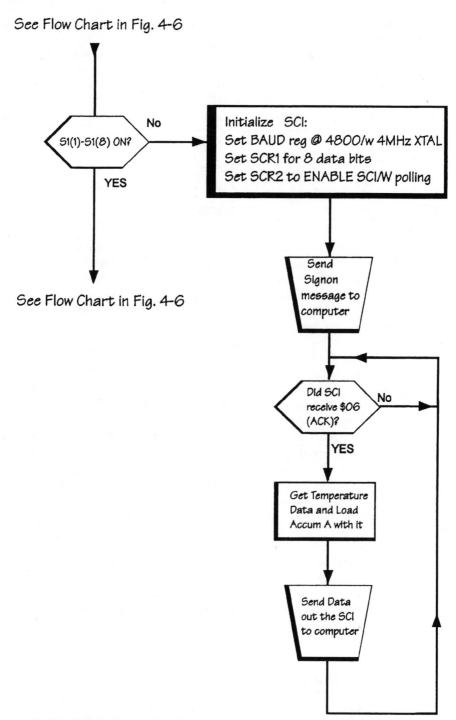

See Flow Chart in Fig. 4-6

S1(1)-S1(8) ON?

No

YES

Initialize SCI:
Set BAUD reg @ 4800/w 4MHz XTAL
Set SCR1 for 8 data bits
Set SCR2 to ENABLE SCI/W polling

See Flow Chart in Fig. 4-6

Send Signon message to computer

Did SCI receive $06 (ACK)?

No

YES

Get Temperature Data and Load Accum A with it

Send Data out the SCI to computer

FIGURE 5-5 Flowchart for M11DIAG2's communication program.

Since this book is about the HC11 and not QBasic programming, I won't go step by step here. Rather, I will simply present the tested program. As an aid, I will include extensive comments in the program shown in Fig. 5-6. However, before I present the program, let's take a more detailed look at the OPEN COM statement, which is QBasic's statement that opens and initiates a communication channel for I/O.

```
REM M11_LITE.BAS by Tom Fox
    'QBASIC Program (v1.0) That Inputs data from
    'Mag-11's temperature measuring circuit and
    'displays it on screen.
    'This program assumes Mag-11 is connected to COM1
    'If you only have COM2 available use M11_COM2.EXE included
    'on the CD-ROM.
    'this program is simplified for instructional purposes
    'and doesn't include run-time choices for ports
    'or for baud rates.
    'ACK (06) sent to Mag-11 causes temperature data to be sent back
ON ERROR GOTO CommunicationProblem
    'if error detected correct it
    COLOR 7, 1                              ' Set screen color.
    CLS
    QUIT$ = CHR$ (27)        'Value returned by INKEY$ when ESC is
pressed
    ACKNOW$ = CHR$ (6)       'causes Mag-11 to send one byte of temp data
    CR$ = CHR$ (13)         'ASCII code for Carriage Return (enter)
    GOSUB quityet           'Check if ESC key was pressed, if so quit
    'setup COM1 for 4800 baud, no parity, 8 data bits, 1 stop bit
    'no hardware handshaking used
    'if you are using different port or different baud change the
    'following statement accordingly before running
    OPEN "COM1:4800,N,8,1, CD0,OP0,DS0,RS" FOR RANDOM AS #1
            GOSUB quityet
            LOCATE 22, 28
            PRINT " <RESET MAG-11 NOW>";
            SLEEP 1
            GOSUB signon "Displ on screen M11DIAG2's sign on message
    DO                              ' Main communications loop.
            GOSUB quityet
            CLS
            LOCATE 15, 18, 1
            PRINT "PLEASE WAIT........RECEIVING DATA NOW. . . ";
    'now send ACK to Mag-11 and receive and displ temperature data
            GOSUB CheckAndDisplayTemp
            CLS
    LOOP
    'The signon subroutine receives signon message from Mag-11's
    'M11DIAG2 firmware and displays the message that was sent.
    signon:
        DO
            IF NOT EOF (1) THEN  'if file is empty then jump to END IF
    'LOC (1) gives the number of characters waiting:
                    Modeminput$ = INPUT$ (LOC (1), #1)
                    SLEEP 1
                    PRINT Modeminput$; 'display signon message
                    SLEEP 3
    'Quit DO LOOP if carriage return detected at end
```

FIGURE 5-6 QBasic program listing.

```
                        IF RIGHT$ (Modeminput$, 1) = CR$ THEN EXIT DO
                        END IF
                        SLEEP 1
                        GOSUB quityet
LOOP
RETURN
'The CheckAndDisplayTemp subrt sends an ACK (06) character to
'Mag-11. M11DIAG2 firmware interprets this as a signal to take 5
'consecutive temperature readings, avg them and send the average
'temperature data back to the computer. The computer then displ
'this temperature along with a descriptive message.
CheckAndDisplayTemp:
                    GOSUB quityet
                    PRINT #1, ACKNOW$;
                    TF$ = INPUT$ (1, #1)
                    TF = ASC (TF$)
'Now TF has data
                    CLS
                    GOSUB quityet
                    LOCATE 15, 14
PRINT "The Present Temperature is "; TF; "degrees Fahrenheit";
SLEEP 5
RETURN
stophere: CLOSE                              ' End communications.
CLS
END
'The quityet subroutine displ heading for the program near the
'top of the screen. It also displ a "press ESC to Quit" message
'near the bottom. It checks the keyboard to see if the ESC key
'was pressed. If it was the program jumps to the routine
'stophere which closes communication and then ends program.
quityet:
LOCATE 4, 6
PRINT "*** MAG-11 COMMUNICATION PROGRAM FOR M11DIAG2 FIRMWARE
VERSION 0.9 ***";
        LOCATE 24, 1, 1
        PRINT TAB(30); "press ESC to Quit";
  KeyInput$ = INKEY$            ' Check the keyboard.
' Quit if the user presses ESC key
  IF KeyInput$ = QUIT$ THEN GOTO stophere
        RETURN
'Communicationproblem is an error handler routine. If a file i/o
'error is detected the routine displays an error message
'and instructs the user what actions to take that will
'(hopefully) correct them. The program then goes back to where
'the problem was encounterd and continues execution.
'If a different error is detected, error trapping is disabled.
CommunicationProblem:
SELECT CASE ERR
                CASE 57  'file i/o error detected
                CLS
                LOCATE 10, 10
PRINT "M11QB has detected a communication problem. . ."
PRINT TAB(5); "Most likely you started M11QB before you ";
PRINT "connected/powered up Mag-11."
PRINT TAB(5); "Make sure Mag-11 is connected and powered "
PRINT " up and then press RESET."
PRINT
```

FIGURE 5-6 (Continued) QBasic program listing.

```
PRINT TAB(20); "Press Enter when you are ready. . ."
   DO
   LOOP UNTIL INKEY$ = CR$   'when enter key is pressed
      CLS                          'clear screen and resume program
      RESUME
        CASE ELSE
              ON ERROR GOTO 0      'disables error trapping
END SELECT
```

FIGURE 5-6 (*Continued*) QBasic program listing.

The statement

```
OPEN "COM2:4800,N,8,1,CD0,OP0,DS0,CS0,RS" FOR RANDOM AS #1
```

sets up the COM2 serial interface port for 4800 baud, no parity, 8-bit data, and 1 stop bit. The options CD0…RS also allows it to operate without any hardware handshaking. "AS #1" refers to the file number that will be referred to with PRINT # and INPUT # statements. This statement assumes that you will be using the COM2 port and 4800 baud. If you use the COM1 port, change the program accordingly. Changing the baud rate is not recommended.

This program is QBasic-compatible. It can be run directly from the QBasic interpreter. QBasic is included in the DOS directory. If you are one of those unfortunates who have only Windows 95/98 and do not have access to one of Microsoft's best programs ever (QuickBasic and Visual Basic are tied here), you can still run this program from the CD-ROM. M11_LITE.EXE is simply the executable form of this program.

To run QBasic from a Windows 95/98 computer, click the Start button, and then click Shut Down. From the Shut Down window screen choose "Restart the Computer in MSDOS mode," and click the "Yes" button. Finally, type QBASIC. Of course, this assumes that your computer originally had a version of MS-DOS before you installed Windows 95/98. If it didn't, you may want to obtain a copy of MS-DOS 6.x.

If you haven't used QBasic before, you are really missing out on one of the really pleasant aspects of an electronics/computer enthusiast's existence. Try QBasic. You'll like it!

In the next section of this chapter we will use the very versatile and handy BUFFALO monitor. However, before changing the firmware (switching EPROMs at U7) from M11DIAG2 diagnostic/data logger to the BUFFALO monitor, spend some time in using the M11DIAG2 firmware. For instance, set up the dipswitches at S1 to check out the CONFIG register again, recheck internal and external RAM (especially interesting is the backup test), check the A/D converter, and try to achieve a runaway program to see if the illegal opcode trap works. Also, try out the binary thermometer again. To use this feature, remember to switch S1(5) and S1(9) on, all others off.

Hint:

If LED6 lights but Mag-11 seems to be operating properly, the HC11 detected an illegal opcode, but it, along with Mag-11 firmware, was able to skip over it by automatically going back to the start of the program.

Finally, make sure that you have some fun with the firmware's data logger feature. *After you examine and reexamine all of M11DIAG2's many features, you are then allowed to install the BUFFALO monitor.* This monitor will be used extensively in the remainder of this book.

The BUFFALO Monitor

This monitor is exceptionally sophisticated. If you have a doubt about its innate complexity, take one glance at its 85+ pages of the source listing (BUF34.asm) that is included on the CD-ROM. In addition to its utility, BUFFALO has two neat characteristics: Its source code is made available for no fee, and it is exceptionally well documented.

INSTALLING AND RUNNING THE BUFFALO MONITOR

In order to use the BUFFALO monitor, an EPROM with BUFFALO firmware must be installed at U7. Also make sure that shorting jumpers are installed as follows:

JP1—between pins 1 and 2

JP4—between pins 1 and 2

JP12—between pins 2 and 3

JP14—install jumper

If you want to purchase a preprogrammed EPROM with BUFFALO firmware, refer to parts and source availability in Appendix C. If you are going to program the EPROM yourself using the BUFFALO.S19 file on the CD-ROM, keep in mind that the addresses in the MOT S (.S19) file seldom correspond to the actual addresses in an EPROM, especially EPROMs used with MOTO's chips. With the BUFFALO monitor, the starting address is $E000. If a 27C512 EPROM is used, the addresses would coincide exactly, but with 27C256 EPROMs, $E000 corresponds to $6000 in the EPROM and $F000 in the MOTS corresponds to $7000. As can be seen, address translation is necessary. This translation can be done directly to the .S19 file or more painlessly (and even invisibly) by most EPROM burners themselves.

In order to communicate at exactly 4800 baud, a 4-MHz crystal must be used. Also, set S2(1) on [S2(1) means position 1 of S2], which results in pin 17 of U1 being grounded. When pin 17 is "high," version 3.400 of the BUFFALO monitor jumps to address $B600, which is the first address of the internal EEPROM.

In order to make use of the BUFFALO monitor, you will require either a terminal or a computer with communication software and an RS-232 interface. Although most users employ a computer with appropriate communication software, I will use the expression "terminal" because this expression is more appropriate here. The terminal should be set for 4800 baud, 8 data bits, 1 stop bit, and no parity or handshaking.

Using a cable with suitable connectors, connect Mag-11's DB-9 connector to the serial connector of the terminal. If you were successful in using M11DIAG2's data logger on your computer, there is no need to change your physical hookup to your computer. Simply skip the next section and continue at the section entitled, "Using the BUFFALO Monitor."

If you did not try out the data logger or, for some reason, did not communicate successfully with your computer, continue reading, and good luck.

Physically Connecting Mag-11 to Your Computer

You will need a three-conductor cable to connect Mag-11's J7 (a DB-9 male connector) to your computer. The connection to Mag-11 will require either a DB-9 female connector or you can tack-solder the cable's wires directly to the Mag-11 board.

The connections at the computer depend on the exact type of connector the serial interface on your computer uses. I will assume here for concreteness that it uses a typical DB-25 (DTE) connector. Pin 2 of DB-9 is Mag-11's receive data (RD) line and should be connected to the terminal's transmit data (TD) line. Pin 3 is Mag-11's TD line and should be connected to the terminal's RD line. Also make sure that Mag-11's ground (pin 7) is connected to the computer's serial interface ground (pin 7 with DB-25 connectors). There is no handshaking used by either Mag-11 or BUFFALO.

Mag-11's BUFFALO firmware was tested on four different computers with four different communications programs. One of these programs, Windows 3.1 Terminal, encountered a slight problem when communicating with Mag-11's BUFFALO monitor. After filling up a screen, new characters were displayed over other characters. No problem was encountered with Windows 95's HyperTerminal or with Hilgraeves HyperAccess on Windows 3.1. BUFFALO even worked well on an old CP/M machine running ACCESS!

Using the BUFFALO Monitor

After properly connecting Mag-11's DB-9 connector to the terminal, turn the terminal on. Connect power to Mag-11. The following message should be seen:

```
BUFFALO v.3.4(ext) Bit User Fast Friendly Aid to . . .
```

If this message isn't seen, use a logic probe to discover the source of the problem. Connect the probe's ground (negative) connection to a ground on Mag-11 and its positive lead to a +5-V power source on Mag-11. Touch the point of the probe to pin 11 of U3. The probe should indicate "high." Now press Mag-11's RESET switch while watching the probe's indicators. If both "high" and "low" indicators briefly light, the problem is likely with the terminal or connections. Make sure that the terminal is set for 4800 and that you connected the pins correctly. Once you do receive the proper message, press the terminal's RETURN key for the prompt:

```
>
```

Now enter "?" or "HELP" for a list of BUFFALO commands (also see Fig. 5-7).

```
>?
   ASM [<addr>]  Line asm/disasm
     [/,=]  Same addr,    [^,-]  Prev addr,    [+,CTLJ] Next addr
     [CR]  Next opcode,                        [CTLA,.]  Quit
   BF <addr1> <addr2> [<data>]  Block fill memory
   BR [-] [<addr>] Set up bkpt table
   BULK  Erase EEPROM,          BULKALL   Erase EEPROM and CONFIG
   CALL [<addr>] Call subroutine
   GO [<addr>] Execute code at addr,   PROCEED  Continue execution
   EEMOD [<addr> [<addr>]]  Modify EEPROM range
   LOAD, VERIFY [T] <host dwnld command>  Load or verify S-records
   MD [<addr1> [<addr2>]]  Memory dump
   MM [<addr>] or [<addr>]/ Memory Modify
     [/,=]  Same addr,   [^,-,CTLH] Prev addr,   [+,CTLJ,SPACE] Next addr
       <addr>O Compute offset,
   MOVE <s1> <s2> [<d>]  Block move
   OFFSET [-]<arg>  Offset for download
   RM [P,Y,X,A,B,C,S]  Register modify
   STOPAT <addr>  Trace until addr
   T [<n>]  Trace n instructions
   TM Transparent mode (CTLA = exit, CTLB = send brk)
   [CTLW]  Wait,     [CTLX,DEL] Abort      [CR] Repeat last cmd


(A)  BUFFALO MONITOR'S HELP SCREEN
```

FIGURE 5-7 BUFFALO's help screen.

BUFFALO expects the following format for commands:

    <cmd>[<wsp><arg><wsp><arg>...]<cr>

[] implies that contents are optional.

<wsp> means whitespace character (space, comma, tab).

<cmd> is a command string of 1 to 8 characters.

<arg> is an argument particular to the command.

<cr> is a carriage return signifying end of input string.

As can be seen from Fig. 5-7, the list of commands is extensive. There will be no attempt here to be thorough in the description of how to use the BUFFALO monitor. One precaution. Unless you know exactly what you are doing or happen to be a personal friend of Motorola's CEO, avoid using the BULKALL command when Mag-11 is placed in special test mode. This command is designed to erase the EEPROM (usually no problem) *and* the CONFIG register (which can be a real pain if you do not know precisely what you are doing).

The ASM (line assembler/disassembler) will be the primary command used in our discussion with Mag-11. This line assembler makes it a snap to try out simple programs and can even be used for the most complex programs known to CPUs. However, when you run out of fingers and toes to count your lines of source code, a sophisticated cross-assembler, such as AS11, which is included on the CD-ROM, will save you time and can brighten your life by simplifying it.

One of the most pleasant features of BUFFALO's ASM is that it allows you to place a program into internal EEPROM as quickly and cleanly as in RAM! Except for the fact that the program in EEPROM remains long after power has been removed, it is impossible to detect, while using ASM, that you are using EEPROM memory rather than RAM. (Of course, internal EEPROM is located at $B600 to $B7FF, so when writing to these addresses, *you know* that you are writing to EEPROM—you just can't tell it from working just with BUFFALO.)

Keep in mind here another feature of BUFFALO when used with Mag-11; when S2(1) is off, BUFFALO jumps directly to EEPROM after reset. This makes it simple to run programs in EEPROM. More on this later.

Don't be a worry wart about EEPROMs, fearing they'll be worn out before you wear out. While the HC11's internal EEPROM is expected to degrade someday (or maybe some century), it is guaranteed to last at least 10,000 erase-write cycles. Apparently, this figure of 10,000 is conservative and is the estimated lifetime at extremely high temperatures. At normal temperatures, this internal EEPROM should last closer to 100,000 cycles. (If one erases and writes once a day, this 10,000-cycle lifetime would last until A.D. 2018; a 100,000-cycle lifetime would end in the year 2265.) More on EEPROM (and EPROM) reliability in Chapter 9.

Using the BUFFALO Line Assembler/Disassembler

The source code for the line assembler is extensive; it takes up about 35 pages. However, the effort by its author, Tony Fourcroy, is well worth it. The following code starts the line assembler, disassembles opcode at <addr>, and then allows the user to enter a line for assembly:

```
ASM <ADDRESS><CR>
```

RULES FOR ASSEMBLY USING BUFFALO'S LINE ASSEMBLER

1 All arguments are in hexadecimal. Do not use the prefix "$."
2 The prefix "#" indicates immediate addressing.
3 A comma indicates indexed addressing, and the next character must be "X" or "Y."
4 Arguments should be separated by one or more spaces or tabs.
5 Any input after the required number of arguments is ignored.
6 Upper- and lowercase are treated identically.

To signify the end of the input line, the following commands are available while using ASM:

<cr>   Finds the next opcode for assembly. If there was no assembly input, the next opcode disassembled is retrieved from the disassembler.

<line feed> *or* <+>   Similar to a carriage return, except that if there was no assembly input, the <addr> is incremented and the next <addr> is disassembled.

<^> or <->   Decrements <addr>, and the preceding address is disassembled.

</> or <=>   Redisassembles the current address.

<.>(period) *or* <Control A>   Exits the assembler/disassembler

Friendly Note:

You may be wondering why I don't start out this discussion using the BUFFALO monitor with a *short flash the MCU check (LED10) LED program*. Well, as you can tell from the schematic, LED10 is connected to pin 5, the PA3/OC5 pin. While it may not be intuitively obvious, you should realize that if you look closely at the schematic, this pin is connected, via JP12, to the HC11's XIRQ input. Normally, this input is inactive after reset (reset sets the CCR's X bit, which makes the HC11 ignore the XIRQ pin). However, BUFFALO's TRACE command uses this pin and thus clears the CCR's X bit. More than this, BUFFALO stores $03 at the TCTL1 register ($1020), which sets the OC5 (pin 5) output line high.

These actions on the part of the BUFFALO monitor make it impossible to "normally" use LED10 when running a program under BUFFALO's control (e.g., using the GO command). Nonetheless, this does not mean that you can't use LED10 if you use BUFFALO's GO command. However, you have to take two actions before you use it properly:

1 The jumper at JP12 must be removed.
2 At the start of the program, add the statement CLR $1020.

I confess that if I were to design a new Mag-11, I probably would control LED10 with the PA4 line (pin 4) instead of PA3 (pin 5). Then, to turn on LED10, you would store $10 at port A instead of $08. This revision was made in the MagPro-11 project in Chapter 9.

Flash Data LEDs with a Program Entered into EEPROM with BUFFALO'S ASM Line Assembler

The best way to learn something is to go ahead and do it. However, start plain and simple. The trivial program in Fig. 5-8 will cause LED1 through LED8 to flash on and off. Here, you want to place the code in EEPROM so that you start at address $B600. Also, the source code is in a format that BUFFALO's line assembler understands. This code is not understood by the AS11 cross-assembler on the CD-ROM, which is used extensively in this book. Please do not get confused here!

BUFFALO's line assembler is started by typing "ASM" followed by a space and then the starting address in hexadecimal. Enter this command with a carriage return.

Examine the listing in Fig. 5-8 and notice several interesting and possibly confusing peculiarities when you actually start to enter the code. After entering "ASM B600" and then carriage return, ASM displays, on the screen (assuming a brand-new HC11), "B600 STX $FFFF." What this means is that ASM looked for an opcode at address $B600, and since the EEPROM is erased, it finds "FF" there. But "FF" is the opcode for storing index

Notes: (1) Line numbers at extreme left are meant for discussion purposes only. (2) BUFFALO's ASM line assembler assumes that all numbers are already in hexadecimal. *Do not add "$" prefix.* (3) Numbers in the second column are addresses. Do *not* type these numbers in. However, these addresses are displayed on the BUFFALO screen. (4) To enter this listing, start BUFFALO and then type "ASM B600 <CR>" and then start entering the text in *third column only.* After you are done typing the line, hit the ENTER key (i.e., carriage return). (5) After entering last line, hit the decimal point key to leave the ASM line assembler.

LINE

1	B600	CLRA
2	B601	STAA 7000
3	B604	STAA 7001
4	B607	STAA 7002
5	B60A	STAA 7003
6	B60D	STAA 7004
7	B610	STAA 7005
8	B613	STAA 7006
9	B616	STAA 7007
10	B619	LDX #3000
11	B61C	DEX
12	B61D	BNE B61C
13	B61F	COMA
14	B620	BRA B601

FIGURE 5-8 Program listing, in BUFFALO's ASM line assembler format, that causes Mag-11's LED1 through LED8 to flash.

register X using extended addressing (mnemonic "STX"). Since the EEPROM is erased, it contains all "ones," which means all "FFs" in hexadecimal. Because of this, ASM assumes that an erased EPROM contains nothing but a whole bunch of STX $FFFF instructions— one right after each other. The opcode listed beneath the STX $FFFF, on the screen, is the new opcode stored there after it is entered in mnemonic form from the terminal.

Notice something else interesting. While it is a BUFFALO dogma that one should not enter the prefix "$" to indicate a hexadecimal number, ASM itself displays the dollar sign for extended addresses. Why it does this is a deep, dark mystery to me. However, dogmas are often nothing but a virtual fence around the real mystery.

There is another thing possibly confusing, and perhaps even a bit weird, about BUFFALO's ASM. If you enter a line of code, at say $A000, and you press the carriage return key, you will see on the next line "rom - xx," where xx is data at the location. This means that the line assembler assumes that you were trying to enter code in read-only memory, since the memory location did not retain the data you tried to enter. In this example you tried to enter a program at $A000. This address accesses EPROM memory space, and ASM was absolutely correct here.

However, this message also will show up if the memory you tried to write to was defective RAM or EEPROM. By the way, the "rom - xx" message also shows up when the address you try to write to has *no memory whatsoever installed.*

While the preceding discussion may frighten you away from using the ASM line assembler, it really shouldn't. Once you get used to its idiosyncracies, it is one of the most pleasurable and useful little programs you will come across.

Running Programs from the BUFFALO Monitor

To execute the program directly from the BUFFALO screen, use the Go command followed by the address. (First, make sure you exited ASM by typing a period.) By the way, simply hitting the "G" key will do the job here.

```
>GO B600<CR>
```

If you entered the program correctly, LED1 through LED8 should flash rapidly on and off. Now stop the flashing by pressing Mag-11's RESET switch. Then turn the power off to Mag-11 and place S2(1) in its off position. This will cause pin 17 of U1 to go "high" when power returns, which causes the BUFFALO firmware to jump immediately to $B600 after reset. Now turn the power on to Mag-11. What happens? Why? You answer these questions yourself. Keep in mind here that the internal EEPROM is located at addresses $B600 through $B7FF and that the BUFFALO monitor jumps to address $B600 if pin 17 of U1 is high.

In order to get back to the BUFFALO prompt, switch position 1 of S2 on, and then press the RESET switch. You can experiment a bit with the frequency of blinking by changing the instruction at address $B619. Loading the index register with a larger number will decrease the frequency.

A Brief Look at Some of BUFFALO's Other Commands

Another simple, but useful, command is MEMORY or just MM or even the undocumented M. Just follow this command with the address you are interested in, and the data contained at that address will appear. Change these data if you like, or simply press the carriage return. This command will be used to enter a program (DAYALERT) in the most basic way possible (see Chapter 8).

If you would like to see a bunch of memory at one time, use the memory dump command MD or DUMP. This command will cause your terminal to display the contents of memory in blocks of 16. The format here is MD <addr1> <addr2><CR>, where addr1 is the starting address and addr2 the ending address. If only one address is listed, 9 blocks of memory are displayed. If only one block is wanted, make addr1 = addr2.

The LOAD T command can be a real time saver. This command will allow you to download S-records (files in MOT-S format and also referred to as .S19 files) via your terminal. We will use this command extensively starting in Chapter 9.

If you have a problem with your program, you might want to use the TRACE command. Here, you place after this command the number (up to $FF in hexadecimal) of instructions you want executed and displayed after each carriage return. If no number is entered, the TRACE command will execute one instruction at a time. The TRACE command starts at the address pointed to by the program counter (the P register). Try the TRACE command out on the blinker program. You can make sure the program counter is pointing to $B600 by using the register modify command RM to change the program counter P, if necessary. The T command is the same as the TRACE command; it merely saves typing four extra letters. (By the way, if you want to use the TRACE or BREAK commands with Mag-11, you must place a jumper on jumper block JP14. In fact, the sole purpose of JP14 is to facilitate the TRACE and BREAK commands.) A carriage return without a previous command on the line repeats the last command.

As shown in Fig. 5-7, there are several other commands you might want to try, although I recommend using BULK sparingly and avoiding BULKALL, at least until you start eating, sleeping, and dreaming about the HC11 and its spooky CONFIG register.

Experiments Using Mag-11 and BUFFALO's Line Assembler

Before discussing the details of the HC11's cool A/D converter, let's look again at representing an 8-bit binary number with Mag-11's LED1 through LED8. Let's do this by examining a short program that turns Mag-11 into a simple binary adder where every time you press the RESET switch, the binary LEDs will increment by 1.

This is not the first time we have looked at representing binary numbers with a bank of eight LEDs. Remember our circa 1965 hobbyist computer way back in Chapter 1? We again looked at this concept in the last chapter when we discussed the binary readout thermometer. Refer again to Fig. 4-13.

As written here, the program in Fig. 5-9 is designed to be entered into the HC11's EEPROM using the BUFFALO monitor's ASM line assembler. The line numbers listed are only meant as an aid in describing the listing and are not entered in the ASM line assembler.

To run the program in Fig. 5-9, all you need to do is place S2(1) in the off position. With S2(1) in this position, the BUFFALO monitor automatically jumps to address $B600 after reset, which is the starting address of this program.

You may wish to leave the binary adder program in EEPROM for a little while. Assuming that you do not erase or program over the instructions that start at $B600, anytime you want to check the operation of Mag-11, switch S2(1) off, and you can test out Mag-11 by trying out the binary adder program.

A few notable comments: First, in keeping with the BUFFALO monitor's line assembler conventions, all numbers are in hexadecimal—do not add a "$" prefix when using the BUFFALO monitor. Second, LED1 is located at hexadecimal 7000, LED2 at 7001,..., and LED8 at 7007. Also, only bit 0 of the data bus (D0) controls the respective LED. This

Note: Start entering program listing by typing "ASM B600 <CR>" at BUFFALO's prompt.

LINE

1	B600	LDS #2D
2	B603	INC 0050
3	B606	LDAB 50
4	B608	LDX #7000
5	B60B	STAB 0,X
6	B60D	LSRB
7	B60E	CPX #7007
8	B611	BEQ B616
9	B613	INX
10	B614	BRA B60B
11	B616	WAI

FIGURE 5-9 Program listing, in BUFFALO's ASM line assembler format, for binary adder experiment using Mag-11. Program turns Mag-11 into a binary adder. Every time the RESET switch is pressed, the data LEDs (1 through 8) indicate a binary number 1 greater than before the switch was pressed.

is the reason for the opcode "LSRB" in line 6. Third, while it is not necessary for proper operation, for simplicity, it is assumed that memory location $50 is clear (all 0s) before this program runs. This clearing of location $50 can be done with the line of code

```
B620   CLR   50
```

A LINE-BY-LINE LOOK AT THE PROGRAM IN FIG. 5-9

Now lets go over, line by line, the listing in Fig. 5-9, since it is so basic.

Line 1   We load the stack pointer immediately with $2D.

Line 2   Adds 1 to data at location $50.

Line 3   Loads accumulator B with data at memory location $50.

Line 4   Loads index register with the address ($7000) of the first LED.

Line 5   Stores accumulator B at the address pointed to by the index register. The respective LED will light now if bit 0 of accumulator B is 1.

Line 6   Shifts all bits of accumulator B to the right. Thus bit 7 goes to bit 6, bit 6 to bit 5,..., and bit 1 to bit 0. This step prepares for lighting the next LED. This step is needed because only bit 0 of the data bus controls the LED.

Line 7 Compares the index register to the address of the last LED (LED8, which has address 7007.)

Line 8 This line causes a branch to the end if the index register contains the address of the last LED. If no match is detected, the program continues to line 9.

Line 9 Adds one to the index register.

Line 10 Branches back to line 5 to turn on the next LED (if bit 0 of accumulator B is 1).

Line 11 The WAI (or wait for interrupt) instruction stops executing programs here until an interrupt occurs. Here, the interrupt will be generated by our old trustworthy RESET switch.

Pay special attention to lines 3 through 10 of this short program, since these lines will be used frequently in the coming experiments. They are used to output data to the binary display consisting of LED1 through LED8.

Now take a wider-angle view of this program. Under normal circumstances, the program will stop after executing the WAI opcode at line 11. However, every time the RESET switch is pressed, 1 will be added to location $50, and the program will proceed as described. Try this program out. It not only is interesting, but a bit of playing with it will help the neurons in your brain become friendlier to the binary numbering system.

A DIGITAL VOLTMETER WITH A BINARY DISPLAY

Since the A/D converter responds to voltages, starting with a digital voltmeter is a great idea for the first experiment that uses HC11's A/D converter. Notice that I use the adjective *digital* here. Do not confuse *digital* with *decimal*—they have different meanings. For instance, it isn't a conflict in terms to state "a binary readout digital voltmeter"—which is what this experiment is all about.

First, if it has not been done yet, adjust R13 on Mag-11 so that pin 22 (VRH) of U1 is as close to 5.12 V as possible. With VRH at 5.12 V and VRL at 0 V, the result register should be at $FF (1111 1111) with an input voltage of 5.12 V and at 00 (0000 0000) with an input of 0 V. As you can see, this means that 0.02 V, or 20 mV, corresponds to a data value of 01 (0000 0001). Thus a data value of 10 decimal ($0A or 0000 1010) corresponds to ten times 20 mV or 0.2 V.

Hook up the circuit shown in Fig. 5-10. Notice, because of its simplicity, that it is not necessary to use a solderless breadboard; the components can be inserted directly into J6's sockets. The capacitor is used here to reduce noise and is optional (see also Fig. 5-11).

This simple circuit, along with the program in Fig. 5-12, forms a digital voltmeter with a binary display. This digital voltmeter demonstrates the use of the HC11's A/D converter. As soon as I show how to actually try this circuit out and give you a chance to play with it, I will get back to theory and look more deeply at the HC11's A/D converter. But first let's keep the fun going.

Digital Voltmeter's Program To create the digital voltmeter, enter the following program into Mag-11 using the BUFFALO monitor's ASM line assembler. Make sure switch S2(1) is on, or you will not be able to get BUFFALO up and running.

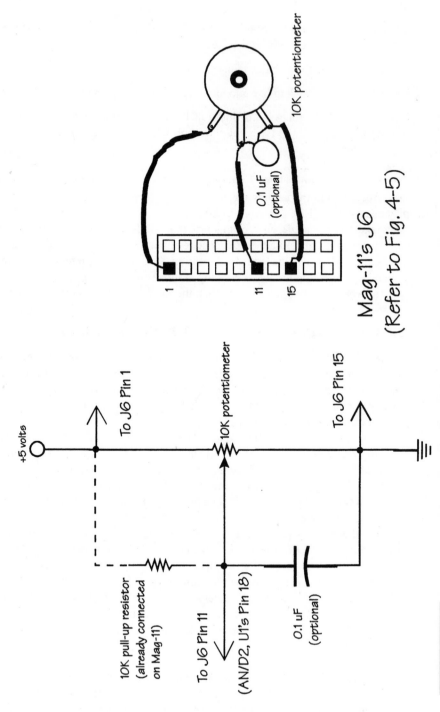

10K potentiometer

0.1 uF (optional)

Mag-11's J6
(Refer to Fig. 4-5)

1

11

15

+5 volts

To J6 Pin 1

10K potentiometer

To J6 Pin 15

10K pull-up resistor (already connected on Mag-11)

To J6 Pin 11

(AN/D2, U1's Pin 18)

0.1 uF (optional)

FIGURE 5-10 Schematic and drawing of potentiometer hooked up for digital voltmeter experiment.

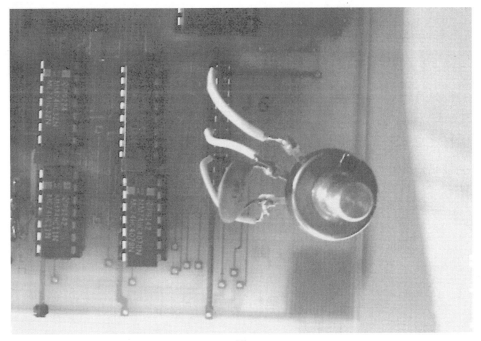

(A)

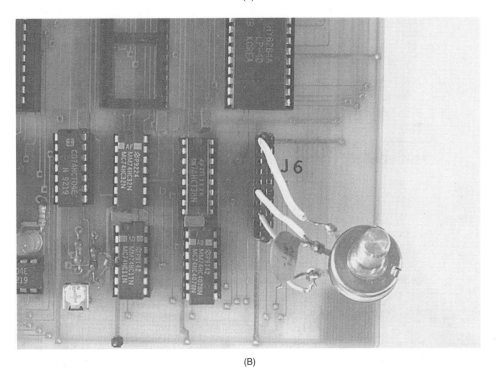

(B)

FIGURE 5-11 Photographs of potentiometer hooked up for digital voltmeter experiment.

Note: Start entering program listing by typing "ASM B700 <CR>" at BUFFALO's prompt.

LINE		
1	B700	LDS #50
2	B703	LDX #1000
3	B706	BSET 39, X CO
4	B709	LDY #3000
5	B70D	DEY
6	B70F	BNE B70D
7	B711	LDAA #01
8	B713	STAA 1030
9	B716	LDAA 1030
10	B719	ANDA #80
11	B71B	BEQ B716
12	B71D	LDAB 1031
13	B720	NOP
14	B721	LDX #7000
15	B724	STAB 0,X
16	B726	LSRB
17	B727	CPX #7007
18	B72A	BEQ B711
19	B72C	INX
20	B72D	BRA B724

FIGURE 5-12 Program listing, in BUFFALO's ASM line assembler format, for binary readout digital voltmeter experiment using Mag-11.

I repeat: The line numbers (at the extreme left) in Fig. 5-12 are not entered. Their purpose is only an aid to discussing the program.

To run this program from the BUFFALO monitor, type "GO B700." Now turn the knob on potentiometer R1. Notice that the LEDs all go out when the voltage is zero (wiper arm is grounded) and the four most significant LEDs (LED8, LED7, LED6, and LED5) light when the voltage is at its maximum. (If the voltage is a bit high, all LEDs may light.) Remove the circuit shown in Fig. 5-10 from Mag-11.

Now try measuring the voltage of a new alkaline D cell by using lengths of no. 22 or 24 hookup wire to connect its negative terminal to ground and its positive terminal to pin 11 of J6. The binary LED display probably will show 01001110, which is 4E in hexadecimal and 78 in decimal. Since this voltmeter has 0.02 V of resolution, the actual voltage is just 78×0.02, or 1.56 V. [Notice that even I, who have used the binary and hexadecimal numbering systems longer than I admit, often revert back to the "finger counting" system (decimal) when my brain attempts to get the "real meaning" of numbers.]

How does this digital voltmeter work? Briefly, all it does is use the HC11's A/D converter to change analog voltage levels into 8-bit digital data. The short program initiates the A/D converter and then interprets the data so as to represent the data on LED1 through LED8. The following is a more detailed representation of what takes place. Only instruction lines that were not used before are explained here.

PROGRAM DESCRIPTION OF LISTING IN FIG. 5-12

Lines 2 and 3   Sets bits 6 and 7 of the OPTION register. This turns on the A/D converter and CSEL.

Lines 4, 5, and 6   Causes a 30-ms delay.

Lines 7 and 8   Sets the ADCTL register for single-conversion, single-channel configuration and selects PE1 (pin 18) as the active input pin. Also starts the conversion process.

Lines 9, 10, and 11   Program waits here until conversion is complete and CCF bit of the ADCTL register is set.

Lines 12, 14, 15, 16, 19, and 20   The display routine. See listing in Fig. 5-9.

Line 17   The NOP instruction saves space for a COMB instruction. Details will be given shortly.

Line 18   When data are finished being displayed, program jumps back to B711 to start the conversion sequence again.

A PRACTICAL APPLICATION—A VCC VOLTAGE MONITOR

Notice with the reference voltages set as described (5.12 and 0 V), 5 V should show a binary reading of 1111 1010 on the binary LED output. This is 250 in decimal. The actual reading in volts would be 0.02 V $\times$ 250 = 5 V.

This simple circuit has a hypothetical yet potentially practical application. For instance, say you connect the +5-V power supply to pin 18 of U1 through a jumper block and write a program that continually monitors S1 for switch closures. See Chapter 4 for details here. If a prearranged combination of switches in S1 is encountered by the HC11, an automatic jump could take place to a subprogram similar to that listed in listing 3. The binary LEDs could then inform the user of the condition of the +5-V power supply and how far off tolerance it may be. If the voltage is off tolerance, the program may then flash one or more LEDs to alert the user that there is a problem.

The HC11's A/D Converter

Now that we have used the A/D converter, let's look a bit more deeply into its capabilities and its workings. An analog-to-digital converter (A/D converter) does just what it says: It converts an analog voltage input into a digital output. Notice that for most measurements, as soon as you convert to digital, you lose accuracy. This is so because in digital parlance an analog input is an infinite-bit measurement with a 0-V resolution. An A/D converter takes this infinite-bit resolution and converts it to an 8-, 10-, 12-, 14-, 16-, or even 22-bit data output. This is a fact of life. Few people mention that you pay this price to convert to digital. However, it is a price well worth paying in many instances. Also, keep this fact of life in mind: Everything is probably digital anyway. The popular and fruitful quantum theory says so. The idea of infinite-bit resolution is only an illusory concept in the minds of theoretical mathematicians. That animal just ain't real!

The HC11's A/D converter is an 8-channel, 8-bit converter with $\pm^1/_2$ least significant bit accuracy over the operating temperature range. Use of 8 bits is still typical with A/D converters. They are economical, and with 8-bit MPU/MCUs, they are the simplest to use because their data lengths are identical.

Eight bits means that you can divide up the input into at most 256 different divisions. For instance, if the VRH (high reference voltage) is 5.12 V and the VRL is 0 V, the A/D converter has a resolution of 5.12/256, or 0.02 V. This means that the HC11 will treat 1.991, 2, and 2.009 V as absolutely the same.

There are many types of A/D converters around. The fastest, the flash converter, is actually the simplest in principle but requires many accurate voltage comparators and many gates. For some idea of how a flash converter works, take a look at the schematic of the trivial 1-bit flash converter in Fig. 5-13. There are only two different outputs of this converter—0 and 1. You may have noticed that the circuit is identical to a voltage comparator with a reference voltage of 2.5 V. As seen in Fig. 5-13, if the input voltage is below 2.5 V, the output is close to 0 V (a binary "0"). If the input exceeds 2.5 V, the output of this circuit is close to +5 V (a binary "1").

While the circuit in Fig. 5-13 is seldom referred to as an A/D converter, it does have uses when working with MPU/MCU circuits. One can get an idea of how a flash A/D converter works by imagining hundreds of voltage converters and a number of gates in a circuit that function similarly in principle to Fig. 5-13.

The HC11's A/D converter is a moderate-speed successive-approximation A/D converter using charge redistribution. While the theory of this A/D converter is not exceptionally difficult to understand, no attempt will be made here to explain it. Chapter 12 of the HC11 bible provides details of the theory in an easy-to-understand language. If you wish optimal accuracy from your design, it may be wise to study this chapter before designing circuits using the HC11's A/D converter.

The HC11—The Author Speaks Frankly

Choices are often the seeds of confusion. Why are Macintosh computers almost universally thought of as easier than PC-compatible computers to set up and use? While there are many valid reasons given, the "real life," down-to-earth reason is simple: There are fewer choices!

The HC11 A/D converter system is basically a snap to use. However, there are so many choices available that it is a confusing system to set up. Depending on which bit is set in the OPTION and ADPU registers, there are eight different ways of operating this A/D converter. In addition to these eight different ways, eight different channels are used. More

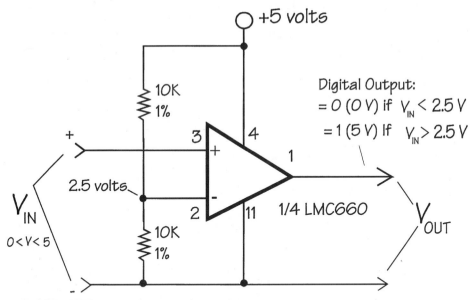

FIGURE 5-13 Schematic for 1-bit flash converter.

than this, there are four other, seldom talked about, "pseudo-channels" that can be used for testing purposes.

There is another facet of the HC11's A/D converter system that leads to this perplexity—the documentation. Both the HC11 bible and the data sheet booklet's descriptions of the A/D system are confusing. Perhaps one of the problems with the documentation is the inherently confusing design of the A/D itself. Whatever it is, I will attempt to make the A/D system easy to understand by using it solely in only one mode. However, before I do this, let's take a brief look at the available choices on setting up the A/D converter. I will do this by looking primarily at the OPTION and ADCTL registers.

The OPTION register is located at $1039 (Fig. 5-14). Only bit 7 (ADPU) and bit 6 (CSEL) are associated intimately with the A/D system, so these are the only ones I will discuss here.

ADPU (bit 7)   Set this bit to turn on the A/D system. Wait at least 100 μs for the A/D system to stabilize before using it.

CSEL (bit 6)   Set this bit to turn on the R-C oscillator, which should be used if a crystal below 4-MHz is used. If you use a 4-MHz crystal, selecting this bit is optional. I recommend setting this bit when using Mag-11. You must wait 10 ms after setting for the A/D system to stabilize. The following discussion refers to the ADCTL register which is shown in Fig. 5-15.

Bit 7/CCF (conversions complete flag)   This bit is not normally used when in the continuous scan mode (SCAN = 1). It is cleared by a write to the ADCTL register (starts a conversion sequence) and set when all four A/D result registers contain valid conversion results. This flag is used in M11DIAG2 firmware as well as in experiments because the firmware and experiments clear the SCAN bit.

REGISTER'S NAME: OPTION					Register Address: $1039			
BIT#	7	6	5	4	3	2	1	0
BIT NAME	ADPU	CSEL	IRQE	DLY	CME	0	CR1	CR0
BIT'S STATUS AFTER RESET	*0*	*0*	*0*	*0*	*0*	*0*	*0*	*0*

FIGURE 5-14 The OPTION register.

Bit 6 Not presently used. Always zero.

Bit 5/SCAN (*continuous scan control*) When this bit is set, conversions are continuous, with the result registers being updated continuously. When this bit is clear, four consecutive conversions are performed on the single channel specified by CD-CA (see Table 5-1). I recommend clearing this bit.

Bit 4/MULT (*multiple-channel/single-channel control*) When set, a conversion is done on each of four channels. The result is placed in the respective result register (see Table 5-1). Note that bits 3 and 2 specify which set of channel signals is associated with the respective result registers. Bits 1 and 0 have no meaning in the multiple-channel mode. When this bit is clear, only one input channel is used, and all conversions are performed on this same analog channel, which is specified by bits 3 through 0 (again see Table 5-1). I recommend clearing this bit.

Bit 3/CD Channel select D.

Bit 2/CC Channel select C.

Bit 1/CB Channel select B.

Bit 0/CA Channel select A.

Table 5-1 gives the A/D's channel assignments. Pay special attention to the channel signal. For instance, Mag-11's experiments that use the A/D converter make use of channel PE1/AN1. With the MC68HC11A1P package, this channel is connected to pin 18. (In Mag-11, pin 18 of the HC11 is connected to pin 11 of J6.)

The last column on the right of Table 5-1 only is pertinent if the ADCTL's MULT bit is set. *Otherwise, ignore it! Things can get even more bewildering if you do not.*

Important stuff to remember about the A/D system:

1 As configured here, you must write to the ADCTL register to start conversion operations.
2 The data in the result registers are valid when the conversion complete flag (CCF) bit is set.
3 A delay of 10 ms is required after turning on CSEL.
4 With a slowly changing input voltage and in the single-conversion, single-channel configuration (SCAN = MULT = 0), result registers ADR1 through ADR4 will contain essentially the same data.

REGISTER'S NAME: ADCTL					REGISTER'S ADDRESS: $1030			
BIT#	7	6	5	4	3	2	1	0
BIT NAME	CCF	—	SCAN	MULT	CD	CC	CB	CA
BIT'S STATUS AFTER RESET	0	0	U	U	U	U	U	U

Note: U means "undeterminate" and can vary from batch to batch, generally being meaningless.

FIGURE 5-15 The ADCTL register.

TABLE 5-1   ANALOG TO DIGITAL CHANNEL ASSIGNMENTS

CD BIT 3	CC BIT 2	CB BIT 1	CA BIT 0	CHANNEL SIGNAL	ADDRESS IN ADDRX IF MULT = 1
0	0	0	0	PE0/AN0	ADR1
0	0	0	1	PE1/AN1	ADR2
0	0	1	0	PE2/AN2	ADR3
0	0	1	1	PE3/AN3	ADR4
0	1	0	0	PE4/AN4*	ADR1
0	1	0	1	PE5/AN5*	ADR2
0	1	1	1	PE6/AN6*	ADR3
0	1	1	1	PE7/AN7*	ADR4
1	0	0	0	Reserved	ADR1
1	0	0	1	Reserved	ADR2
1	0	1	0	Reserved	ADR3
1	0	1	1	Reserved	ADR4
1	1	0	0	VRH pin†	ADR1
1	1	0	1	VRL pin†	ADR2
1	1	1	0	(VRH)/2†	ADR3
1	1	1	1	Reserved for factory testing	ADR4

Note: IF MULT = 0, ADR1 through ADR4 contain data of channel specified by bit 0 through bit 3.
*Not bonded in 48-pin DIP version that is used by Mag-11.
†Internally connected used by Mag-11 HC11DIAG firmware.

The A/D Converter Documentation's Fine Print

Although the HC11 apparently has the capability to convert 16 different channels, 4 are used for testing and 4 are not used at present. Also, while the A/D converter is listed as an 8-channel device, there are only four result registers. The particular set of channels whose data are stored in the result registers is determined by bits 2 and 3 (CD and CC). Notice that if bit 3 and bit 2 are both "1," the A/D system monitors the reference voltages. M11DIAG2 firmware uses this built-in feature to test the A/D converter. (Also notice that since only 48 pins are used in the DIP package, Motorola left out 4 channels, which means that the MC68HC11A1P A/D converter is a 4-channel *not* an 8-channel peripheral. This reduction in channels is seldom a hindrance and does simplify using the A/D converter, since you can be assured that a particular result register corresponds to a particular channel.)

The MCU Controlled Photometer Experiment

By now you should have a practical and theoretical feel for the A/D converter. It is time to try something really interesting and potentially practical, like an MCU-based photometer.

To try this experiment out, first adjust Mag-11's R13 for a reading of 5 V at pin 22 of U1. Then hook up the simple phototransistor circuit shown in Fig. 5-16. Notice that all you have to do to construct this circuit is insert the phototransistor's collector lead in pin 11 of J6 and the emitter in pin 19. The 10-kΩ resistor shown in the circuit is already contained in Mag-11's RN3. Also make sure that switch 2 of S2(2) (i.e., position 2 of S2) is in the off position. To be on the safe side, switch S2(4) should be on.

I will first demonstrate what happens using the program in Fig. 5-12, which should still be in the EEPROM from the last experiment. (If it is not or you skipped the last experiment, go back to Fig. 5-12 and enter it now.) Start the program with BUFFALO's GO command (i.e., GO B700 <cr>).

Shine a flashlight on the phototransistor. Most, if not all, LEDs will be off. Why? Look at the circuit in Fig. 5-16 carefully. At high light levels, the phototransistor conducts heavily, which causes its emitter voltage to drop to near zero. Now put your hand over the phototransistor. What happens? Most LEDs will light, indicating an emitter voltage near +5 V. When dark, the phototransistor is cut off, effectively disconnecting it from the circuit. Now vary the light level by passing your hand in front of the phototransistor in a brightly lit room. Notice how the LEDs change.

The more light, the lower is the binary data displayed on LED1 through LED8. This is not much help if you want to make a photometer. However, we can change a single instruction in the listing in Fig. 5-12 and wind up with a binary display that increases in value with increasing light levels. Simply replace the NOP instruction in line 13 of the listing in Fig. 5-12 with a COMB instruction, and then run the program again. Now the LEDs make sense, huh!

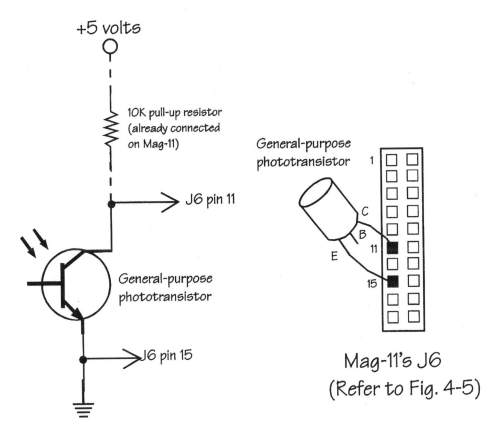

FIGURE 5-16 Schematic and drawing of basic light-measuring experiment using Mag-11.

While the experiment now seems to make sense, the reading on the binary display is only good for relative measurements. This fact allows me to demonstrate one of the most useful features of an MPU/MCU: its ability to change its apparent operation just by revising the instructions used to control it. For instance, it is possible to turn the simple circuit shown in Fig. 5-16 into a fairly accurate photometer that will measure light levels up to 75 footcandles with a resolution of 10 footcandles. The program in the listing in Fig. 5-16 accomplishes this. Run this program by entering "GO B750" on the BUFFALO monitor. Here LED1 through LED8 operate as a bar-type readout where each LED corresponds to 10 footcandles. For instance, if LED1 and LED2 are on, the reading is 20 footcandles. If LED 1 through LED5 are on, this indicates that the light intensity is about 50 footcandles. If all LEDs are on, the light intensity is more than 75 footcandles.

Notice that the readout is nearly linear (except for LED8). This brings up an interesting point: Before digital readouts were used, little importance was placed on linear sensors. All one had to do was mark the faceplate of an analog-type meter where the numbers get all squished up at one end. Who needed a linear sensor? Nobody back then! Then along came the first non-MPU-based digital instrument (e.g., a digital thermometer), where linearity was paramount. Absolutely vital!

Things have again changed. Thanks to the built-in intelligence of MCUs! Now, with MPU/MCU-based instruments, linearity may be nice, but it is *not* mandatory—not at all. One can simply write a program that takes all those sensors with squished outputs and "desquish" them so that the display will be nice and linear. We are back to where we were before the digital revolution! Things are really changing! Or are they?

How does one arrive at a program such as the one in Fig. 5-17? You must start with facts, just the facts, and nothing but the facts!

Fact: A calibrated photometer was used to determine several light levels. *Fact:* As shown in Fig. 5-16, a phototransistor was connected to Mag-11 exactly as shown. *Fact:* The program in Fig. 5-12 was run, and the corresponding binary data on the LEDs were written down. *Fact:* A graph was drawn to determine what binary data corresponded with footcandles in multiples of 10 (see Fig. 5-18). *Fact:* An appropriate table (Table 5-2) was created with data, from Fig. 5-18, obtained by interpolation.

These facts are then used to create the program. For instance, the program loads the data from the A/D converter, and if it finds the data over 5, it causes LED1 to light; if the data are over 8, LED2 also will light; and if the data are over $FE (254), all LEDs will light.

If you use the phototransistor listed in the schematic, the accuracy of this photometer is roughly 12 percent of full scale (about ±10 footcandles). This is accurate enough for many practical uses. See Table 5-3 for a listing of various recommended footcandles for various settings.

As before, run this program from the BUFFALO monitor with the GO B750 command. When it is absolutely dark, all LEDs should be off, and when you shine a flashlight on the phototransistor, all LEDs should light. Under bright light, wave your hand in front of the phototransistor. Notice the interesting movement of the LEDs. This circuit functions as a photometer with a bargraph-type display. When only LED1 is on, the light intensity is roughly 10 footcandles; when LED1 and LED2 are both on, it is roughly 20 footcandles, etc.

Notice that while most LEDs indicate a resolution of 10 footcandles, the difference between LED8 and LED7 is only 5 footcandles. When all LEDs are on, this means a light level exceeding 75 footcandles—it actually could be higher than 1000 footcandles! The reason for this is that the phototransistor becomes saturated at around 75 footcandles and appears as an effective short circuit above this value.

Obviously, a phototransistor does not make a good light sensor at high light levels, although a filter could help here a bit. A silicon photovoltaic cell makes a better sensor, although a selenium cell is inherently better, however, because its sensitivity to light more closely matches that of the human eye.

An Improved Photometer with a Binary Output that Measures Light Intensities from 0 to 255 Footcandles

A circuit that you can use to accurately measure light levels of from 0 to 255 footcandles is shown in Fig. 5-19. Again, the readout will be the binary-data LEDs (LED1 through LED8). However, here the binary-data readout is directly in footcandles. The output of this

Note: Start entering program listing by typing "ASM B750 <CR>" at BUFFALO's prompt.

LINE		
1	B750	LDS #50
2	B753	LDX #1000
3	B756	BSET 39,X C0
4	B759	LDY #3000
5	B75D	DEY
6	B75F	BNE B75D
7	B761	CLRB
8	B762	CLR 7000
9	B765	CLR 7001
10	B768	CLR 7002
11	B76B	CLR 7003
12	B76E	CLR 7004
13	B771	CLR 7005
14	B774	CLR 7006
15	B777	CLR 7007
16	B77A	LDAA #01
17	B77C	STAA 1030
18	B77F	LDAA 1030
19	B782	ANDA #80
20	B784	BEQ B780
21	B786	LDAA #01
22	B788	LDAB 1031
23	B78B	COMB
24	B78C	CMPB #5
25	B78E	BLO B77A
26	B790	STAA 7000
27	B793	CMPB #8
28	B795	BLO B77A
29	B797	STAA 7001
30	B79A	CMPB #0F

FIGURE 5-17 Program listing, in BUFFALO's ASM line assembler format, that turns Mag-11, along with Fig. 5-16, into a bargraph-type photometer that reads from 0 to 75 footcandles.

LINE		
31	B79C	BLO B77A
32	B79E	STAA 7002
33	B7A1	CMPB #28
34	B7A3	BLO B77A
35	B7A5	STAA 7003
36	B7A8	CMPB #55
37	B7AA	BLO B77A
38	B7AC	STAA 7004
39	B7AF	CMPB #B4
40	B7B1	BLO B77A
42	B7B3	STAA 7005
43	B7B6	CMPB #EB
44	B7B8	BLO B77A
45	B7BA	STAA 7006
46	B7BD	CMPB #FE
47	B7BF	BLO B77A
48	B7C1	STAA 7007
49	B7C4	BRA B77A

FIGURE 5-17   (*Continued*) Program listing, in BUFFALO's ASM line assembler format, that turns Mag-11, along with Fig. 5-16, into a bargraph-type photometer that reads from 0 to 75 footcandles.

circuit is quite linear, so the program in Fig. 5-12 can be used. Make sure that line 13 is listed with the inclusion of the NOP instruction. (Note that the program in Fig. 5-12 should still be in EEPROM starting at $B700.)

A BRIEF LOOK AT THE PHOTOMETER CIRCUIT IN FIG. 5-19

Notice the circuit's simplicity. One reason for this simplicity is that this circuit uses the LMC660 quad CMOS op amp. The main reason for the use of this neat IC here is that even with a single +5-V power supply, its output voltage range is from 0 to +5 V as long as its output is connected to a relatively high impedance. This is "fantabulous" for the simplification of circuits such as these. The values chosen for R1, R2, R3, and R4 assume that the silicon photovoltaic cell used is Edmund Scientific's catalog no. T37,332 or a near equivalent. If a larger cell is used, reduce R3. If a smaller cell is used, increase R3. In this circuit, the photovoltaic cell is connected directly across the op amp's input. This type of circuit is known as a *current-to-voltage converter*. In an op amp circuit with a supply of +5 V and GND, pins

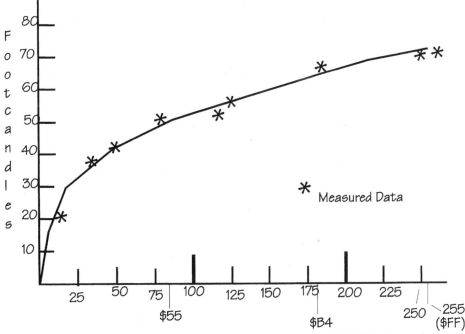

Number in decimal and hexadecimal displayed on Mag-11's LED1 through LED8.

Best fit curve drawn through measured data points.

This graph was used to obtain the data in Table 5-2.

FIGURE 5-18 Graph that uses raw data from Table 5-2 to determine interpolated data for program used in Fig. 5-17.

2 and 3 can be thought of as shorted together as long as the op amp's output voltage is greater than 0 V but less than +5 V. Thus pin 2 is at ground potential. Nonetheless, the op amp's input current is nil, and PC1's output current can only flow through R4, resulting in an output voltage, at pin 1, of $I \times 100$. This output voltage is amplified by U1(b), which is connected as a noninverting amplifier whose gain is adjustable from roughly 5 to 48.

A solderless breadboard can be used to actually construct this circuit. I have recently solved the problem of using potentiometers with solderless breadboards. Many PC pots can be used with solderless breadboards; few are entirely satisfactory, however. Standard potentiometers can be used with solderless breadboards as well. Short lengths of hookup wires are attached to the pot's terminals. The wire's other end is inserted in the breadboard. However, this is awkward because the pots are usually "just hanging around." Now enters Panasonic's Cermet series of pots. Specifically, the laydown version: Panasonic part no. EVN-36CA00Bxx. These pots are ideal for working with solderless breadboards. (The 10-kΩ R4 can be purchased from Digi-Key: part no. 36C14.)

There are only three connections required from the breadboard to Mag-11: ground, +5 V, and a wire from pin 7 of the op amp (U1 on the breadboard) to pin 11 of Mag-11's J6 (which is connected to pin 18 of Mag-11's U1). After wiring the circuit, run the program in Fig. 5-12. If you already entered this program and the BUFFALO EPROM is still

TABLE 5-2　RAW DATA TABLE USING FIGS. 5-10 AND 5-12

Binary Data (shown on LED1—LED8) Hexadecimal	Decimal	Measured Footcandles
0	0	0
5	5	10
8	8	20
F	15	30
28	40	40
55	85	50
B4	180	60
EB	235	70
FE	254	75

Note: Data given were determined by interpolating actual measured data from graph in Fig. 5-18.

installed, power the system up. After receiving BUFFALO's prompt, type "GO B700" and press ENTER.

To calibrate the photometer, place PC1 two *horizontal* feet from a new GE 100-W crystal-clear bulb with an average lumen rating of 1750. The light intensity should be close to 50 footcandles. Adjust R4 so that 50 decimal (0011 0010 binary) shows on LED1 through LED8. Now you have a photometer that is remarkably accurate between 0 and 255 footcandles.

Advantages of an HC11-Based Photometer

The design of this improved photometer spotlights the advantage of MCU-based instruments over more conventional designs—they can be adapted easily for accurate measurements even if the sensor is far from ideal.

MCUs also make it simple to add a digital display. For instance, when in special bootstrap mode or single-chip mode you can connect the HC11's port B and/or port C to a seven-segment LED display. Another easy way to hook up an LED or an LCD display to the HC11 is with use of the SPI (serial peripheral interface) system. See Chapter 7 for details on the SPI.

Say you want to build an accurate photometer that you want to display light readings in both footcandles and lux. (One footcandle is the illumination produced by one standard candle, now referred to as a *candela,* at a distance of 1 foot. A lux, also sometimes referred to as a *metercandle,* is the illumination measured at a distance of 1 meter from the standard candle.) It is simple to have the MCU automatically make the conversion between lux and footcandles.

TABLE 5-3 RECOMMENDED ILLUMINATION LEVELS FOR SEVERAL COMMON ACTIVITIES

FootCandle Level	Activity
20–50	Sufficient light to read newspapers
50–75	Kitchen work
75–100	Regular office work area
100–200	Designing and detailed drafting
Over 250	Fine machine shop work

Another point: The HC11 has a whole bunch of built-in gizmos that make it a snap to input physical data parameters such as light levels (A/D converter, input capture, etc.). This really helps in making for a cheap design.

Another point: The program for the photometer can be built right into the MCU—either in its EEPROM or its ROM (for large production runs). This brings up one of the neatest features of not only the HC11-based products but most MCU/MPU-based products—their inherent "virtual intelligence." (The adjective *virtual* is used here because the *real* intelligence comes ultimately from the designer.) This intelligence allows these products to do seemingly amazing things—and do them cheaply. It also allows for rapid product enhancement—all you need to do is change the program in the HC11. [If the program is in EEPROM and the product has an RS-232 interface built in, you do not even have to replace the HC11—just hook it up to a suitable programmer (even a PC) and reburn the EEPROM with that new, "super-duper" program you just came up with! You can then sell your new advanced photometer for 50 percent more than the old, standard model. Well, too bad they don't make $500 bills anymore; that roll of $100 bills that will be in your pocket might get a bit uncomfortable!]

Another point: As mentioned earlier, MCU-based instruments do not need linear sensors. All they need for accuracy are sensors that do not change significantly over time. (This statement is not completely true. If the sensor's parameters change over time and the changes are documented, this information can be used in the program so that the HC11 "knows" what the new parameters of the sensor are at any time. Of course, this assumes that the project has a built-in clock that runs all the time.)

Oh, did I mention about the HC11's output ports and/or SPI and how easy and cheap it is to hook up a LED/LCD display to it? Oh, that's right. I did! But this cheap and simple display deal is a thing to keep in mind and deserves to be repeated and repeated.

One final thing: Think about the letters "HC" in HC11. Recall what they stand for? (*Hint:* It isn't "How Come.") They stand for "high-speed CMOS." HC is also an acronym for "heap low electrical consumption." This means that the HC11-based photometer can be powered by a tiny, cheap 9-V battery. If you want to reduce power way, way down so that the battery lasts real, real long, simply run your gadget with a slow clock—even a 32.768-kHz crystal will work in many designs.

Do you get the impression that I think highly of MCUs, especially the HC11? You're right! In the next chapter we will continue to use Mag-11 to look into the depths of the HC11 silicon. We will look at, and play with, the numerous features of HC11's versatile programmable timer.

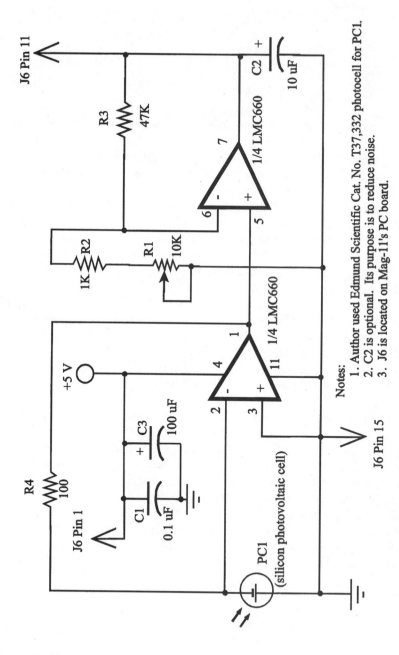

Notes:
1. Author used Edmund Scientific Cat. No. T37,332 photocell for PC1.
2. C2 is optional. Its purpose is to reduce noise.
3. J6 is located on Mag-11's PC board.

A more practical photometer circuit than that shown in Fig. 5-16. This circuit can be wired on a solderless breadboard and connected to Mag-11's J6 as indicated.

FIGURE 5-19 Much improved photometer circuit that uses silicon cell and LMC660CN. This circuit, along with Mag-11 and program listing in Fig. 5-12 forms a light-measuring device that is accurate from 0 to 255 footcandles.

6

PROGRAMMING AND CUSTOMIZING
THE HC11 MICROCONTROLLER

Chapter 4 introduced the programmable timer during a discussion of the HC11's input capture system. You should by now have at least a hint of the enormous versatility of the HC11's design. For instance, the programmable timer is just one of the chip's multitude of subsystems, and the input capture function is just one of the several peripherals that make use of this programmable timer. More than this, there are many ways to use the input capture pins— even more than have already been mentioned. Now let's look at another peripheral that makes use of the timer: the *output compare function*. I will keep it simple here, since it can get quite complex if I include information on the fifth output compare (OC1). Relax, only one more time will I bring up the concept of the OC1—and that time only briefly.

Most concepts, when dealing with MCUs, are trivial. The fine details, however, are often confusing. The output compare function does not conflict with these assertions. As in the input compare function, output compare makes use of the free-running timer. Here, however, you set up a 16-bit "number" (up to 65,535 in decimal). When a match is made between this "number" and the free-running timer, a status flag is set (OCxF), an interrupt may be generated (if enabled), and timer output pins are automatically changed according to software-accessible control bits.

Instead of going directly into boring and bewildering theoretical details, let's look at a simple, yet practical example. Let's use the output compare function to generate an accurate 1000-Hz squarewave signal. If you would like to hear this signal, breadboard the schematic shown in Fig. 6-1. This circuit uses an optoisolator input and an LM386 audio power output IC to drive a small speaker. Connect a wire from pin 4 of J3 (which is connected to the HC11's OC2 pin) to the circuit's input. You also can use an oscilloscope to "see" the signal. Use the BUFFALO monitor's ASM line assembler to enter the program's listing in Fig. 6-2. Run this program with the GO B650 command. (By actual measurement with a Data Precision 5740 digital frequency meter, the output frequency was measured at 1000.1 Hz when "E" was measured at 1.000087 MHz.)

Before going into a line-by-line discussion of the 1000-Hz squarewave generator program, there are two registers used by this example that we should look at. The first, the timer output compare register 2, or TOC2 for short, is a 16-bit (2-byte) output compare register that simply holds the "number" with which we wish to compare the free-running timer. It is not shown here pictorially because there is no need—it basically is just a plain old memory location that is intimately connected with the timer. Since the TOC2 register is a 16-bit register, it has two memory addresses: $1018 (MSB) and $1019 (LSB). By the way, there are four other similar registers: TOC1, TOC3, TOC4, and TOC5. Obviously, TOC2 is associated with the OC2 pin, TOC3 with the OC3 pin, etc.

The timer control register 1, or TCTL1, is shown in Fig. 6-3, since it has several important control bits. Figure 6-3 also shows how the bits control output compare 2.

Line-by-Line Explanation on How the Squarewave Generator Program Works

This discussion refers to Fig. 6-2.

Line 1 loads the stack pointer with $50. This is a good habit.

Line 2 merely loads the index register with $1000 so that reads or writes to HC11's registers can use index addressing, if desired.

Lines 3 and 4 configure TCTL1 so that pin 2 of the 68HC11A1P toggles when a successful compare is made.

Lines 5 and 6 clear the double-byte register, TOC2.

Lines 7 and 8 clear the output compare 2 flag in the timer interrupt flag register 1.

Line 9 loads the double accumulator with 500 decimal ($1F4). Assuming a 4-MHz crystal is used, the free-running timer's clock has a period of 1 μs. Thus, if we want a 1000-Hz output signal (its period is 1 ms or 1000 μs), we must toggle pin 2 every 500 μs—thus we simply load D with decimal 500!

In line 10 we add the data in the TOC2 to D, and then (line 11) we store the result back in the TOC2 register.

We wait at line 12 until the timer register (TCNT) matches TOC2. After the compare is successful, we branch back to line 7 and we clear the output compare 2 flag and continue.

Notice that in the program listing in Fig. 6-2 we use polling of the TFLG1 register. Before we look at an interrupt-driven program, which is more versatile, let's experiment here a bit and change the frequency. To do this, we modify line 9 by loading D with a different number. There are limits to the frequency we can generate. Unless we change bit 0 and bit 1 of the TMSK2 register ($1024), the longest wavelength signal that can be generated (assuming

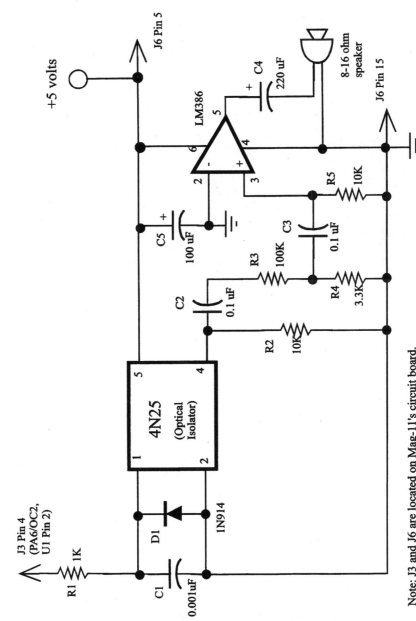

Note: J3 and J6 are located on Mag-11's circuit board.
Also refer to Fig. 4-5.

FIGURE 6-1 (A) Schematic of circuit that when connected to Mag-11 provides sound output. Circuit uses HC11's output compare feature.

135

FIGURE 6-1 *(Continued)* **(B) Photograph of circuit wired according to schematic.** *Note:* **Circuit used in various output compare experiments in Chap. 6.**

E = 1 MHz) is simply $2 \times 65{,}535$ µs or 0.13107 seconds, which is 7.6295 Hz. This is done by loading D, in line 9, with 65,535 decimal or FFFF in hexadecimal (1111 1111 1111 1111 in binary). (Again, keep in mind here that the BUFFALO line assembler requires hexadecimal numbers.) What if you load D with 1 in line 9? Is a 500-kHz signal produced? Remember, the shortest instruction (e.g., NOP) takes at least 2 cycles, or 2 µs, to complete. Thus it should seem obvious that it is not possible to generate a 500-kHz signal. (For instance, the instruction on line 12 alone takes 7 µs to complete.) Practically speaking, do not try to create a signal shorter than 70 µs ($f = 14{,}286$ Hz) using a 4-MHz crystal.

While there are definite restrictions on generating high frequencies, very low frequency squarewaves can be created by changing the contents of the TMSK2 register. If both bits 0 and 1 of this register are set, a $\frac{1}{2}$-Hz squarewave can be generated, i.e., a squarewave with a wavelength slightly longer than 2 seconds (16×0.13107).

Creating a Squarewave Using the Output Compare Interrupt

In polling-driven software, one wastes CPU time. In many practical applications this waste of time cannot be tolerated. The designers of the HC11 must hate the mere thought of wasting CPU time because they included a myriad of interrupts. We have looked at several

LINE		
1	B650	LDS #0050
2	B653	LDX #1000
3	B656	LDAA #40
4	B658	STAA 1020
5	B65B	CLR 1018
6	B65E	CLR 1019
7	B661	LDAA #40
8	B663	STAA 23,X
9	B665	LDD #01F4
10	B668	ADDD 1018
11	B66B	STD 1018
12	B66E	BRCLR 23,X 40 B66E
13	B672	BRA B661

Notes: (1) Line numbers at extreme left are meant for discussion purposes only. (2) BUFFALO's ASM line assembler assumes that all numbers are already in hexadecimal. *Do not add "$" prefix.* (3) Numbers in the second column are addresses. Do *not* type these numbers in. However, these addresses are displayed on the BUFFALO screen. (4) To enter this listing, start BUFFALO, then type "ASM B650 <CR>," and then start entering the text in *third column only.* After you are done typing the line, hit the ENTER key (i.e., carriage return). (5) After entering last line, hit the decimal point key to leave the ASM line assembler.

FIGURE 6-2 Program listing, in BUFFALO's ASM line assembler format, that creates a 1000-Hz squarewave using output capture and the polling method.

REGISTER'S NAME: TCTL1							Register Address: $1020	
BIT#	7	6	5	4	3	2	1	0
BIT NAME	OM2	OL2	0M3	OL3	OM4	OL4	OM5	OL5
BIT'S STATUS AFTER RESET	0	0	0	0	0	0	0	0

OM2	OL2	Action Taken
0	0	OC2 does not affect pin (OC1 still may)
0	1	Toggle pin 2 of 68HC11A1P on successful compare
1	0	Clear pin 2 of 68HC11A1P on successful compare
1	1	Set pin 2 of 68HC11A1P on successful compare

FIGURE 6-3 THE TCTL1 register.

already, including the reset interrupt, the illegal opcode interrupt, the clock monitor interrupt, and the general-purpose input capture interrupts. We will look at the timer output compare 2 interrupt next.

The HC11's interrupt vector assignments are shown in the left column of Table 6-1. However, if one is going to use the BUFFALO monitor, these assignments are not very useful because they are occupied by BUFFALO's brain space. Luckily, BUFFALO provides pseudovectors that are located in internal RAM space. (By the way, these pseudovectors are identical to those used by the HC11's bootstrap ROM. The only difference here is that at reset the bootstrap ROM jumps to $BF40, whereas the BUFFALO monitor jumps to $E000.)

TABLE 6-1   HC11'S INTERRUPT VECTOR ASSIGNMENTS

ON-CHIP NORMAL VECTOR ADDRESS	INTERRUPT SOURCE	STARTING ADDRESS OF PSEUDOVECTOR USED BY BUFFALO MONITOR
FFD6,D7	SCI	00C4
FFD8,D9	SPI	00C7
FFDA,DB (input edge)	Pulse accumulator	00CA
FFDC,DD (overflow)	Pulse accumulator	00CD
FFDE,DF	Timer overflow	00D0
FFE0,E1	Timer output compare 5	00D3
FFE2,E3	Timer output compare 4	00D6
FFE4,E5	Timer output compare 3	00D9
FFE6,E7	Timer output compare 2	00DC
FFE8,E9	Timer output compare 1	00DF
FFEA,EB	Timer input capture 3	00E2
FFEC,ED	Timer input capture 2	00E5
FFEE,EF	Timer input capture 1	00E8
FFF0,F1	Real-time interrupt	00EB
FFF2,F3	IRQ	00EE
FFF4,F5	XIRQ	00F1
FFF6,F7	SWI	00F4
FFF8,F9	Illegal opcode trap	00F7
FFFA,FB	COP fail	00FA
FFFC,FD	Clock monitor	00FD
FFFE,FF	Reset	E000 (in ROM)

Note: All addresses are in hexadecimal.

Refer to Table 6-1's right column for the starting addresses of these pseudovectors. These pseudovectors differ slightly from normal vector assignments in that they require a JMP opcode (7E) placed before the address of the interrupt service routine.

Using the Interrupts

The first step in using the OC2 interrupt is to bring it to life by clearing the I bit in the CCR (this is easily accomplished with the CLI opcode) and setting the OC2I bit (bit 6) in the TMSK1 register (this register was discussed earlier). The program listing in Fig. 6-4 provides a 1000-Hz signal as did Fig. 6-2; only this time the interrupt method, instead of polling, is used. As before, it is written here in a form that allows direct entry using the BUFFALO monitor's ASM line assembler, and the program resides in the HC11's EEPROM.

To run this short program enter "GO B6B0." If you have the circuit hooked up as shown in Fig. 6-1, you should hear a loud, moderate-pitched sound from the speaker.

Line-by-Line Explanation of the Program Listing in Fig. 6-4

Lines 2 and 3 load the JMP opcode at location $00DC, which is the pseudovector for output compare 2.

The next two lines (lines 4 and 5) load the starting address ($B6A0) at addresses $00DD and $00DE. Thus after an OC2 interrupt is detected, the program will jump to $B6A0, the starting address of the interrupt service routine.

Line 7 sets OC2 for a toggle of pin 2 on a compare match.

Line 8 clears the OC2 flag bit in the TFLG1 register.

Line 9 enables OC2 interrupts and the CLI.

Line 10 enables interrupts.

Lines 11 through 14 are a trivial program that merely functions as a place to wait for interrupts. (A "JMP xxxx" instruction can take the place of the three NOPs. This JMP instruction can be used to lead into a more meaningful program. This will be done shortly.)

Line 15 starts the interrupt service routine. It loads the double accumulator with the data (prestored in internal RAM at $60) for a delay time for 1/2 cycle.

Lines 16 and 17 add this value to the last compare value and store it back in the OC2 compare register.

Before returning from the interrupt (line 20), lines 18 and 19 clear the OC2 flag in the TFLG1 register.

Note: Start entering program listing by typing "ASM B680 <CR>" at BUFFALO's prompt.

LINE	INSTRUCTION	
1	B680	LDS #50
2	B683	LDAA #7E
3	B685	STAA DC
4	B687	LDX #B6A0
5	B68A	STX DD
6	B68C	LDAA #40
7	B68E	STAA 1020
8	B691	STAA 1023
9	B69A	STAA 1022
10	B697	CLI
11	B698	NOP
12	B699	NOP
13	B69A	NOP

Note: After entering line 14, press the line feed or + key and not the ENTER key.

14	B69B	BRA B698

The next line starts the interrupt service routine. Press the line feed or + key until address B6A0 shows up.

15	B6A0	LDD 60
16	B6A2	ADDD 1018
17	B6A5	STD 1018
18	B6A8	LDAA #40
19	B6AA	STAA 1023
20	B6AE	RTI

Again press the line feed or + key instead of ENTER until address B6B0 shows up.

The next short program segment places the frequency/period data at addresses 60 and 61 and then jumps to the start of the main program.

21	B6B0	LDX #1F4
22	B6B3	STX 60
23	B6B5	JMP B680

Important note: To run this program, type "GO B6B0."

FIGURE 6-4 Program listing similar to Fig. 6-2 only using the interrupt method.

Lines 21 and 22 are a short program that stores 500 decimal, which is $1F4 in hexadecimal, at location $60.

Line 23 causes the program to jump to $B680, back to where the squarewave program starts, and the cycle repeats itself until you hit the RESET switch or turn the power off.

Some Other Interesting Programs

The program listing in Fig. 6-5 will automatically sweep the scale from about 93 Hz to nearly 9 kHz. The rate of climb of frequency is determined in the third line (the LDY #300 instruction). Change the data here if you want to experiment a bit. If you want to try out this program, line 11 of the program listing in Fig. 6-4 must be changed to jump to the start of the program listing in Fig. 6-5 as follows:

```
LINE          NEW INSTRUCTION
11            B698 JMP B6C0
```

To run this program from the BUFFALO monitor, type "GO B680."

A Fun Program

The program in Fig. 6-6 is fun to listen to. Try it. (As in the preceding example, line 11 of the program listing in Fig. 6-4 must be altered to run the program in Fig. 6-6. Here, change line 11 of listing 7 to "JMP B6D8." Again, use "GO B680" to run this program.)

The description of this listing will be brief and fragmentary. The first three lines of this listing just load D with a pseudorandom number. (The original number comes from the free-running timer register TCNT.) Lines 4, 5, 6, and 7 reject this number if it is below 56 decimal ($38) or above 4608 decimal ($1200). If it passes this test, the number is stored at $0060 and $0061 by line 8. This number determines the frequency produced by listing 7. Lines 9, 10, and 11 cause a $1/3$-s delay for the "note" to be heard, and line 12 merely causes the program to get another number.

Mag-11's Musical Talents

After listening to the "alien" sound produced by the program listing in Fig. 6-6 for a few minutes, it becomes obvious that it is not difficult to have Mag-11 produce computer music. Keep in mind that the length of time a note is held is determined by line 9. If D is loaded with $B000, the length of a note would be about $1/3$ s. The basic program to produce music is given in Fig. 6-7. The data required to play "Old MacDonald" are given in Fig. 6-8. Use the MM 1100 <CR> command to enter these data. Use the space bar instead of the ENTER key to step to the next address for loading the data. Data "FF" cause a delay

Note: This program also uses the program in Fig. 6-4. In order to run this program successfully, line 11 of the program listing in Fig. 6-4 must be changed to jump to the start of the program listing in Fig. 6-5 as follows:

```
Line            NEW INSTRUCTION
11              B698 JMP B6C0
```

If this wasn't done already, do it now!
To run this program, type "ASM B680 <CR>" at BUFFALO's prompt.

LINE		
1	B6C0	LDX #1500
2	B6C3	STX 60
3	B6C5	LDY #300
4	B6C9	DEY
5	B6CB	BNE B6C9
6	B6CD	DEX
7	B6CE	DEX
8	B6CF	CPX #38
9	B6D2	BLO B6C0
10	B6D4	BRA B6C3

FIGURE 6-5 Program listing, in BUFFALO's ASM line assembler format, that automatically sweeps the scale. Uses the HC11's output-capture subsystem, along with interrupts, to create a sweep generator from about 93 Hz to nearly 9 kHz.

(lines 6, 7, and 19 through 24), and data "00" cause the program to halt (lines 5, 16, 17, and 18). The larger the number in data, the lower is the note. The pitch of the note is also determined by lines 3 and 9 of listing 10. The length of the note is determined by line 11, and the relative length of a pause is determined by line 20.

Notice that the data listing in Fig. 6-8 assumes that memory is installed starting at address $1100. If you have not installed U9 in its socket yet, this must be done before you try to enter the data in Fig. 6-8. U9 is an 8-kB static CMOS RAM, such as the 6264LP. Other similar memory ICs can be used. Just make sure that they have 28 pins and identical pinouts. Remember, these data are stored in CMOS static RAM—not EEPROM—so they are subject to "evaporation" shortly after power is removed.

Important:

If you are going to run the program in Fig. 6-7, you must first change line 11 in the program listing in Fig. 6-4 to "JMP B7C7." Again, run the program with "GO B680."

Note: This program also uses the program in Fig. 6-4. In order to run this program successfully, line 11 of the program listing in Fig. 6-4 must be changed to jump to the start of the program listing in Fig. 6-6 as follows:

```
Line           NEW INSTRUCTION
11             B698 JMP B6D8
```

If this wasn't done already, do it now!
To run this program, type "ASM B680 <CR>" at BUFFALO's prompt.

LINE

1	B6D8	LDD 100E
2	B6DB	MUL
3	B6DC	MUL
4	B6DD	CPD #38
5	B6E1	BLO B6D8
6	B6E3	CPD #1200
7	B6E7	BHI B6D8
8	B6E9	STD 60
9	B6EB	LDY #B000
10	B6EF	DEY
11	B6F1	BNE B6EF
12	B6F3	BRA B6D8

FIGURE 6-6 **Program listing, in BUFFALO's ASM line assembler format, that produces weird music.**

A real sharpie might see another possibility for a similar program—voice synthesis. To experiment here, try to load D in line 11 of listing 10 with $150 or even $100 instead of $B000. With this modification, it will take several hundred of these short "notes" to voice a single "word." While I have only experimented briefly with using Mag-11 in voice synthesis, it seems to be possible. A few tips. Vary the notes sharply. For instance, a fairly accurate "s" sound can be made with the following data: F0 01 E0 05 F0 01 D0 80 01 F0 01 E0….Notice that with 100 to 300 distinct notes a second, the human ear interprets such a combination of notes as the "pink noise" of the "s" sound. Such a rapid alteration of frequencies is necessary to obtain the sound of the human voice. However, some sounds, such as long "e," do not require such abrupt changes in frequency as the "s" sound. It is obvious that it will take considerable memory for Mag-11 to say a single word, and there are no guarantees here on the quality of speech. Today there are better ways of implementing artificial speech in products. However, if you just want your HC11 gadget to say a word or two, it may be most cost-effective to use the HC11's output compare system to do the job.

Note: This program also uses the program in Fig. 6-4. In order to run this program successfully, line 11 of the program listing in Fig. 6-4 must be changed to jump to the start of the program listing in Fig. 6-7 as follows:

```
Line                NEW INSTRUCTION
11                  B698 JMP B7C7
```

If this wasn't done already, do it now!
To run this program, type "ASM B680 <CR>" at BUFFALO's prompt.

LINE

1	B7C7	LDX #1100
2	B7CA	CLI
3	B7CB	LDAB #6
4	B7CD	LDAA 0,X
5	B7CF	BEQ B7F0
6	B7D1	CMPA #FF
7	B7D3	BEQ B7F4
8	B7D5	MUL
9	B7D6	ADDD #A7
10	B7D9	STD 60
11	B7DB	LDY #B000
12	B7DF	DEY
13	B7E1	BNE B7DF
14	B7E3	INX
15	B7E4	BRA B7CA

......
......
...... Press line feed or + key.
......

16	B7F0 SEI
17	B7F1 NOP
18	B7F2 WAI

......
...... Press line feed or + key.
......

FIGURE 6-7 Program listing, in BUFFALO's ASM line assembler format, that can produce "player piano" music.

LINE		
19	B7F4	SEI
20	B7F5	LDY #3000
21	B7F9	DEY
22	B7FB	BNE B7F9
23	B7FD	INX
24	B7FE	BRA B7CA

FIGURE 6-7 (*Continued*) **Program listing, in BUFFALO's ASM line assembler format, that can produce "player piano" music.**

A Brief Look at Advanced I/O Pin Control

My examples of the output compare control of pin 2 are labeled "normal I/O pin control" in the M68HC11 Reference Manual. Also available are advanced methods using the OC1. The OC1 allows one output compare to control up to five pins or two output compares to control one pin. The details of using OC1 in this manner are beyond the scope of this book.

Another, simpler feature of the HC11's output compare system is *forced output compares.* This feature is actually used in some automotive spark timing systems. In simple terms, all it does is enable the program to "force" a compare before a compare would take place in the normal way. This force mechanism uses the CFORC register ($100B) (see Fig. 6-9).

To force one or more output compare channels, write to this register with ones in the bit positions corresponding to the channels to be forced. Thus storing $40 (0100 0000 binary) at $100B will force a successful compare of OC2 at the next timer counter clock cycle. (In the examples, this would occur at the next E clock cycle, since the prescale factor for the timer system is the default value of 1. If the prescale factor were set to 16, the forced compare would require 16 E clock cycles to occur.)

The Timer's Other Life

As mentioned earlier, the real-time interrupt (RTI), pulse accumulator, and COP systems are also associated with the programmable timer. We will look briefly at these now.

THE RTI (REAL-TIME INTERRUPT)

The RTI function is used to generate periodic interrupts. (*Caution:* Do not confuse the acronym of the real-time interrupt with the mnemonic for the instruction return from interrupt, which is also RTI. It is unfortunate that these two separate concepts have the same acronym.) These periodic interrupts can be used in a real-time clock, or they can be used to perform program switching in a multitasking operating system. I will demonstrate its use with a little experiment that will flash the red LED8 at exactly 1-s, 5-s, and 1-min intervals.

Notes: (1) Use BUFFALO's MM (memory modify) command to enter the data at the starting address $B600 (e.g., MM B600). (2) Data reads across; press space bar for next address.

```
77 FF 77 FF 77 FF 90 FF 87 FF 87 FF 90 90 FF FF FF FF FF FF 65 FF 65 FF 70 FF 70 FF
77 77 FF FF FF FF FF FF 77 FF 77 FF 77 FF 90 FF 87 FF 87 FF 90 90 FF FF FF FF FF FF
65 FF 65 FF 70 FF 70 FF 77 77 FF FF FF 77 FF 77 FF 77 FF FF FF FF 77 FF 77 FF 77 FF
FF FF FF 77 FF 77 FF 77 FF 77 FF 77 77 FF FF FF FF FF FF FF FF 77 FF 77 FF 77 FF 90
FF 87 FF 87 FF 90 90 FF FF FF FF FF FF 65 FF 65 FF 70 FF 70 FF 77 77 00 00
```

FIGURE 6-8 **Data for "Old MacDonald" music using Figs. 6-7 and 6-1.**

REGISTER'S NAME: CFORC					Register Address: $100B		
BIT#	7	6	5	4	3	2	1	0
BIT NAME	FOC1	FOC2	FOC3	FOC4	FOC5	0	0	
BIT'S STATUS AFTER RESET	0	0	0	0	0	0	0	0

FIGURE 6-9 The CFORC register.

RTI experiment In order to try this experiment out exactly as described, you will need to use the BUFFALO monitor's line assembler again. Figure 6-10 provides the program listing. As described earlier, in order to enter this program, start out, at BUFFALO's prompt, by typing "ASM" followed by the hexadecimal start of your program, which here is B600:

```
>ASM b600
```

As before, enter the data in Fig. 6-10, but do not enter the line numbers at the extreme left. Recall that these line numbers are for discussion purposes only.

Of course, to run this program, type "GO B600" at the prompt. If everything is OK, LED8 should flash on briefly every 1 s. Later we will change one line to have it flash at exactly 5-s intervals.

First I will explain, in nice, simple terms, what this program does. Then I will discuss, line by line, the fine details. After initializing the stack pointer, store 61 ($3D) at a temporary storage register. This number (61) is chosen because we set up the RTI rate at 16.384 ms (0.016384 × 61 = 0.999424 s), which means that we get a real-time interrupt to occur every 16.384 ms. The program keeps track of the number of interrupts, and after every 61 of them, which occurs after 1 s (actually, 0.999424 s, but this is close enough, it lights LED8 for a fraction of a second and then shuts it off and waits for another second to light it again.

A couple of points should be mentioned before I go into a line-by-line discussion of the program. First, while the RTI can be used with the polling method, we restrict ourselves to the interrupt method because of its superiority—but this leaves us open to possible confusion. Remember earlier when I discussed pseudovectors? Pseudovectors are needed here for two intertwined reasons:

1. The BUFFALO monitor already uses the normal vectors to do its "thing."
2. Since we are sticking our program in EEPROM, which is located from $B600 through $B7FF, we cannot use the normal vectors anyway because they are located from $FFC0 through $FFFF. As mentioned earlier in this chapter, pseudovectors are located in RAM space, and unlike true vectors, they require a jump (JMP) instruction. For instance, the pseudovector associated with the RTI is located at RAM memory locations $EB, $EC, and $ED. At these locations we place the instruction "JMP $B700." $B700, by the way, is the start of our real-time interrupt routine.

Note: Start entering program listing by typing "ASM B600 <CR>" at BUFFALO's prompt.

LINE

1	B600	LDS #0050
2	B603	LDX #1000
3	B606	LDD #003D
4	B609	STD #0000
5	B60C	BSET 26,X,01
6	B60F	LDAA #40
7	B611	STAA 1025
8	B614	STAA 1024
9	B617	CLI
10	B618	LDD #0000
11	B61B	BNE B618
12	B61D	LDAA #01
13	B61F	STAA #7007
14	B622	LDAB #FF
15	B624	DECB
16	B625	NOP
17	B626	NOP
18	B627	BNE B624
19	B629	CLRA
20	B62A	STAA 7007
21	B62D	BRA B600

Now type a period to get to BUFFALO's prompt (>); then type "ASM B700."

30	B700	LDAA #40
31	B702	STAA 1025
32	B705	LDX 0000
33	B708	DEX
34	B709	STX 0000
35	B70C	RTI

Again type a period, and then type "ASM 00EB."

40	00EB	JMP B700

FIGURE 6-10 Program listing, in BUFFALO's ASM line assembler format, for real-time interrupt experiment that uses the Mag-11.

Before I actually start the line-by-line discussion, notice that there are actually three different sets of lines: (set 1) lines 1 through 21, (set 2) lines 30 through 35, and (set 3) line 40. Set 1 is the main program—the one that you start with the BUFFALO's GO command. Set 2 is the interrupt routine. It is in the interrupt routine that the "work" associated with the interrupt gets done. Notice that this routine, like most interrupt routines, ends with a return from interrupt (RTI) instruction. Set 3, which is simply the instruction "JMP B700" tells the CPU where to go to start its interrupt routine.

Finally, the line-by-line discussion:

Line 1 initializes the stack by storing $50 in the stack pointer. (Remember that with the ASM line assembler, all numbers are assumed by the line assembler to be hexadecimal.)

Line 2 sets up the register base address by storing $1000 in the X index register. The primary reason that this is done is that we have to use index addressing in line 5. Notice that I have limited the use of index addressing. While index addressing is often the neatest way to accomplish things, it often is also the most confusing, and thus I try to avoid it when possible—at least when writing a program that is supposed to be easy-to-understand.

Lines 3 and 4 store the "count" number, which is $003D or 61 in decimal at $0000 and $0001. (Note that the $3D winds up being stored at $0001.)

Line 5 uses the BSET instruction to set bit 0 of the pulse accumulator control register (PACTL) at $1026 (see Fig. 6-11). Bits 0 and 1 of this register determine the RTI rate. After reset, both bits are clear, which sets the RTI rate at 8.19 ms with a 4-MHz crystal. Here, we set bit 0 and clear bit 1, which results in a rate of 16.384 ms. With both bits set, the rate is 65.54 ms. (With an 8-MHz crystal, all rates are halved.)

Lines 6, 7, and 8 store $40 at $1025 (the TFLG2 register; see Fig. 6-12), which clears the real-time interrupt flag (RTIF), and also at $1024 (the TMSK2 register; see Fig. 6-13), which configures the RTI system for interrupt-driven operation.

Line 9 uses the CLI instruction to clear the interrupt mask in the CCR. This allows normal interrupt operation.

Lines 10 and 11 are rather interesting. Keep in mind here that the interrupt routine starting with line 30 subtracts 1 from the data at 0000 and 0001 every time it is called on.

REGISTER'S NAME: PACTL				REGISTER'S ADDRESS: $1026			
BIT#	7	6	5	4	3	2	1	0
BIT NAME	DDRA7	PAEN	PAMOD	PEDGE	0	0	RTR1	RTR0
BIT'S STATUS AFTER RESET	0	0	0	0	0	0	0	0

FIGURE 6-11 The PACTL register.

REGISTER'S NAME: TFLG2						REGISTER'S ADDRESS: $1025	
BIT#	7	6	5	4	3	2	1	0
BIT NAME	TOF	RTIF	PAOVF	PAIF	0	0	0	0
BIT'S STATUS AFTER RESET	0	0	0	0	0	0	0	0

FIGURE 6-12 The TFLG2 register.

REGISTER'S NAME: TMSK2						REGISTER'S ADDRESS: $1024	
BIT#	7	6	5	4	3	2	1	0
BIT NAME	TOI	RTII	PAOVI	PAII	0	0	PR1	PR0
BIT'S STATUS AFTER RESET	0	0	0	0	0	0	0	0

FIGURE 6-13 The TMSK2 register.

Line 10 loads the double accumulator D with data at 0000 and 0001. Line 11 checks if these data are zero. If not, the program again loads D with the data at the same memory locations. The program keeps looping here until the data at 0000 and 0001 are zero. This happens after 61 real-time interrupts have occurred. Also see the interrupt routine starting on line 40.

Once the data at 0000 and 0001 are both zero, the program continues to line 12, where accumulator A is loaded with 1.

In line 13, LED8 lights because accumulator A is stored at memory location $7007, which is the location of LED8.

Lines 14 through 18 merely provide a short delay so that LED8 stays lighted long enough to notice.

The next two lines (lines 19 and 20) turn off LED8, and then the next line (line 21) sends the program back to the start at $B600.

The interrupt routine starts on line 30. Here, the first two lines clear the RTIF in the TFLG2 register by storing $40 at $1025. The next three lines subtract one from the counter at 0000 and 0001. The RTI instruction causes the CPU to start executing instructions where it left off when it was so rudely interrupted by the real-time interrupt.

Notice the last line of the listing—line 40. This line sets up the address in the pseudovector at 00EB through 00EC. Notice that the JMP B700 instruction here results in a jump to $B700 every time the CPU receives a real-time interrupt.

Having More Fun with the RTI

Now that you are better acquainted with the RTI, think how you can increase the time between LED8's flashes to 5 s and to 1 min. I will show you, inch by inch, foot by foot, how to decrease the flash rate to one every 5 s, and then you should be able to figure, all on your own, how to decrease this rate to one every 1 min.

First, recall that we found that for a 1-s rate, we needed 61 interrupts. Also recall that 61 in decimal is 3D in hexadecimal. This hexadecimal number is important because BUFFALO's line assembler assumes that all numbers are in hexadecimal. Since we want a 5-s rate, we have to multiply 61 by 5, which is 305.

What is 305 in hexadecimal? While there are many rote formulas to convert decimal to hexadecimal, all are easily forgotten. However, by getting down to the real meaning of hexadecimal and decimal, I have come up with a method—not a formula, just a method—that does the job. First, let's look at the *finger-counting format,* which is also referred to as the *decimal* or *base-10 format.* Here, the number on the right is in the one's place, the next number to the left is in the ten's place, the next number to the left is in the hundred's place, and so on. In hexadecimal or base-16 format, the number on the right is also in the one's place, but the number to the left is in the sixteen's place, the next number to the left is in the 256's place, and so on. Our problem is the number 305. How many 256's are in 305? One, of course. So our hexadecimal number starts with a "1." Now we subtract 256 from 305 to get 49, which is what we have to represent next. How many 16's are in 49? Three! So our next number is 3, and so far we have the number 13?, with a question mark for the number in the one's place. Now we subtract 48 from 49 to get 1, which is our last "mystery" number. So we now know 305 in decimal is 131 in hexadecimal!

While it is possible to use BUFFALO's memory modify command (MM) here to enter the data, the simplest way is to go back to the BUFFALO's ASM command:

```
>ASM b606
B606 LDD 0131
```

Run the program again ("GO B600"), and you should have LED8 now light every 5 s instead of every 1 s.

Experiment   Change the value of the argument at line 3 so that the LED flashes at 1:00-min intervals.

Another quick brain teaser   With only changing line 3, what is the slowest rate you can flash LED8?

Another quickie   What is the slowest rate you can flash LED8 if you are allowed to change both lines 3 and 5?

Hint:

Go back to the discussion on lines 3 and 5. Also refer to the pictorial of the PACTL register in Fig. 6-11.

Introducing the Pulse Accumulator

Another system associated with the HC11's timer is the pulse accumulator (PA.) The PA can operate in two different modes: event counting and time measurement. I will discuss only its first use—event counting. At the end of the discussion, I will briefly mention its other uses.

To demonstrate the PA's event-counting feature, we will use a phototransistor to count (1) the number of times a light beam hits the phototransistor and (2) the number of times darkness descends on the phototransistor.

Before I get to the real thrills, however, let's look at a few practical matters. First, the input to the PA is the PA7/PAI pin, which is pin 1 of the 68HC11A1P. For those who want to get their hands on this pin, notice that this pin is connected to pin 2 of Mag-11's female header J3.

If you are wondering where the real work of the PA is done, wonder no longer. The pulse accumulator counter (PACNT) register is the one you seek. The PACNT, located at $1027 and shown in Fig. 6-14, however, is a laborer and not the boss with the brains. While the real boss is the human or human-like user, the status and control registers can be looked on as "supervisors." All three of these registers (TMSK2, TFLG2, and PACTL) were used by the RTI. However, the RTI used different bits in these registers. For instance, in the PACTL, we used only bits 0 and 1 to set the real-time interrupt rate. We do not use these bits here, although we use the four most significant bits (bits 4 through 7). The only bit we change here after reset is bit 6, which is called the *pulse accumulator enable bit.* In order to operate the PA, this bit must be set—it is cleared after reset.

Instead of a dreary discussion of the other two registers, let's get right into another exciting experiment. All you will need here is the same general-purpose phototransistor you used in the experiment in Chapter 5 on the photometer. The other part—a 10-kΩ resistor— is already connected from the HC11's PA7/PAI pin to +5 V.

To wire the *real complicated circuit,* insert the phototransistor's collector in pin 2 and its emitter in pin 20 of Mag-11's J3.

Caution:

Make sure the power is off before inserting the phototransistor's leads in J3.

REGISTER'S NAME: PACNT					REGISTER'S ADDRESS: $1027			
BIT#	7	6	5	4	3	2	1	0
BIT NAME	BIT 7	BIT 6	BIT 5	BIT 4	BIT 3	BIT 2	BIT 1	BIT 0
BIT'S STATUS AFTER RESET	—	—	—	—	—	—	—	—

FIGURE 6-14 The PACNT register.

Now enter the program listing in Fig. 6-15. To run the program, first leave the line assembler by typing a period, and then start the program by typing "GO B650."

LED1 through LED8 should shut off. Take a flashlight and place it within a few inches of the phototransistor. LED1 should go on. Now move the flashlight back and forth. The bank of LEDs should indicate, in binary, the number of times the flashlight beam struck the phototransistor. Notice the LEDs seem to respond only to the rapid increase in light intensity (i.e., the change from dim light to bright light). At first glance, this may be confusing because bit 5 of the PACTL register (located at $1026) is clear, which means that the pulse accumulator only responds to falling edges of pulses. What's going on? Well, this

---

Note: Start entering program listing by typing "ASM B650 <CR>" at BUFFALO's prompt.

LINE		
1	B650	LDS #50
2	B653	LDAA #40
3	B655	STAA 1026
4	B658	CLR 1027
5	B65B	LDAA 1027
6	B65E	STAA 7000
7	B661	RORA
8	B662	STAA 7001
9	B665	RORA
10	B666	STAA 7002
11	B669	RORA
12	B66A	STAA 7003
13	B66D	RORA
14	B66E	STAA 7004
15	B671	RORA
16	B672	STAA 7005
17	B675	RORA
18	B676	STAA 7006
19	B679	RORA
20	B67A	STAA 7007
21	B67D	BRA B65B

FIGURE 6-15 Program listing, in BUFFALO's ASM line assembler format, that uses Mag-11 along with the circuit in Fig. 5-16 to demonstrate one use of the HC11's pulse accumulator subsystem.

is where a little knowledge of circuit theory can take some of the mystery out of life. Notice from the simple circuit that the phototransistor's emitter is connected directly to ground, while its collector is connected to +5 V through a 10-kΩ resistor. Well, when light hits the phototransistor, its current increases, which simply means its resistance decreases, which makes its collector voltage drop. *Voilà!* There is our negative edge.

Now try something else interesting. Change the instruction on line 2 to "LDAA #50," and run the program again. Note that what you did was set bit 4 of the PACTL register, which makes the PA respond to rising edges—remember, before it responded to falling edges.

You probably have noticed the simplicity of the PA. However, I have showed only one of the ways of using the PA. To be honest, I showed not only the simplest use of the PA—counting events—but also the simplest example of counting possible. The PA also can be used to measure the frequency of an unknown signal as well as its duration. There are a multitude of similar uses. Another feature, which is related to the PA, is use of the PA7/PAI pin as a general-purpose interrupt somewhat similar to, but in various ways superior to, the well-known IRQ pin.

What Is the Computer Operating Properly (COP) Watchdog System?

The COP watchdog timer system also uses the HC11's timer. The COP system differs from most HC11 systems and subsystems in that its purpose is simple: to keep the HC11-based system running normally. In this respect its purpose is the same as the illegal opcode trap and the clock monitor system, both of which have been discussed already. While the illegal opcode trap is ingeniously simple, it is possible to have a system lock up without ever encountering an illegal opcode. Most of the time a properly designed COP watchdog system will solve this problem. To be honest, however, the COP watchdog system is not only used infrequently, but it does not seem to be encouraged by Motorola! Why do I say this? Well, the HC11's CONFIG register's NOCOP bit (bit 2) must be clear for the COP system to be enabled, and as shipped, most derivatives of the HC11 have this bit set. By the way, the only HC11s that are shipped with COP enabled are labeled *HCP11*. While the user can clear this bit and enable the COP system, it is obvious that Motorola is not shoving the use of the COP system down the users' throat. The people at Motorola believe, as I do, in *freedom*. Ah, freedom—it's the only way to treat people and MCUs.

WHAT DOES THE COP SYSTEM DO?

When the COP system is enabled, the user's software is responsible for keeping a free-running watchdog timer from timing out. If the watchdog timer times out, this indicates trouble with execution of the software (most likely a stray pulse-type glitch), and a system reset is initiated. While power-on reset, clock monitor reset, and COP reset all have different reset vectors, it is typical to direct these vectors to the same initialization software (i.e., send the program to its start).

WHAT DOES SOFTWARE NEED TO DO TO KEEP THE COP SYSTEM FROM CREATING A SYSTEM RESET?

A system reset is the medicine that cures the sickness that I have labeled the "wild, wild program syndrome." You have to watch it, however. The medicine can have terrible side effects, especially if the "reset medicine" becomes addictive. It is possible for the medicine to become worse than the disease—and this is one reason (other reasons may include pure slothfulness) that many designers are hesitant about using the COP watchdog timer.

Once the COP watchdog system is enabled (remember, this is done by clearing bit 2 of the CONFIG register), the COP timer must be reset by a specified software sequence before a specified time runs out. If the software fails to do this in time, a COP reset is initiated, and the "medicine" is administered, whether the system needs it or not. What is the specified software sequence? Here it is:

1. Write $55 to address $103A (COPRST register). This arms the COP timer-clearing mechanism.
2. Then write $AA to the same address, which actually clears the COP timer.

While both writes must occur in correct order, other instructions can be placed between the writes. Also, the time between the reset sequence must be shorter than the chosen COP timeout period. This reset sequence is referred to as *servicing the COP timer.*

A question: How patient is the COP system? In other words, how long is the chosen COP timeout period? Well, again, Motorola gives us freedom in the form of choices. Assuming the use of a 4-MHz crystal, the timeout period is 32.768 ms right out of reset, which means that the COP timer must be serviced more frequently than 32.768 ms or the system will experience a forced reset (i.e., the "medicine"). Other timeout period choices depend on bits 0 and 1 of the OPTION register (see Table 6-2). Recall that the OPTION register is located at $1039 and is shown in Fig. 5-14.

There is one precaution here concerning the OPTION register. While this register can be written as is normally done (e.g., with a simple STAA or STAB instruction), it can only be written to once, and that "write" must be within 64 bus cycles after reset. Because of this restriction, it is often wise simply to use the shortest timeout in the table in your COP software servicing design. Before leaving the COP timer, I will show a subroutine lifted from the BUF-FALO monitor's own source code (Fig. 6-16). Notice how this subroutine services the COP.

TABLE 6-2 COP TIMEOUT CHANGES

CR1 (BIT 1 OF OPTION REGISTER)	CR0 (BIT 0 OF OPTION REGISTER)	NOMINAL COP TIMEOUT FOR	
		4-MHZ CRYSTAL (E = 1 MHZ)	8-MHZ CRYSTAL (E = 2 MHZ)
0	0	32.768 ms	16.384 ms
0	1	131.07 ms	65.536 ms
1	0	524.29 ms	262.14 ms
1	1	2.1 s	1.049 s

```
*********
*INPUT() - Read device. Returns a=char or 0.
* This routine also disarms the cop.
*********
INPUT    EQU *
         PSHX
         LDAA #$55  reset cop
         STAA COPRST
         LDAA #$AA
         STAA COPRST
         LDAA IODEV
         BNE INPUT1  jump not sci
         JSR INSCI  read sci
         BRA INPUT4
INPUT1   CMPA #$01
         BNE INPUT2  jump not aci
         JSR INACIA  read acia
         BRA INPUT4
INPUT2   LDX #PORTA
         CMPA #$02
         BEQ INPUT3  jump if duart a
         LDX #PORTB
INPUT3   JSR INUART  read uart
INPUT4   PULX
         RTS
```

FIGURE 6-16 COP service routine lifted from BUFFALO's source code.

The next chapter is devoted to the HC11's SPI. While not really hard to use or understand, the SPI peripheral is probably the least understood of all the HC11's systems. However, it has enormous possibilities. I will again attempt to stick to the concrete (dry, not wet) and avoid the abstract (where possible). Chapter 7 demonstrates how to interface the SPI with two different inexpensive real-world chips. It also demonstrates how to add two seven-segment LED displays to Mag-11 with only five or six wires connecting them. Chapter 8 then goes into detail on the joys of using the AS11 cross-assembler.

PROGRAMMING AND CUSTOMIZING

THE HC11 MICROCONTROLLER

HC11's SPI Interface

We already examined, in some detail, the HC11's serial communication interface (SCI). Like the SCI, the serial peripheral interface (SPI) exchanges data serially. A minimal SCI system can consist of just two wires (one data transmission wire and one ground) for a one-way exchange or three wires (two data transmission and one ground) for a two-way exchange. However, this paucity of connecting wires is only possible because the SCI is asynchronous, which means that the receiver and transmitter must be set up, before the "real" communication begins, for roughly the same bit rate. The maximum data rate of the standard HC11's SCI is in the neighborhood of 10 kb/s. At rates substantially higher than this, problems arise. (New derivatives of the HC11 can operate at substantially higher data rates; see Chapter 13.) One analogous problem you probably have experienced a number of times if you have watched enough movies is that which occurs when an actor's voice becomes unsynchronized with his or her lip movements. If the difference is less than a second, this problem will be just annoying, since you still can understand the actor correctly. However, if sound and video are out of sync for longer periods (which is very rare), it can become difficult to determine which actor says what. The same thing can happen with asynchronous communications. This type of communication is *always* out of perfect sync, but if the error is small, the system is smart enough to still get the data right. However, a point is reached (and this point gets farther and farther out as the years go by) where the system gets so far out of sync that the data stream becomes gibberish to the receiver. One solution is to provide another line that carries a clock signal so that the receiver is always in near-perfect sync with the transmitter. The HC11's SPI system does just this.

In the HC11's SPI, the line that carries the stable clock signal is called the *SCK*. The SCK "synchronizes" the data exchange between the transmitter and receiver. Since the data exchange is synchronized, the maximum data rate can be increased more than a hundredfold over the HC11's SCI system. This is fast enough to compete with a parallel bus. The reduction in wires and resulting circuit simplicity often make the SPI the preferred method of hooking up external devices such as latches and analog-to-digital (A/D) and digital-to-analog (D/A) converters, real-time clocks, frequency synthesizers, light-emitting diode (LED) and liquid-crystal display (LCD) drivers, additional MCU/MPUs, etc.

It has been stated that the SPI subsystem has the steepest learning curve of any of the HC11's built-in goodies. While this "steep learning curve" jargon is overused and thus boring, it is unusually descriptive when talking about the SPI. Why? Probably because once one gets used to the terms and acronyms used when talking about the SPI, the warmth of the sun feels better on your shoulders, the robin's song is more melodious, and the odor of hot solder becomes even more alluring.

Master Versus Slave

Two types of devices are involved in SPI data transfer: a master and one or more slaves. As the name indicates, the master is the "boss." The master can initiate a data transfer, but a slave can only respond to it. Do not get confused here, though, and jump to the conclusion that only a master can transmit data and thus can only function as a transmitter. A slave also can transmit data. Nonetheless, the master must tell the slave that it wants data— the slave is never allowed to tell the master what it wants or does not want.

While the HC11 SPI subsystem can be configured either as a master or a slave, I will restrict my discussion here in this book to the HC11's SPI being configured as a master. This not only simplifies things but is the "normal" way of doing things.

Using HC11'S SPI to Add Another Output Port to MAG-11

Although this is a simple experiment, it is a practical one. While an HC11 in single-chip mode has a good supply of output ports, an output port in an HC11 expanded multiplex system is precious because of its scarcity. The purpose of this experiment is not only to lessen your fear of the SPI but also to show you one way to add an additional 8-bit output port to an HC11 system. In this first experiment we will add one output port that will control eight LEDs. In the second experiment, which is actually even more useful, we will hook up two seven-segment LED displays to the HC11 with only five wires. In the third experiment we will add an input port.

The schematic for the first experiment is shown in Fig. 7-1. A solderless breadboard is recommended for wiring up this circuit (see also Fig. 7-2). On the breadboard, eight LEDs are used to display binary/hexadecimal numbers. To make "sense" out of these LEDs,

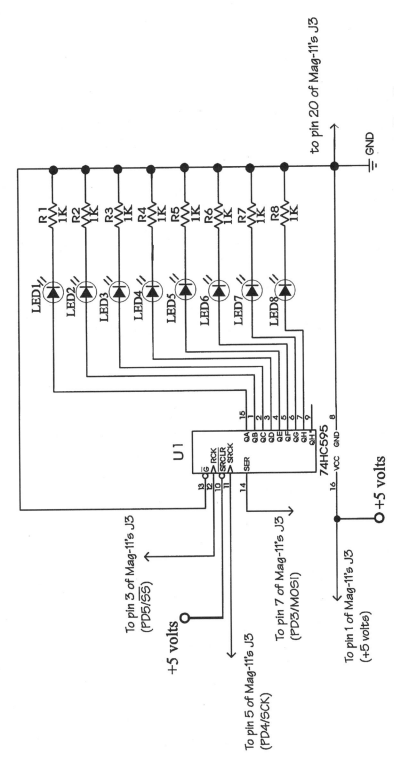

FIGURE 7-1 Schematic showing use of the HC595 and the SPI module to add an additional output port to Mag-11.

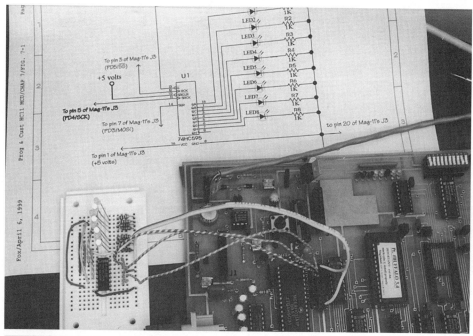

FIGURE 7-2 The HC595.

locate them in a single row with LED8 (the MSB) on the left, as you normally look at the board. As in most experiments, it is recommended that you use color-coded, no. 22 hookup wire to connect the breadboard to Mag-11.

You may have noticed that there are five wires connecting the circuit on the breadboard to Mag-11. If you have a separate +5-V power source, you can eliminate the wire going to pin 20 (+5 V) of J3. Now you are left with only four wires connecting Mag-11 with the circuit. As you will see in the next experiment, it is possible to connect two seven-segment LED displays to Mag-11 with only five wires.

Of course, just as in any experiment with the HC11, you need a program. Not just any program, but a program that is specifically designed for the application. Here again, we will use the nearly ideal programming tool that makes these experiments so easy to do and so much fun—BUFFALO's ASM line assembler.

After typing "ASM B7B0" (at the BUFFALO's > prompt), type in the appropriate information in the program listing in Fig. 7-3. *Again, do not type in the line numbers at the extreme left.*

Before running this program, look at line 6. Here accumulator A is loaded with $55, which is 01010101 in binary. This is the number that will be displayed on the breadboard's LEDs. You can change this number at will to display different hexadecimal/binary numbers. Why not try $AA (10101010)?

To test this program, wire it up as shown in Fig 7-1, enter the program in Fig. 7-3, and then run it with Go B7B0. If everything is OK, the LEDs on the breadboard should indicate 01010101, which is $55. Make sure you go back and change line 6 as indicated so that the LEDs will indicate a different number.

Notes: (1) Line numbers at extreme left are meant for discussion purposes only. (2) BUFFALO's ASM line assembler assumes that all numbers are already in hexadecimal. *Do not add "$" prefix.* (3) Numbers in the second column are addresses. *Do not type these numbers in.* However, these addresses are displayed on the BUFFALO screen. (4) To enter this listing, start BUFFALO and then type "ASM B7B0 <CR>" and then start entering the text in *third column only.* After you are done typing the line, hit the ENTER key (i.e., carriage return). (5) After entering last line, hit the decimal point key to leave the ASM line assembler.

```
line #
1      B7B0   LDX   #1000
2      B7B3   LDAA  #50
3      B7B5   STAA  1028
4      B7B8   LDAA  #38
5      B7BA   STAA  1009
6      B7BD   LDAA  #55
7      B7BF   STAA  102A
8      B7C2   LDAB  1029
9      B7C5   BITB  #80
10     B7C7   BEQ   B7C2
11     B7C9   BCLR  08,X 20
12     B7CC   BSET  08,X 20
13     B7CF   BRA   B7B0
```

FIGURE 7-3 **Program listing, in BUFFALO's ASM line assembler format, that demonstrates use of the HC595 to add an output port to Mag-11.**

Now that wasn't hard, was it? To see what was actually taking place, let's look at the HC595, the HC11's SPI-related signal pins, and the SPI registers before we undergo a line-by-line explanation of the program listing itself.

A Look at the 74HC595

The HC595 is an 8-bit shift register with output latches. Figure 7-4 shows a simplified logic diagram for this chip. Both the shift register and latch (storage register) use positive-edge triggered clocks. Because of this, bits CPOL and CPHA in the SPCR register should be equal. Shortly, I will undertake a longer discussion on the CPOL and CPHA bits than you (and I) will like.

As Fig. 7-4 shows, you connect your serial data stream to pin 14 and the clock to pin 11. Simple. However, no matter how many bits of data you enter, nothing will register at the latches' outputs (pins 1 to 7 and 15) until the leading edge of a pulse is received at the latch clock pin. This latch clock input is also referred to as the *RCK pin* in some data sheets, as well as in the schematic in Fig. 7-8. In our experiment, this pin is connected to port D's pin PD5/_SS, and the data in the HC595's shift register are transferred to the latch with two instructions (BCLR $08,X $20 and BSET $08,X $20) that send out a negative pulse. *Remember: The last edge of a negative pulse is a positive edge.*

Pin 13 is connected to ground because it is the *output enable pin.* This pin is also referred to as the _G pin in some data books, as well as in Figs. 7-1 and 7-8. A positive

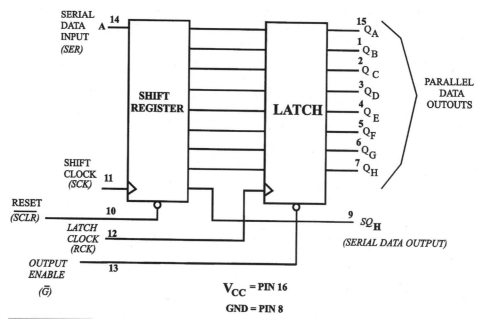

FIGURE 7-4 Simplified logic diagram for the 74HC595. *Note:* **Pin description in italics is used by some chip manufacturers.**

level at this pin disconnects the output latches, which will completely extinguish the display.

Pin 10, the reset pin, which clears the shift register, is connected to +5 V because we do not really use it here. This pin is also referred to as the _SCLR pin in some data books, as well as in Figs. 7-1 and 7-8.

HC11'S SPI-Related Signal Pins

Four of HC11's port D ($1008) related pins are used by the SPI: PD2/MISO, PD3/MOS1, PD4/SCK, and PD5/_SS. Figure 7-5 gives a look at port D, whereas Fig. 7-7*A* provides Motorola's own block diagram of the SPI subsystem.

PD2/MISO. MISO stands for "master in, slave out." Since we restrict ourselves to configuring the HC11's SPI as a master, we can think of this as the *serial input* to the HC11. By the way, the most significant bit is received first. This is opposite to the way SCI typically sends out data. In the MC68HC11A1P, the PD2/MISO pin is pin 44. This pin is connected to pin 9 of Mag-11's J3.

PD3/MOSI. MOSI stands for "master out, slave in." Again, since we will only discuss the HC11's SPI being configured as a master, we consider this pin as the *serial output* from the HC11. As before, the most significant bit is sent first. This is pin 45 on the MC68HC11A1P. This pin is connected to pin 7 of Mag-11's J3.

Register Name: PORT D				Register Address: $1008			
BIT#	7	6	5	4	3	2	1	0
BIT NAME	0	0	BIT 5	BIT 4	BIT 3	BIT 2	BIT 1	BIT 0
BIT'S STATUS AFTER RESET	—	—	—	—	—	—	—	—

FIGURE 7-5 The PORTD register.

PD4/SCK.   SCK stands for "serial clock." This is *the* clock. It is used to synchronize data exchange between the master and slave. As you might guess, eight SCK cycles are needed to transmit 1 byte of data. The fastest possible rate of the SCK is one-half the E clock. Thus, for a 4-MHz crystal, the fastest rate is 500 kHz, whereas with an 8-MHz crystal, the fastest SCK is 1 MHz. All experiments in this book will use the fastest SCK. Since we are restricting our discussion to the HC11's SPI being configured as the master device, the PD4/SCK pin will be configured as an output. Pin 46 of the 68HC11A1P is the PD4/SCK pin. This pin is connected to pin 5 of Mag-11's J3.

PD5/_SS.   _SS stands for "slave select." As its descriptive name suggests, this pin is used to select a slave device. However, since the SPI looks on the _SS as an input pin (the "_" prefix means it is active when a 0), it will not be used here directly by the SPI because we are restricting ourselves to configuring the HC11's SPI as a master, and master's are not selected as slaves. Nonetheless, this pin is sometimes used indirectly by the HC11's SPI as an output pin to "select the slave device." Read on.

Important Point:

PD5 must be configured here as a general-purpose output. This is necessary to disable the mode fault circuit. In order to configure PD5 as a general-purpose output, set bit 5 (make it a 1) of the DDRD register ($1009).

Now that we have looked at the HC11's pins used by the SPI, we have to look deep inside the HC11 if we want to actually use the SPI.

HC11'S SPI-Related Registers

As in all HC11's subsystems, the related registers provide control and status information. Four registers affect, more or less, the SPI system. These registers are as follows:

Port D data direction register (DDRD).   This register, which we looked at briefly earlier, is located at $1009. To make things simple, we will store $38 at this register. We can do this because we are only discussing using the HC11's SPI system as a master, What does this $38 do? It enables the SCK output, configures the MOSI and MISO lines properly, and turns bit 5 (_SS) of port D into a general-purpose output so that it is not affected directly by the SPI system. [*Note:* We will use bit 5 of port D and the related pin in the first SPI experiment to fire a pulse at the 74HC595's RCK (register clock) input. This pulse transfers the contents of the HC595's shift register to its output latches.]

Serial peripheral status register (*SPSR*).   This register, which is located at $1029, is another simple register to use. Here, only bit 7, the SPI transfer complete flag (SPIF) will be used by us. This flag is set (SPIF = 1) on completion of data transfer between the processor and external device. In order to clear the SPIF bit, you read the SPSR and then access the serial peripheral data register (SPDR). With SPIF = 1, attempts to write to the SPDR are futile unless the SPSR is first read.

Since SPI interrupts will not be enabled in our experiments (see the section on the SPCR), our software will use the SPIF to keep track of when data transfer has been complete between the SPI and an external device. In effect, our program waits until the transfer is complete before continuing. The following code snippet accomplishes this:

```
**
            STAA SPDR
LOOP1       LDAB SPSR
            BITB #$80
            BEQ LOOP1
***
```

The workings of this above code snippet should be obvious. However, a somewhat less obvious but more elegant way of accomplishing the same thing is

```
**
            STAA SPDR
LOOP2       LDAB SPSR
            BPL  LOOP2
***
```

The trick here is to realize the difference between signed and unsigned numbers. With signed numbers, the most significant bit (bit 7) determines whether a number is positive (plus) or negative. If bit 7 = 0, the CPU thinks of the number as positive, so the BPL instruction will only branch if bit 7 = 0. This means that the serial data transfer between the processor and external device is not complete.

Serial peripheral data register (*SPDR*).   This register is reminiscent of the SCI's system SCDR register because it is the one that does the real work. This input-output (I/O) register is used to transmit and receive data on the serial bus. Only a *write* to this register initiates transmission/reception of another byte. Recall, however, that with SPIF = 1, attempts to write to the SPDR are futile unless the SPSR is first read.

Writes to the SPDR are not buffered, and data are placed directly into the shift register and transmitted. However, when the SPDR is "read," a buffer is actually being read and not the shift register itself. An overrun condition will exist if the SPIF is not cleared before a second transfer of data from the shift register to the read buffer is initiated. If this happens, the data byte causing the overrun will be lost.

SPI control register (SPCR).   This register located at $1028, along with the DDRD, must be configured properly before the SPI system can be used. This register is shown in Fig. 7-6, and this register does the following: Turns on (enables) the SPI system itself, enables SPI interrupts (we will not use these interrupts, so we keep them disabled), configures the SPI as the master, sets clock polarity and phase (the most confusing aspect of the HC11's SPI subsystem), and sets the SPI bit rate (we will use only the fastest available). In other words, it does a whole bunch.

Because of the importance of the SPCR, I will provide a bit-by-bit explanation of its operation here:

Bit 7: SPI interrupt enable.   Since we do not want to use the SPI interrupt system we make sure this bit is 0.

Bit 6: SPI system enable.   Set this bit to 1 to turn on the SPI.

Bit 5.   Make sure this bit is 0. A 1 here causes port D outputs to be open drain, which only has use in rather complex systems that we are purposely avoiding here.

Bit 4: Master/slave mode select.   Since we want the HC11's SPI to be master, we set this bit to 1.

Bit 3: Clock polarity select (CPOL).   A 1 here makes the SCK idle low (which means an active "high"). This is one of the confusing fine details about the SPI system. The confusion here does not arise from the uncertainty of what Motorola means by an SCK idling low but rather from the lack of detailed data from the manufacturers of the devices used with the HC11 (e.g., serial A/D converters, shift registers, etc.). More on this potential problem and solutions shortly.

Bit 2: Clock phase select (CPHA).   This bit works in conjunction with the CPOL bit to set the clock-data relationship between the master and slave. This bit selects one of two fundamentally different clocking protocols. In general, Motorola seems to favor setting this bit (i.e., CPHA = 1), since it often simplifies things. This bit is set at RESET. Much more, actually too much more, on the CPOL and CPHA bits shortly.

Register Name: SPCR				Register Address: $1028			
BIT #	7	6	5	4	3	2	1	0
BIT NAME	SPIE	SPE	DWOM	MSTR	CPOL	CPHA	SPR1	SPR0
BIT'S STATUS AFTER RESET	*0*	*0*	*0*	*0*	*0*	*1*	*U*	*U*

Note: U means "undeterminate" and can vary from batch to batch and is generally meaningless.

FIGURE 7-6 The SPCR register.

Bits 1 and 0. These bits, referred to as SPR1 and SPR0, respectively, select the clock rate. Since we usually want the fastest rate possible (which is $^1/_2$E or $^1/_8$ the crystal frequency), normally both bits are set to 0. The slowest rate ($^1/_{32}$E) is achieved by setting both bits to 1.

CPOL and CPHA

Here we go! A time to perk up. Or, perhaps, a time to take a snooze. It all depends on which particular chips you are trying to connect to the HC11's SPI.

CPOL and CPHA are bits 3 and 2 (respectively) of the SPCR ($1028). Since these bits are frequently the cause of problems in using the HC11's SPI, we will look into them more deeply than most other bits. Before we do this, however, be alerted to the fact that several manufacturers, notably Maxim and Dallas Semiconductor, frequently spell out in simple, concrete terms in their data sheets what to set CPOL and CPHA at. If you use only these parts, you can, if you like, skip the following discussion on these two bits and use the manufacturer's recommendation.

First off, the SPI clock (available at the SCK pin) does not run continually like the HC11's own clock (E). It normally is idling—either high or low (+5 V or 0 V, respectively) depending on how the CPOL bit is set. If CPOL = 0, it idles low; if CPOL = 1, it idles high. Once data are written to the shift register, eight clock cycles are generated to shift the 8 bits of data. This makes sense, right? After it has done its job, the SCK goes idle again, waiting for another write to the SPDR to send out another eight clock cycles.

The other bit, which is actually the more confusing of the two, is the CPHA, which sets the phase you want to use to mark valid data. The setting of this bit depends (and is really the confusing aspect here) on exactly which device or devices [usually chip(s)] you want to connect to your SPI. This usually requires a careful study of the chip's data sheets and sometimes even its block diagram and timing sheets. When CPHA = 0, the data are valid on the leading edge of SCK; when CPHA = 1, the data are valid on the trailing edge. However, many data sheets do not mention leading and trailing edges but rather rising and falling edges. For instance, the data sheet (at least the data sheet published by National Semiconductor) mentions that the popular 74HC595 8-bit shift register with output latches is positive-triggered (i.e., rising-edge-triggered.) Do not jump to the conclusion that a leading edge and a rising edge are the same animal. This is true only when SCK is idling at 0 V (idling low)—which occurs when CPOL is 0. When CPOL is 1 and the SCK is idling at +5 V, a leading edge is the start of a negatively going pulse. The trailing edge can be explained similarly. For more details, see Fig. 7-7B and *C*.

In general, when CPOL = CPHA, the SPI is configured for rising edges, and when CPOL and CPHA are different, the SPI is set up for falling edges. Also, the master device (which is always the HC11's SPI in our somewhat limited discussion) always places data on the MOSI line a half-cycle (i.e., right in the middle of the data bit) before the clock edge (SCK). It chooses this timing so that the slave device can reliably latch the device.

A rule of thumb that I hesitate to mention because it only is pertinent when the HC11 SPI is set up as a slave device is this: Things are simplified when CPHA = 1. The reason for this is that when CPHA = 0, the _SS input line must go high between successive char-

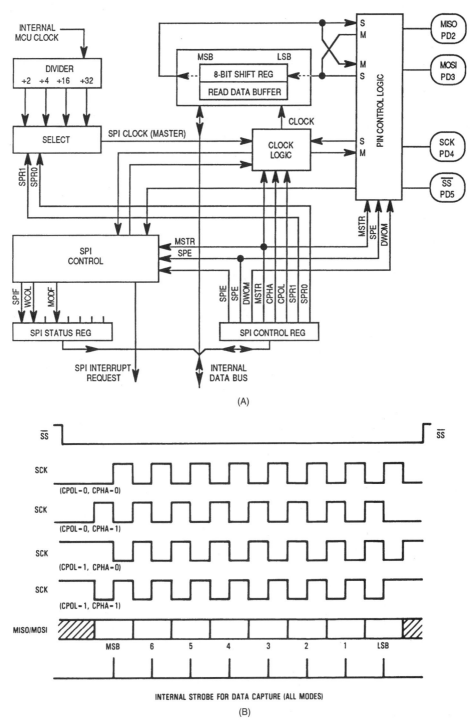

FIGURE 7-7 (*A*) SPI's block diagram. (*B*) SPI's data clock timing diagram.

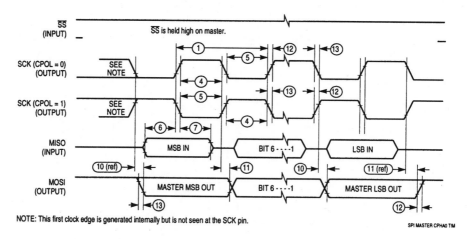

NOTE: This first clock edge is generated internally but is not seen at the SCK pin.

SPI MASTER CPHA0 TIM

SPI Master Timing (CPHA = 0)

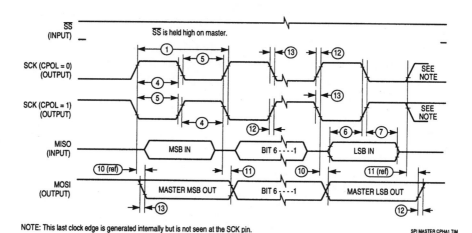

NOTE: This last clock edge is generated internally but is not seen at the SCK pin.

SPI MASTER CPHA1 TIM

SPI Master Timing (CPHA = 1)

FIGURE 7-7 (*Continued*) (*C*) SPI's mater timing diagram.

acters in an SPI message. With CPHA = 1, this is not required. As already mentioned, the HC11 comes out of reset with CP0L = 0 and CPHA = 1.

SETTING CPOL AND CPHA—A PRACTICAL LOOK

Setting the CPOL and CPHA correctly, if the manufacturer of the device does not spell it out in *big block letters* in the data sheet, may sound as if you need to be an experienced computer designer to do it right. This just is not true. Your best bet here, though, is to examine carefully the timing charts for both the HC11's SPI and the device you want to interface with. If your design is similar to the one already mentioned that uses the HC595, where you are using the SPI as only an output device (i.e., only using the MOSI pin), figuring out the timing charts is relatively simple. Here, you know that the data at pin MOSI are perfectly valid at the SCK's rising edge if CPOL = CPHA = 0 or 1 and are valid at the trailing edge if CPOL and CPHA are different. Now all you have to do is discover exactly when the interface device accepts the data on the MOSI line. If it, as with the HC595, accepts the data at the rising edge, then you simply want CPOL = CPHA, and you are done. You will see shortly with the discussion of the last experiment dealing with the HC597 that things are slightly more complicated when the data are traveling to the HC11 (i.e., when the MISO is used). As always, do not rely only on the timing charts. A good hands-on trial by breadboarding a simple circuit that uses the device is invariably wise. This can be done by connecting a "test circuit" to Mag-11 or similar HC11 prototyping board. Here, use BUFFALO's line assembler to write a test program that includes the CPOL and CPHA you think are the best ones.

A LINE-BY-LINE LOOK AT THE LISTING IN FIG. 7-3

Line 1. We load index register X with $1000. $1000 is the base address of the HC11's non-CPU registers. The index register (a CPU register) will only be used in addressing port D at $1008. Notice that we left out loading the stack pointer. The BUFFALO monitor did this for us.

Lines 2 and 3. These lines set the bits in the SPCR at $1028 so that the SPI is enabled, the SPI interrupt is disabled, the master mode is selected, the clock idles low, the very first edge is rising and occurs in the middle of the transmission of the MSB, and the SCK is operating as fast as possible.

Lines 4 and 5. These lines set the bits in the DDRD at $1009 so that pin PD5 is a general-purpose output, PD4 becomes the SCK output, PD3 becomes the MOSI output, and PD2, PD1, and PD0 become inputs *for the heck of it since they are not used here.*

Lines 6 and 7. These two lines do the work. Accumulator A is loaded with some data (here $55) and then stores them at the SPDR at $102A. This last instruction causes the SPI system to send the 8-bit data, one bit after another starting with the MSB, out the MOSI pin. This instruction also sends out eight clock cycles from the SCK pin.

Lines 8, 9, and 10. These lines are nearly identical to the code snippet described earlier. Note they could be replaced by the two lines: B7C2 LDAB $1029 and BPL B7C2. In short, these three lines of code are included so that the program waits here until the SPI is through sending the byte of data out the MOSI pin.

Lines 11 and 12. These two instructions, both of which use index addressing (refer to the discussion about line B7B0), cause a pulse with a rising edge to go out the PD5 pin. This rising edge causes the 74HC595's data to be transferred from its shift register to its latches so that they can be seen on the LEDs.

Line 13. This branch instruction should be self-explanatory.

NOTES CONCERNING THIS EXPERIMENT AND THE CPOL AND CPHA BITS

Notice that the instructions at lines B7B3 and B7B5 set bits CPOL and CPHA to 0. This sets up the SPI's clock so that data at the MOSI pin is valid at SCK's rising edge. It seems logical to think that if CPOL = 1 and CPHA = 1, the same results will happen because the MOSI is also valid at SCK's rising edge. I have checked this all out, and as expected, it works perfectly according to theory. Now, with CPOL = 0 and CPHA = 1, the MOSI is valid at the falling edge, which is not compatible with the HC595. Again, I have tried it and found that the experiment does not work right when $54 (CPOL = 0, CPHA = 1) is stored at $1009, which is the SPCR. This is great news. No surprises.

The unpleasant, pleasant surprise came when CPOL was set to 1 and CPHA to 0. The experiment worked perfectly again. But this time it did not seem like it should. This is bad. This is real, real bad because it implies that the designer of the experiment (me) does not really and truly know what's going on. To say this is "troubling" reminds me of the man who was looking at a tornado through his front window that was heading right for his house and saying to his wife who was at the kitchen table, "Honey, I think something's coming." To verify that this terrifying "success" was not an anomaly reminiscent of those which have occurred in various chemically induced nuclear effect experiments, I replaced the HC595 chip with another one and even changed the 4-MHz crystal to an 8-MHz one— getting the SCK right up near its rated maximum. Still, the experiment worked well.

I then again carefully, this time very, very carefully, examined the data clock timing diagram and noticed something potentially interesting. I imagined, "What if the SCK is just a few nanoseconds ahead of the data stream coming out of the MOSI pin?" Well, if this were true, the experiment would work exactly as discussed, with it not working when CPOL = 0 and CPHA = 1 but working OK when CPOL = 1 and CPHA = 0. My initial conclusion: The SCK is likely at least a few nanoseconds ahead of the data stream. While I did not use a scope to try to check this out, it seems logical to assume that this is what is going on. Nonetheless, logic is often the refuge of a slothful person.

Driving Two Seven-Segment LED Displays with the SPI

In this next experiment we will use the HC11's SPI and two HC595's to drive two seven-segment LED displays. This experiment has many applications.

Seven-segment LED displays are really neat. They also are easy to use and seem to be longer lived than LCD displays. While generally they are only thought of as useful to display decimal numbers, they are of use in displaying many more characters such as A, b, C, d, E, and F. In other words, they also can be used to display hexadecimal numbers. In addition, they can display the letters H, I, J, L, O, P, r, S, and U. In fact, it is possible to display 128 unique "symbols" (256, if you include the decimal point) on every seven-segment display. The fact that most of these symbols are meaningless to us is our fault, and not the display's fault. For instance, why does the Z have to be like it is and not just three parallel lines over each other? The point is that seven-segment displays are more versatile than many of us give them credit for. It is our lack of versatility that is the problem, not theirs.

The schematic in Fig. 7-8 shows how to wire up a two-digit display circuit, using Mag-11, to the HC11's SPI (also see Fig. 7-9). The program listing in Fig. 7-10 converts binary data (Mag-11's S1 is the input) into hexadecimal. This hexadecimal number is shown on the seven-segment LEDs inserted into the breadboard. DIS2, which is located above DIS1 in Fig. 7-9, displays the 4 most significant bits of the 8-bit number. These 4 bits are used so often that they have acquired a name: a *nibble*. No kidding! Well, what is an 8-bit number called? A *byte* of course! Well, you know, *byte* sounds just like *bite*, right? And a little bite is often called a *nibble*, right? So why shouldn't a 4-bit number be called a *nibble*?

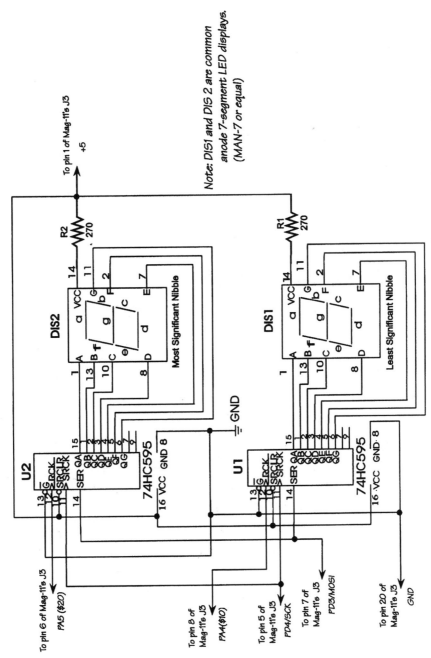

FIGURE 7-8 Schematic for a two-digit, seven-segment LED display for Mag-11 that uses the HC11's SPI module along with two HC595s.

171

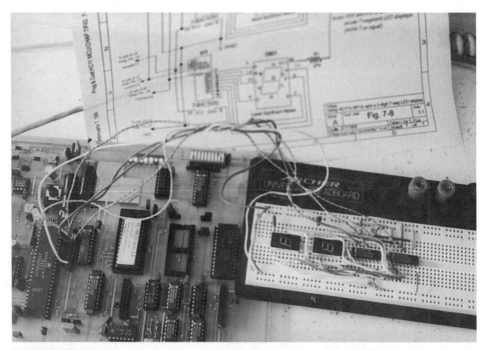

FIGURE 7-9 Photograph of LED display for MAG-11 shown in Fig. 7-8.

This experiment is really nice and practical for two reasons: (1) as mentioned earlier, it demonstrates how to easily connect up a two-digit display to Mag-11/HC11, and (2) it gives you excellent hands-on practice in converting from binary to its shorthand notation called *hexadecimal*.

After wiring up the experiment as shown in Fig. 7-8, enter the program listing in Fig. 7-10. Notice that this listing is rather long. However, even if you are rather slow at typing, it should take you less than 5 minutes to enter this listing into the HC11's EEPROM. Do not worry if you make a mistake. Thanks to the HC11's EEPROM and BUFFALO's ASM, there is no need to start over—just correct your mistake and continue typing.

RUNNING THE EXPERIMENT

Once the wiring and programming are done, the real fun starts—the anticipation of throwing the big one. Yes, the big switch! Since the program in this experiment starts at B600, it is possible to get it up and running immediately after reset by first switching S2(1) off. Doing this avoids having to type "Go B600" at BUFFALO's prompt. Also, before applying power, switch positions 1 through 8 of S1 [i.e., S1(1), S1(2),…, S1(8)] to their on positions. [Make sure S1(9) is on as well.] Recall that Mag-11's S1 switches are grounded when they are on. By turning all eight switches on, the program will load the "00" byte (2 nibbles) into HC11's accumulator A. The LEDs, LED1 through LED8, are instructed, by the program, to display the binary number indicated by S1(1) through S1(8). As in earlier experiments, LED1 displays the least significant bit, and LED8 displays the most significant bit. Also recall that S1(1) controls the least significant bit and S1(8) the most significant bit.

FIGURE 7-10 Program listing, in BUFFALO's ASM line assembler format, for XXX.

Note: Start entering program listing by typing "ASM B600 <CR>" at BUFFALO's prompt.

```
Line #
0       B600    LDS     #2D
1       B603    LDX     #1000
2       B606    LDAB    #50
3       B608    STAB    1028
4       B60B    LDAA    #38
5       B60D    STAA    1009
6       B610    LDAA    7000
7       B613    LDY     #7000
8       B617    STAA    0,Y
9       B61A    LSRA
10      B61B    CPY     #7007
11      B61F    BEQ     B62F
12      B621    INY
13      B623    BRA     B617
14      B625    NOP
15      B626    NOP
16      B627    NOP
17      B628    NOP
18      B629    NOP
19      B62A    NOP
20      B62B    NOP
21      B62C    NOP
22      B62D    NOP
23      B62E    NOP
24      B62F    LDAA    7000
25      B632    LSRA
26      B633    LSRA
27      B634    LSRA
28      B635    LSRA
29      B636    BSR     B660
30      B638    STAB    102A
31      B63B    LDAA    1029
32      B63E    BPL     B63B
33      B640    BCLR    0,X 20
34      B643    BSET    0,X 20
35      B646    LDAA    7000
36      B649    ANDA    #0F
37      B64B    BSR     B660
38      B64D    STAB    102A
39      B650    LDAA    1029
40      B653    BPL     B650
41      B655    BCLR    0,X 10
42      B658    BSET    0,X 10
43      B65B    BRA     B600
44      B65D    NOP
45      B65E    NOP
46      B65F    NOP
47      B660    CMPA    #0
48      B662    BEQ     B6A0
49      B664    CMPA    #1
50      B666    BEQ     B6A3
51      B668    CMPA    #2
52      B66A    BEQ     B6A6
53      B66C    CMPA    #3
54      B66E    BEQ     B6A9
```

FIGURE 7-10 *(Continued)* **Program listing, in BUFFALO's ASM line assembler format, for XXX.**

Note: Start entering program listing by typing "ASM B600 <CR>" at BUFFALO's prompt.

```
Line #
55    B670    CMPA    #4
56    B672    BEQ     B6AC
57    B674    CMPA    #5
58    B676    BEQ     B6AF
59    B678    CMPA    #6
60    B67A    BEQ     B6B2
61    B67C    CMPA    #7
62    B67E    BEQ     B6B5
63    B680    CMPA    #8
64    B682    BEQ     B6B8
65    B684    CMPA    #9
66    B686    BEQ     B6BB
67    B688    CMPA    #A
68    B68A    BEQ     B6BE
69    B68C    CMPA    #B
70    B68E    BEQ     B6C1
71    B690    CMPA    #C
72    B692    BEQ     B6C4
73    B694    CMPA    #D
74    B696    BEQ     B6C7
75    B698    CMPA    #E
76    B69A    BEQ     B6CA
77    B69C    LDAB    #E
78    B69E    RTS
79    B69F    NOP
80    B6A0    LDAB    #40
81    B6A2    RTS
82    B6A3    LDAB    #F9
83    B6A5    RTS
84    B6A6    LDAB    #24
85    B6A8    RTS
86    B6A9    LDAB    #30
87    B6AB    RTS
88    B6AC    LDAB    #19
89    B6AE    RTS
90    B6AF    LDAB    #12
91    B6B1    RTS
92    B6B2    LDAB    #2
93    B6B4    RTS
94    B6B5    LDAB    #F8
95    B6B7    RTS
96    B6B8    LDAB    #0
97    B6BA    RTS
98    B6BB    LDAB    #18
99    B6BD    RTS
100   B6BE    LDAB    #8
101   B6C0    RTS
102   B6C1    LDAB    #3
103   B6C3    RTS
104   B6C4    LDAB    #46
105   B6C6    RTS
106   B6C7    LDAB    #21
107   B6C9    RTS
108   B6CA    LDAB    #6
109   B6CC    RTS
```

Throw the switch—*now*! The bank of data LEDs (i.e., LED1 through LED8) should all be off, and both digits of the LED display should show "0." Now turn off S1(1). LED1 should light, and DIS1 should show "1." Now also turn off S1(2) through S1(4). LED1 through LED4 should light, and DIS1 should show "F." To check out DIS2, turn off S1(5). LED5 should light, and DIS2 should show "1," while "F" should still show on DIS1. If everything so far appears OK, start experimenting. Set positions of S1 so that different binary numbers show on the data LEDs, and then look at the seven-segment display to see the hexadecimal representation for these numbers.

Quick question How should S1's switches be set so that "BE" shows on the seven-segment display? *Hint:* Remember that setting the position *off* sets the bit (i.e., makes it a 1).

I strongly recommend that you spend some time trying out this experiment. Try out a number of switch settings, and try to figure out, before setting the switches, what will show on the display. This is an excellent hands-on refresher course for reviewing the hexadecimal numbering system. For some reason, experiments like this are more effective than a similar computer program and much more effective than simply reading about such programs.

A BRIEF LOOK AT THE CIRCUIT IN FIG. 7-8

Experienced but critical designers may look at the circuit in Fig. 7-8 and shake their heads. "That Fox guy," they might say, "fell into the dabbler designer's trap of only using a current-limiting resistor in series with the display's common anode!" My answer: "Take it easy guys, give me a break—and a moment to explain." There is no doubt that a better display will be achieved by connecting 680-Ω resistors in series with each of the LEDs' cathodes—this will eliminate a "1" being brighter than an "8," for instance. Nonetheless, things are simpler by replacing the 14 resistors with just 2. And all we are doing is experimenting, for goodness sakes! Besides, no damage is done. OK? (Of course, another alternative is to use a resistor network.)

The rest of the circuit is quite similar to Fig. 7-1 except that bits 4 and 5 (PA4 and PA5) of port A are used to toggle the HC595s latch clock pins (RCK). In the earlier experiment, bit 5 from port D was used.

A Quick Look at the Program

Relax! I will not do a line-by-line discussion here. Rather, let's look at the program listing in Fig. 7-10 from a broader, more general perspective.

The first line loads the stack pointer with $2D. This instruction is only necessary if you are going to run the program automatically at power up. [S2(1) is off.] The next five lines are identical to the first five lines in the listing in Fig. 7-3 and are basically initialization instructions. *Notice that the LDS instruction was not used in the program in Fig. 7-3. Since this program starts at B7B0, not B600, it is not possible to start it at power up. It must be run directly through BUFFALO's Go command, and BUFFALO has already set up the stack pointer for us.*

The next eight lines take the switch data from S1 (located at $7000; hopefully, you already knew this) and then turns on the respective LEDs. This routine is similar to the one in Fig. 5-9. The next ten lines are NOP instructions put there by me because I made changes in the program and was *too darn lazy* to completely write the program over. In addition to showing my vice of sloth, it also demonstrates concretely one of the uses for the NOP instruction.

Lines 24 through 28 shift the most significant nibble (4 bits) of data in accumulator A into the least significant nibble—this is done so that we can use a single subroutine that decodes the raw binary data into a code that will turn on the appropriate segments of the LED display. This subroutine starts at location B660 (line 46). Its purpose is simple: to turn on the appropriate segments of the display so that the display informs the user, in standard hexadecimal graphic notation, the binary value of the Mag-11's data LEDs (LED1 through LED8). For instance, line 46 compares accumulator A's contents to "00." If A contains zero, we want to turn on all segments of the display except "g," the center segment, in order to create a "0" on the display (also refer to Fig. 7-8).

Before continuing with this discussion, I want to refresh your mind with the fact that the displays are common anode. This means that the display's segment are *turned on* when the respective bit is 0 (0 V). Thus, it is obvious that we want to send 0100 0000 ($40) out the SPI. Notice that if accumulator A is "00," then the instruction on line 48 (BEQ B6A0) branches to B6A0. But look at the instruction at B6A0 (i.e., line 80). It is an LDAB 40 instruction. Wouldn't you expect this? After accumulator B is loaded with the right code, the program returns from the subroutine and starts again at line 29, where the next five instructions send out this "$40" code to U2—which then lights DIS2.

Stop everything! You may have a question here.

READER: "Apparently, this same $40 code is sent to both U1 and U2. Right?"

AUTHOR'S TERSE RESPONSE: "Yeahess."

READER: "Then why is only one DIS2 affected by this code?"

Great question. It is true. This $40 code is in both HC595's shift registers. But look at pin 12, connected to U2's latch clock input. It is connected to bit 5 of HC11's port A. The program toggles this pin. The result is that the data in U2's shift register are transferred to U2's latches, which turn on the displays' segments. U1's pin 12 is connected to bit 4 of port A and is controlled by toggling this pin—and this pin wasn't toggled. Only one display is controlled at a time, and it does not matter to us in the outside world what is in the HC595's shift register—the only thing the "users" (us guys in the outside world) care about is what is in the HC595's latches.

It may seem to some of you that the display will not work right because so many instructions are required to set up DIS2, and then a whole bunch more instructions are necessary to set up DIS1. These critics may say, "There would be too much flickering." However, all these instructions take place in a fraction of a millisecond—in about the same time it takes a speeding bullet to travel a foot. In other words, it "appears" to happen instantaneously—and this is with a 4-MHz crystal. Speed is everything and is the thing that provides MCUs with virtual intelligence.

Using the HC11'S SPI to Input Data

I just showed how to use the SPI to add an 8-bit output port to an HC11 system as well as how to add a two-digit seven-segment LED display. Now I will show how to add an 8-bit input port.

This experiment will use a 74HC597 8-bit parallel-input/serial-output shift register with input latches. This experiment will accomplish two objectives: (1) explain, with a hands-on project, how to use the SPI to receive serial input data (the preceding experiments showed how to send serial input) and (2) demonstrate one easy way to add an additional 8-bit input port to an HC11-based system such as Mag-11.

Wiring It Up

Follow the schematic in Fig. 7-11 to wire the HC597 experiment on a solderless bread-board (see also Fig. 7-12). The phototransistor called for in the parts list and shown on the schematic is the inexpensive PN128 NPN-type phototransistor. However, any general-purpose NPN phototransistor (e.g., TIL 78) also can be used. In place of the phototransistors, a 16-pin, eight-position dipswitch can be substituted. Nonetheless, phototransistors are preferred because they are more intriguing and perhaps even a bit titillating.

Entering the Program

Again, use BUFFALO's ASM line assembler to enter the program listing in Fig. 7-13 starting at $B700.

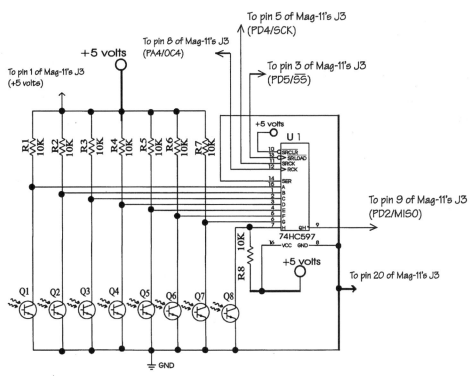

FIGURE 7-11 Schematic that shows how to use an HC597 and the HC11's SPI module to add an input port to Mag-11.

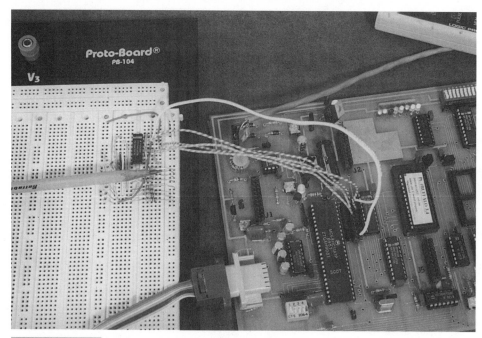

FIGURE 7-12 Using an HC597 and the HC11's SPI module to add an input port to MAG-11 (See Fig. 7-11).

Assuming that you have wired up the circuit in Fig. 7-11 correctly and entered the program listing in Fig. 7-13, run the program by typing "Go B700 <cr>" at BUFFALO's prompt. [Since this program does not start at B600, you must make sure that S2(1) is on and run the program from BUFFALO's prompt.] With the phototransistors in virtual darkness, LED1 through LED8 should light. When the phototransistors are in a bright light, all LEDs should be off. If this seems confusing, take a look at the schematic and think about a phototransistor's operation. When an NPN phototransistor has its collector connected to a positive source with respect to its emitter, the current flowing through its collector will increase when the light intensity rises. In effect, the phototransistor acts as a light-dependent resistor whose resistance drops when light hits it. Notice from the schematic that when the phototransistor's resistance drops, the voltage at its collector also drops. If this voltage drops below 1 V, the HC597 treats it as a 0; when it rises above about 3.5 V, the HC597 treats it as a 1. Between 1 and 3.5 V, the manufacturer has no guarantees. However, practically speaking, generally anything above 2.7 V is treated as a 1 and anything below 2 V is treated as a 0. However, voltages between these levels (2 and 2.7 V) can cause problems (e.g., increased supply current) and should be avoided. This can be done here simply by using only bright light and near darkness.

If you prefer to have the LED turn on when light hits the respective phototransistor, simply replace the NOP instruction in line B71F with the COMA instruction. For details on this instruction, see the next chapter.

Now, working in a brightly lit room, put your finger over one or more of the phototransistors. Notice the response on LED1 through LED8. While this may appear at first glance to be an interesting but useless experiment, the circuit and program have at least one prac-

Note: Start entering program listing by typing "ASM B700 <CR>" at BUFFALO's prompt.

```
line #
1       B700    LDX     #1000
2       B703    LDAA    #58
3       B705    STAA    28,X
4       B707    LDAA    #38
5       B709    STAA    09,X
6       B70B    BCLR    0,X 10
7       B70E    BSET    0,X 10
8       B711    BCLR    08,X 20
9       B714    BSET    08,X 20
10      B717    STAA    2A,X
11      B719    LDAA    29,X
12      B71B    BPL     B719
13      B71D    LDAA    2A,X
14      B71F    NOP
15      B720    LDY     #7000
16      B724    STAA    0,Y
17      B727    LSRA
18      B728    INY
19      B72A    CPY     #7008
20      B72E    BNE     B724
21      B730    BRA     B70B
```

FIGURE 7-13 **Program listing, in BUFFALO's ASM line assembler format, that is used to add an input port to Mag-11.**

tical purpose—reading punch cards. Details on this possible application are left to you. Now, let's take a look at how it all works.

The Circuit

In order to understand this circuit, we will first look at the HC597 chip. The simplified logic diagram for this chip is given in Fig. 7-14. Notice what it does—it takes in parallel data (pins 15 and 1 through 7) and then sends them out serially through pin 9. It also takes in serial data through pin 14, the purpose of which is to allow simple cascading of chips so that multiple bytes can be accessed through one circuit. We will ignore this here. (Pin 14 will be connected to ground.) We also will ignore the reset pin, pin 10. (Pin 10 is connected to +5 V.)

Looking at Fig. 7-14, it is quite obvious that the SPI's SCK line should be connected to the HC597's shift clock pin (pin 11). Other than the parallel data inputs, which are connected to the phototransistors' collectors, there are only two control inputs left—the latch clock (pin 12) and the serial shift_parallel load input (pin 13). Let's look at these pins in some detail:

Latch clock (pin 12). Abbreviated LC, the low-to-high transition on this input loads the parallel data from inputs A through H (pins 15 and 1 through 7) into the input latch. This loading of input data to latches must precede the action initiated by the next control input.

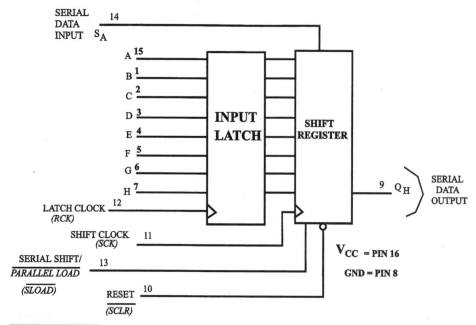

FIGURE 7-14 Simplified logic diagram for the 74HC597.

Serial shift/_parallel load (pin 13).   Abbreviated SS/_PL, this is the shift register control mode. Unlike the LC control pin, this one is level-dependent and *not* edge-dependent. When a low level is applied to this pin, the shift register receives data from the input latch. *With a low level at this pin, no serial shifting takes place.* When the level of this pin (SS/_PL) goes high, the shift register shifts serial data with every low-to-high transition of the shift clock input (pin 11). Here, data in stage H are shifted out QH (pin 9), and the data that were in stage H are replaced by the data that were stored in stage G, the data that were in stage G are replaced by the data that were in stage F, and the data in stage F are replaced.

The operation of the HC597 is fairly simple—send a positive pulse to lock in data from the input pins to the input latch, send a negative pulse to get the data into the shift register, and then connect a clock with eight pulses to get the data back into the SPI's MISO pin. However, notice that we are dealing here with a synchronous serial interface, and timing is everything—everything. Timing, as before, has to do with the CPOL and CPHA bits. As you can tell from lines 2 and 3 of the program listing, $58 is stored at the SPCR, which means that CP0L = 1 and CPHA = 0. This results in SCK idling low and the SPI being configured for falling edges.

Note:

When SCK idles low, leading and rising edges are synonymous, but when SCK idles high, leading and falling edges are synonymous.

The Mystery of the Wrong CPOL and CPHA Bits

The design choice of choosing CPOL = 1 and CPHA = 0 is apparently correct because the experiment works right. But so does CPOL = CPHA = 0 and CPOL = CPHA = 1, but not CPOL = 0 and CPHA = 1. This is an important point. No matter how much went into the design of a project, the design is only successful if the project works as it was planned to work. However, it is also important to investigate why setups that should not work do work. For instance, I spent a considerable amount of time poring over the HC597 and the SPI's timing charts. I came to the conclusion that the best choice was CPOL = 1 and CPHA = 0, and that's why $58 (%01011000) was stored at SPCR. But why did the two other choices for CPOL and CPHA also work? Back to the timing charts and (especially) the HC11's SPI and HC597's timing specifications. I pored over these charts and specs and discovered something interesting—if the SPI's data input hold time was at least 100 ns, then it was apparent that the SPI would be able to get the right data from the HC597 when it was configured for data capture on a rising edge (i.e., CPOL = CPHA = 0). Then going back to the HC597's timing charts and specs, I found out that at a supply voltage of 2 V, the maximum propagation delay from the rising edge to QH was 175 ns at 25 C. This delay would be plenty of time for the SPI to capture the right data. However, at a supply voltage of 4.5 V, this delay was only 35 ns. This was not enough time—or was it? Going back again to the SPI timing specifications, I took note that the 100 ns was the *minimum* setup time required. Apparently, 35 ns was enough for this particular chip's production run. However, slight variations may allow errors to creep into the data transfer, and while the data may be close to being accurate, close only counts when playing horseshoes—it ain't good enough when working with MCUs.

Before leaving this tale of mystery and suspense, let's delve into the details of why I chose CP0L = 1 and CPHA = 0. First, notice one important point: With CPOL = 0 and CPHA = 1, the experiment produces a result where the input data read by the SPI were out of phase. (You can try this out by changing line B703 to read LDAA #54.) For instance, bit 7 (MSB) was interpreted by the SPI as bit 6, bit 6 as bit 5,..., and bit 0, well, it wasn't there at all. In other words, this choice, apparently, put the system perfectly out of phase by a full SCK cycle. Doesn't it seem logical to think that if we are off by one cycle, all we need do is change the phase? Well, switching to CPOL = 1 and CPHA = 0 does just this.

While it seems obvious that the choice of CPOL = 1 and CPOL = 0 is correct here, let's look even more closely. This choice of bits causes the SPI to capture the data on its MISO line when there is a falling edge on the SCK. Remember that the HC597 shifts data when it receives a rising edge on the SCK. Before thinking of what happens here, think about what would happen if we chose CPOL = CPHA = 0. Here, the SPI would capture the data at a rising edge of SCK—which is the same theoretical instant that the HC597's shift register is shifting data. In other words, the data would theoretically, at least, be indeterminate. Since it takes about 30 or 40 ns for the shift register to change data, usually there will not be a problem—but sometimes there could be. Now, if we set CPOL and CPHA so that data will be captured by the SPI at the falling edge of SCK, we have a good half cycle (e.g., 500 ns with a 4-MHz crystal) to get the data. In other words, the SPI will capture the data right in the middle of the cycle, which provides plenty of time to get the correct data.

AVOIDING THE TIMING CHART NIGHTMARE

To be honest, timing charts can get confusing. What do you think the preceding discussion was all about? Clear? Come on, get real! Happily, many newer chips that interface with the SPI system have data sheets that tell you, straight on, the best choice for CPOL and CPHA—there is no need with these chips for detailed timing charts.

Examine the data sheets carefully. Sometimes the recommended setting for CPOL and CPHA are spelled out in *huge block letters* right in them. Table 7-1 lists a few chips, their description, and the recommended settings for CPOL and CPHA. Now that we have gotten most of the fine print explained, let's take a line-by-line look at the program that does the job.

A Line-by-Line Look at the Listing in Fig. 7-13

Line 1. This first line loads the X index register with $1000, which is the address for port A. $1000 is also the base address for the registers. Loading the X register with $1000 makes it simple to use index addressing when dealing with the registers.

Lines 2 and 3. These lines of code were mentioned already. In addition to setting CP0L = 1 and CPHA = 0, which idle SCK low and provide for data capture at the

TABLE 7-1 SEVERAL CHIPS AND RECOMMENDED CPOL AND CPHA FOR USE WITH THE HC11'S SPI

CHIP	DESCRIPTION	CPOL	CPHA	SPCR (AUTHOR'S RECOMMENDATION)	RELATIVE PRICE (1 = LOWEST)
74HC595 (generic)	8-bit serial latch	0	0	$50	1
74HC589 (generic)	8-bit parallel-input/serial-output shift register with input latch (3-state output)	1	0	$58	1
74HC597 (generic)	8-bit parallel-input/serial output shift register with input latch	1	0	$58	1
MAX512 (Maxim)	Triple 8-bit DAC	0	0	$50	2
MC145157-1 (Motorola)	Serial in frequency synthesizer	0	0	$50	3
DS1305* (DALLAS)	Serial real time clock	0	0	$50	2

*The DS1305 does not care about clock polarity. The chip samples the inactive clock when CE becomes active. CPHA must equal CPOL.

falling edge of SCK, storing $58 at SPCR enables the SPI system, configures the HC11 as master, and sets the SCK at the highest rate.

Lines 4 and 5.   These lines cause $38 (%00111000) to be stored at the data direction register for port D. This sets up the _SS, SCK, and MOSI pins as outputs and the MISO pin, which will be used here, as an input.

Lines 6 and 7.   These lines use the BIT CLEAR instruction to pull PA4 (bit 4 of port A connected to J3) low, and then the next line (line 7) uses the BIT SET instruction to pull the same pin high. This low-to-high transition latches phototransistor data into the HC597's latch.

Line 8.   This line then pulls the _SS pin (connected to J3's pin 3) low, which causes the shift register, in the HC597, to get loaded with the data from the latches.

Line 9.   This line pushes the _SS pin high, which selects the serial shift mode. At this point the HC597 is waiting for the SCK to start.

Line 10.   Here, the STAA 2A,X instruction writes to the SPDR register, which starts the SCK clock and initiates reception of the serial data from the HC597.

Lines 11 and 12.   These lines form a polling loop, causing the program to wait here until the SPI transfer complete flag (bit 7) in the SPI status register is set. Once this flag is set, the data are valid in the SPDR register.

Line 13.   This line uses the LDAA 2A,X instruction to load accumulator A with the data. Now accumulator A has the data that was at the eight input pins of the HC597 a couple of microseconds previously—actually 8 cycles previously, which is 8 μs with a 4-MHz crystal.

The rest of the program is rather routine, although I will continue the discussion anyway.

Line 14.   This line consists of an NOP instruction that simply leaves room for you to replace it with the COMA, as described earlier.

Lines 15 through 20.   These lines are the display routine where LED1 through LED8 display the contents of accumulator A, after it received the data from the HC597.

The first step in the display routine is to load index register Y with the address of LED1, which is $7000. You should recall that Mag-11 uses *only bit 0* in memory locations $7000 through $7007 to light LED1 through LED8. Thus, if accumulator A has a 1 at bit 0, storing it to $7000 will light LED1. A 0 at bit 1 will turn off the respective LED. After using index addressing to store accumulator A at location $7000, bit 1 is shifted to bit 0 by the next instruction LSRA. Of course, LSRA not only shifts bit 1 to the right but *all* bits to the right. The INY instruction on line B728 adds 1 to the Y index register, i.e., $7000 + 1 = $7001, so the index register is now pointing to LED2. The CPY #7008 instruction compares the Y index register with $7008, which is 1 more than the address of the last LED. If Y does not equal this yet, the program branches back to line B724 where the accumulator A is stored at $7001, the address of LED2. The loop keeps up, shifting A right and adding

1 to Y until it runs out of LEDs, and then the program goes on to line 20, where the BRA B70B instruction prepares to take another reading of the eight input pins of the HC597. The program continues until a power failure or you get tired of looking at the LEDs flash, whichever comes first.

We have completed our look at the HC11's SPI. Obviously, it is an extremely useful peripheral. Hopefully, the discussion on the CPOL and CPHA bits has not scared you from using it in your design. Keep this in mind: If timing charts and detailed block diagrams frighten you, simply stick to using chips where the data sheets spell out the recommended CPOL and CPHA settings.

In the next chapter I go into detail about using the AS11 cross-assembler, which is included on the CD-ROM.

THE AS11 CROSS-ASSEMBLER

Most experiments so far have used the BUFFALO monitor's ASM line assembler. As you probably know by now, this line assembler is nearly perfect for entering short programs. However, for long programs, especially long programs that use lots and lots of different memory locations and conditional branching instructions, writing and (especially) reading programs in ASM line assembler format often become bewildering. For instance, it is easier for an average mind to associate the location in memory with the label MINUTES with minutes than it would be the simple address $01. This advantage with labels is especially pertinent when it is necessary to use many memory locations. For instance, with most real-time clock (RTC) chips, there are registers for seconds, minutes, hours, days, date, month, year, minute alarm, *ad semi-infinitum*. Programming is much easier when labels are used rather than the memory address of the register. Isn't MINALARM easier to remember and refer to, when talking about the minute alarm register, than $0C?

Sadly, the BUFFALO line assembler does not recognize labels, just hexadecimal memory locations. However, most standard assemblers make good use of labels. Regular assemblers have many other neat points that make programming simpler and quicker. However, a regular assembler requires the use of an HC11 as the CPU. As far as I know, there are no inexpensive general-purpose, personal computers (PCs) that use an HC11 as their CPU. Most PCs are based on either Intel's 80xx6 series of CPUs (remember, the original Pentium should really have been labeled 80586) or Motorola's 60xxx series. However, properly written software can do amazing things. For instance, it is rather simple to design a program that will run on the 80xx6 series of MPU that will be able to understand 68HC11 instructions. If the object of the program is to act as an assembler for the 68HC11, then such a program is often referred to as a *cross-assembler*. Today, cross-assemblers are used so frequently that they are often simply referred to as *assemblers*. In this book I will use both terms interchangeably when I refer to the AS11—so don't get confused.

From an Idea to the Program

Once an idea is formulated in your head, you should come up with some descriptive name for it. This "naming procedure" helps organize things, especially if you are working on several projects at once. The next concrete thing to do is to put your idea down in a simple flowchart. Then you must decide in what storage medium you are going to place your program. Normally (at least in this book), the program will be stored in either an external EPROM or the HC11's internal EEPROM. Closely related questions are, "At what address is the program to start? Will there be enough space for the program?"

The next step is to sit down with a computer running a text editor that saves in ASCII and start writing the program. The program must be written in a special format that the assembler understands. The file must be saved in ASCII format and is called the *source code.*

After a program has been completed, it must be converted from an easily remembered form called a *mnemonic form* into binary code that the CPU can later execute. This is the job of the assembler program. Here we will use the AS11 cross-assembler. The AS11 cross-assembler reads the source code (also referred to as the *mnemonic version* of the program) and produces an *object code* file with an extension .S19. If the listing option (l) is included, the assembler produces a composite listing showing both source code and the machine-code translation. This listing is useful to discover the source of possible errors in the program and is usually referred to as the *assembler listing.*

After the program has been assembled and the object-code file with the .S19 extension has been created, the next step is to physically program the EPROM or EEPROM. To program EPROMs, you will need an EPROM programmer. The HC11's internal EEPROM can be programmed, with help from the BUFFALO monitor, with either Mag11 (for 48-pin DIP HC11's) or MagPro11 (to be described in the next chapter) for the 52-pin FN series of HC11s and HC811s.

Let's start our discussion of the AS11 assembler with a concrete example and take it step by step, to its hands-on, real-life solution.

THE IDEA

An experiment, using a single phototransistor, that will inform the user of the presence of three different light intensities provides the theme for our idea. The purpose of this idea, in addition to helping us along our path of knowledge concerning the programming and customizing of the HC11 could be:

1. If the phototransistor is placed against a window, it could tell us whether it is dark out by lighting green LED1; if it is dawn or dusk (or possibly stormy), red LED3 will flash; when it is daytime, LED4 through LED8 will light.
2. If the phototransistor is simply inserted in J6, this experiment could be used to determine whether there is the recommended light in the room for reading and general work and flash red LED3 if the light level is questionable.

CIRCUIT

Use the circuit shown in Fig. 5-16.

NAME

Let's name our program and experiment DAYALERT. Cute, huh?

THE FLOWCHART

The flowchart of DAYALERT is given in Fig. 8-1. Notice how general it is, with few specifics. Some programmers use a more detailed flowchart, whereas many skip the flowchart step altogether and go right into writing the program. You will have to choose the best way for yourself. In real life, I use (usually, anyway) a rather crude flowchart when writing nontrivial programs.

WHERE THE PROGRAM'S MACHINE LANGUAGE IS GOING TO BE STORED

The primary reason for this step is to determine what address to choose to start the program. We have several choices here, but we will choose the HC11's own internal 512-byte EEPROM, which resides between $B600 and $B7FF.

WRITING THE SOURCE CODE

As mentioned before, use a suitable text editor/word processor that you are familiar with. The primary criterion for its "suitability" is that the program must be able to save files in ASCII format.

The source code, which was written with a word processor, is given in the program listing in Fig. 8-2. Notice that I have named the disk file, which is in ASCII format, as DAYALERT.ASM. The .ASM extension is not needed by the AS11 cross-assembler, but since AS11 does not mind us using it, we add the .ASM extension in an attempt to keep things straight.

Notice all the comments. You can easily tell the comments from the program itself because, except for the first letter, comments are usually in lowercase. This is not required by the cross-assembler or even noticed by the cross-assembler, it just is the custom.

The cross-assembler can tell comments from the actual code in two different ways. First, if there is an asterisk in the first column, the cross-assembler "knows" that it is a comment and will ignore it when assembling. Second, if the text starts after the operand and is separated from it by a space or tab, the text is treated as a comment. Do not be too concerned here about the placement of comments. If you get it wrong, the cross-assembler will let you know and tell you where you went wrong and give you valuable clues to what you did wrong. This is true with most of the formal rules the cross-assembler requires. Do not sweat it; let the cross-assembler sweat for you.

Comments form part, and sometimes all of, the documentation of a program. Use lots and lots of documentation to describe your basic idea, code, memory layout, etc. This documentation may appear unnecessary when you are working intently with the program.

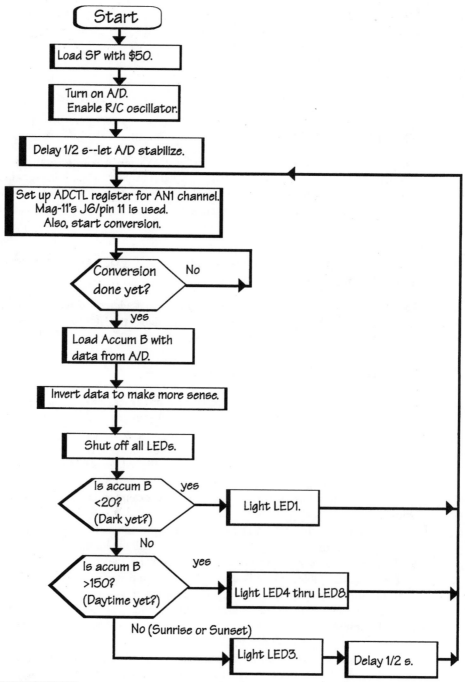

FIGURE 8-1 Flowchart for DAYALERT program.

```
                OPT 1 ;turn on listing option
***
*DAYALERT v1 — Program to be used with Mag-11 and
*collector connected to J6's pin 11
*phototransistor's emitter connected to J6's pin 19
*When dark, green LED1 lights
*At sunset/sunrise or dark day red LED3 flashes
*During day LED4-LED8 light
***
REGBASE    EQU   $1000    ;Hc11's reg base addr
OPTIO      EQU   $39      ;Offset:use index addr
ADCTL      EQU   $1030    ;Use extended addr
ADR2       EQU   $32      ;Offset:use index addr
LED1       EQU   $7000    ;Green LED
LED2       EQU   $7001    ;Green LED
LED3       EQU   $7002    ;Red LED
LED4       EQU   $7003    ;Green LED
LED5       EQU   $7004    ;Green LED
LED6       EQU   $7005    ;Red LED
LED7       EQU   $7006    ;Red LED
LED8       EQU   $7007    ;Red LED
AN1        EQU   01
DARK       EQU   20       ;Set darkness level
DAY        EQU   150      ;Set day level
*Sunrise and sunset are between dark and day
*Place program in eeprom starting at b600
           ORG   $B600
*Set up stack pointer so program
*will still work without BUFFALO
           LDS   #$50
           LDY   #REGBASE   ;base adr in x ind rg
*Turn on the A/D system and enable RC osc
           BSET OPTION,X#$CO
           BSR   DELAY      ;gives A/D time to stabl
*Setup ADCTL for channel AN1: J6's pin 11
NEWLT      LDAA #AN1
           STAA ADCTL
*Get data from A/D and wait until data is done
LOOP2      LDAA ADCTL
           BPL   LOOP2
           LDAB ADR2,X
*Light data is now in accum B
*We now inverse it to be more logical
           COMB
           BSR   CLRLEDS    ;shut off all LEDs
           CMPB #DARK
           BHI   NOTDARK    ;If not dark goto NOTDARK
*If it is dark light green LED1
*and then get next byte of light data from AN1
           STAA LED1
           BRA   NEWLT
*If it's not dark is it daytime?
NOTDARK    CMPB #DAY
           BLO   NOTDAY
           STAA LED4    ;if it's daytime
           STAA LED5    ;light these LEDs
```

FIGURE 8-2 Source code, in AS11 format, for DAYALERT program.

```
               STAA LED6
               STAA LED7
               STAA LED8
               BRA  NEWLT        ;get new light data
***************************
*If neither dark nor day then
*light red LED3 and make it blink
NOTDAY    STAA LED3
          BSR  DELAY
          CLR  LED3
          BSR  DELAY
          BRA  NEWLT
***************************
*Simple subroutine creates about
*a 1/2 sec delay with 4 mHz crystal
DELAY     PSHY
          LDY  #$FFFF
LOOP3     DEY
          BNE  LOOP3
          PULY
          RTS
***************************
*Subroutine turns off all LEDs
CLRLEDS   CLR  LED1
          CLR  LED2
          CLR  LED3
          CLR  LED4
          CLR  LED5
          CLR  LED6
          CLR  LED7
          CLR  LED8
          RTS
          END
```

FIGURE 8-2 (*Continued*) Source code, in AS11 format, for DAYALERT program.

Nonetheless, such documentation is invaluable for anyone reading your source code, and it may be invaluable to you, yourself if three full moons have occurred since you assembled the program's source code for the last time. One other point: The comment field isn't any place to be overly concerned about "prawrper" grammar or spellin'. Just make the comments understandable.

ASSEMBLING THE PROGRAM

This is probably the easiest part, since the AS11.exe program does the work. It is simplest if both the AS11.exe and source code file are on the same directory (i.e., "same folder" for you people who cut their first computer teeth on Windows 95/98.) Assuming that this is the case, we invoke the cross-assembler with the line

```
AS11 DAYALERT.ASM
```

If no errors are reported, AS11 will produce the file DAYALERT.S19 in the same directory/folder that DAYALERT.ASM is located. This .S19 file can be used by EPROM programmers, or it can be used with the BUFFALO monitor's LOAD T command to store the code in selected memory locations.

If desired (and I usually desire it), AS11 also can produce another file called the *assembler listing*. There are two ways to produce this neat file. The first, which is used in DAYALERT.ASM, employs the OPT 1 directive at the start of the listing. The second way to produce the listing file is to add the "l" option right after the command line. For instance, AS11 DAYALERT.ASM -1>DAYALERT.LST will produce the desired assembler listing. Note the addition of ">DAYALERT.LST." This addition is required to produce the actual file. If this addition to the command line is not included, the listing would only show on the monitor, and a listing file would not be produced.

The actual listing for DAYALERT.LST is shown in Fig. 8-3. At a quick glance, this listing may seem confusing. However, it supplies lots of information that can help a whole bunch when you discover that the program does not work as planned. (And when you write your first nontrivial assembler language program, it will *not* work—trust me! This is the reason you should tell AS11 to create a program listing.) However, right now we want to take a look at just one aspect of this listing—the machine-language instructions.

Notice the numbers at the extreme left of the listing (e.g., 0001, 0002,…, 0091, 0092). These numbers are *not* machine language. They are line numbers inserted in the listing by AS11. Their purpose is to merely identify the line of instructions for discussion purposes. They are sort of like the line numbers we used when we wrote programs using BUFFALO's ASM line assembler. However, with BUFFALO's line assembler, I inserted the numbers— here, AS11 does it automatically. Much easier.

The actual machine code starts on line 0023. The first number, b600, is the hexadecimal starting address of the program. The next hexadecimal number, 8E, is a machine-language opcode that instructs the CPU to immediately load the stack pointer with a 2-byte hexadecimal number 0050. The next instruction is on the next line, 0024, and it starts with "ce," which is the "load index register X immediately with the following 2 bytes" instruction.

While the primary use for listing machine-language instructions is debugging a program, we will make use of this machine-code listing to enter the program directly into memory. This is about as basic as you can get and really gives you the feeling of direct control and a strange feeling of *power*. It also takes some of the mystique away from the computer, and you discover that a computer, no matter how complicated it may appear, has a complexity that is virtual, and it is really, deep down in its innards, a simple machine after all. The other reason we are going to enter the program in this rather basic (yes, I could have used the adjective *crude* here instead) way is that for short programs, this is the simplest way to do it. Well, enough discussion. Let's get going.

First, you will be using the BUFFALO monitor, so get it up and running. Then look at Table 8-1. This is the "program," in machine language, you will be entering into the HC11's EEPROM. Since you are entering machine language directly, you will not be using BUFFALO's line assembler, like you did previously, but something even more basic, BUFFALO's MEMORY MODIFY command. Start by entering "MM B600 <cr>," where <cr> means "press the Enter key." Then enter "8E" and press the space bar (I will designate this action with <sp>). Keep entering the machine code 1 byte at a time. After typing each byte of code, press the space bar. When you are all done entering the machine code, press the Enter key.

To run this program, either type "Go B600" at BUFFALO's prompt or switch S2(1) off and it will run automatically at power up or after pressing the Reset switch.

Notice that this is a more time-consuming way of entering a program than using BUFFALO's line assembler. However, it is a more basic way. To be honest, you will not be entering programs like this very frequently. I will show in the next chapter how to take the .S19 file created by AS11 and use BUFFALO's LOAD T command to indirectly program

```
     Freeware assembler ASxx.EXE Ver 1.03.
0001 1000     REGBASE                    EQU   $1000    ;Hc11's reg base addr
0002 0039     OPTION                     EQU   $39      ;Offset:use index addr
0003 1030     ADCTL                      EQU   $1030    ;Use extended addr
0004 0032     ADR2                       EQU   $32      ;Offset:use index addr
0005 7000     LED1                       EQU   $7000    ;Green LED
0006 7001     LED2                       EQU   $7001    ;Green LED
0007 7002     LED3                       EQU   $7002    ;Red LED
0008 7003     LED4                       EQU   $7003    ;Green LED
0009 7004     LED5                       EQU   $7004    ;Green LED
0010 7005     LED6                       EQU   $7005    ;Red LED
0011 7006     LED7                       EQU   $7006    ;Red LED
0012 7007     LED8                       EQU   $7007    ;Red LED
0013 0001     AN1                        EQU   01
0014 0014     DARK                       EQU   20       ;Set darkness level
0015 0096     DAY                        EQU   150      ;Set day level
0016*Sunrise and sunset are between dark and day
0017
0018*Place program in eeprom starting at b600
0019
0020 b600                                ORG   $B600
0021*Set up stack pointer so program
0022*will still work without BUFFALO
0023 b600 8e 00 50              LDS    #$50
0024 b603 ce 10 00              LDX    #REGBASE      ;base adr in x ind rg
0025*Turn on the A/D system and enable RC osc
0026 b606 1c 39 0c              BSET   OPTION,X #$CO
0027 b609 8d 39                 BSR    DELAY         ;gives A/D time to stabl
0028*Setup ADCTL for channel AN1: J6's pin 11
0029 b60b 86 01 NEWLT           LDAA   #AN1
0030 b60d b7 10 30              STAA   ADCTL
0031*Get data from A/D and wait until data is done
0032 b610 b6 10 30 LOOP2        LDAA   ADCTL
0033 b613 2a fb                 BPL    LOOP2
0034 b615 e6 32                 LDAB   ADR2,X
0035*Light data is now in accum B
0036*We now inverse it to be more logical
0037 b617 53                    COMB
0038 b618 8d 37                 BSR    CLRLEDS       ;shut off all LEDs
0039 B61a c1 14                 CMPB   #DARK
0040 b61c 22 05                 BHI    NOTDARK       ;If not dark goto NOTDARK
0041*If it is dark light green LED1
0042*and then get next byte of light data from AN1
0043 b61e b7 70 00              STAA   LED1
0044 b621 20 e8                 BRA    NEWLT
0045*If it's not dark is it daytime?
0046 b623 c1 96 NOTDARK         CMPB   #DAY
0047 b625 25 11                 BLO    NOTDAY
0048 b627 b7 70 03              STAA   LED4          ;If it's daytime
0049 b62a b7 70 04              STAA   LED5          ;light these LEDs
0050 b62d b7 70 05              STAA   LED6
0051 b630 b7 70 06              STAA   LED7
0052 b633 b7 70 07              STAA   LED8
0053 b636 20 d3                 BRA    NEWLT         ;get new light data
0054***********************************
0055*If neither dark nor day then
0056*light red LED3 and make it blink
```

FIGURE 8-3 Program listing, created by the AS11 cross-assembler, for the DAY-ALERT source code.

```
0057 b638 b7 70 02 NOTDAY      STAA   LED3
0058 b63b 8d 07               BSR    DELAY
0059 b63d 7f 70 02            CLR    LED3
0060 b640 8d 02               BSR    DELAY
0061 b642 20 c7               BRA    NEWLT
0062********************************
0063*Simple subroutine creates about
0064*a 1/2 sec delay with 4 mHz crystal
0065 b644 18 3c DELAY         PSHY
0066 b646 18 ce ff ff         LDY    #$FFFF
0067 b64a 18 09    LOOP3              DEY
0068 b64c 26 fc               BNE    LOOP3
0069 b64e 18 38               PULY
0070 b650 39                  RTS
0071********************************
0072*Subroutine turns off all LEDs
0073 b651 7f 70 00 CLRLEDS    CLR    LED1
0074 b654 7f 70 01            CLR    LED2
0075 b657 7f 70 02            CLR    LED3
0076 b65a 7f 70 03            CLR    LED4
0077 b65d 7f 70 04            CLR    LED5
0078 b660 7f 70 05            CLR    LED6
0079 b663 7f 70 06            CLR    LED7
0080 b666 7f 70 07            CLR    LED8
0081 b669 39                  RTS
0082                          END
        Number of errors 0
```

FIGURE 8-3 (*Continued*) Program listing, created by the AS11 cross-assembler, for the DAYALERT source code.

TABLE 8-1 MACHINE LANGUAGE LISTING IN HEXADECIMAL FORMAT FOR DAYALERT PROGRAM

```
8e 00 50 ce 10 00 1c 39 0c 8d 39 86 01 b7 10 30
b6 10 30 2a fb e6 32 53 8d 37 c1 14 22 05 b7 70
00 20 e8 c1 96 25 11 b7 70 03 b7 70 04 b7 70 05
b7 70 06 b7 70 07 20 d3 b7 70 02 8d 07 7f 70 02
8d 02 20 c7 18 3c 18 ce ff ff 18 09 26 fc 18 38
39 7f 70 00 7f 70 01 7f 70 02 7f 70 03 7f 70 04
7f 70 05 7f 70 06 7f 70 07 39
```

Notes: (1) Use BUFFALO's MM (Memory Modify) Command to enter the data at the starting address $B600 (e.g. MM B600) (2) Data reads across; press 'SPACE BAR' for next address (3) Data was excepted from assembler listing (Fig. 8-3).

EEPROM. As will be shown, RAM can be programmed directly with the use of the LOAD T command. For now, however, let's look at using the AS11 assembler again by looking at the source code in Fig. 8-2 with a slightly different, more formal approach.

How AS11 Expects Its Line of Source Code to Look

For this description of AS11 "fields," refer to Fig. 8-4. As indicated in Fig. 8-4, each line of code can include up to four fields. Each field is separated by a space or tab. This space or tab separation is sometimes called a *whitespace*. Usually, one or more of these fields are blank.

Each line of code can include up to 4 fields, each field separated by a space or tab. This space or tab separation is sometimes called a *whitespace*. See below.

(1)	(2)	(3)	(4)
Label	Operation	Operand	Comment

Usually, one or more of these fields is blank.

The description of each field follows:

1. Label field—If the first character in this field is an asterisk (*), AS11 ignores the line and treats the entire text following the asterisk as a comment. The comments are printed on the assembly listing for information purposes only. If the first character in this field is a symbol (e.g., a–z, A–Z, 0–9, etc.), then AS11 treats the set of symbols as a label. By the way, AS11 differentiates uppercase and lowercase letters when they appear in symbols. *Loop, loop,* and *LOOP* are all treated by AS11 as different labels.
Example: In the line of code

```
        REGBASE    EQU  $1000  ;Hc11's reg base addr
```

REGBASE is in the label field and is a label.

2. Operation field—This field, which comes after the label field, must be preceded by at least one whitespace (space or tab). This field must contain either a mnemonic for a legal HC11 opcode (e.g., CLRA) or an assembler directive (e.g., EQU). AS11 does not differentiate here between uppercase and lowercase letters. Thus *EQU, equ,* and *Equ* are all treated the same by AS11.

Example: In the line of code

```
        REGBASE    EQU  $1000  ;Hc11's reg base addr
```

EQU is in the operation field and is an assembler directive.

3. Operand field—This field, which comes after the operation field, must be preceded by at least one whitespace. This field is the most difficult to understand for both the programmer and (especially) the AS11 itself. The AS11 needs both the operation field and the label field to figure out the significance of the operand field. For instance, in the line of code

```
        REGBASE    EQU  $1000  ;Hc11's reg base addr
```

the operand field contains $1000, which is a hexadecimal number. Since the operation field contains the assembler directive EQU and the label field contains REGBASE, the AS11 knows that all it has to do is replace the label REGBASE with the number $1000 everyplace it finds it in the source code.

4. Comment field—This field, which comes after the operand field, must be preceded by at least one whitespace. Unlike the operand field, "there is nothing to this field." In fact, this statement often can be taken as literal, since this field often has nothing in it! AS11 ignores this field except that it repeats this field in the assembler listing—if it is instructed to create one, that is.

FIGURE 8-4 Source code format expected by the AS11 cross-assembler.

> *Quick question:*   In the line of code
>
> ```
> REGBASE    EQU  $1000  ;Hc11's reg base addr,
> ```
>
> Hc11's reg base addr is a comment in what field?
>
> *Answer:*   You've gotta be kidding!
>
> *Hints:*   In our example of lines of source code, the comment starts with a semi-colon (;). This is not at all necessary, but it does help when reading the source code and assembler listing. Try to keep the comment to less than 20 spaces in length. This often helps the looks of the assembler listing when you print it out.

FIGURE 8-4   *(Continued)* **Source code format expected by the AS11 cross-assembler.**

Don't Sweat About AS11 Format Rules

Now that I have gotten you all worried that you will not do it right when you start writing the source code, I have a personal message for you: "Don't sweat it!" If you get something wrong, say a comment in the operand field, the AS11 will inform you that something is wrong, perhaps with a message such as, "Illegal operand at line xx." Well, if this happens, simply go back and try to figure out, with clues given you by AS11, what is wrong and correct the goofs. This "no sweat it" attitude is good when using the AS11, especially when you first get started. Mistakes are an excellent way of learning. To be honest, one wonders if mistakes did not exist, would learning exist? Perhaps this wonderment is best left to more philosophical educators than myself.

Time to Sweat—A Look at Number Formats

Now with all this talk about "no sweat," there *is* one thing to worry about when using the AS11—the number format. It is important here to worry because if the wrong number format is used, AS11 usually will not know it, so it will not spit out any error messages. In short, the error will be passed right through to the CPU, and the program will not work right—if at all. As already discussed, three different number formats are currently used widely in the computer field: binary, hexadecimal, and decimal.

Decimal.   Unlike BUFFALO's ASM line assembler, AS11 treats a number without a prefix (e.g., 56) like the standard garden variety of number that is more technically referred to as *decimal-formatted* or *base-10 numbers* and sometimes nicknamed "the finger-counting format."

Hexadecimal.   If there is a dollar sign prefix (e.g., $38), AS11 treats the number as if it is in hexadecimal format.

Binary. If there is a percent prefix (e.g., %00111000), the number is thought of by AS11 as already in "digital computer" format. Another name for this format is *binary.*

While AS11 also supports octal-formatted numbers (base 8), they are almost as rare as vacuum tube flip-flops, so we ignore them here. One final note: AS11 recognizes the apostrophe prefix (e.g., 'T) to indicate that the character following it is in ASCII.

AS11 Instructions

There are two types of instructions used in assembly source code that appear to be identical—and often are treated as identical by novice programmers—but are fundamentally different. The first types of instructions are the CPU's *machine-language mnemonic instructions* (e.g., LDAA, STAA, CLR, etc.), which we have already used in the listings in BUFFALO's ASM line assembler format, and the others are *assembler-language instructions,* also referred to as *assembler directives* (e.g., EQU, ORG, RMB, etc.). More simply stated: There are two types of instructions in AS11 source code: (1) machine instructions, which as its name implies, are instructions to the CPU itself, and (2) directives, which are instructions to the AS11 program. Next, we will describe *all the AS11 directives.* Later, we will look at common and carefully selected HC11 machine instructions. All the HC11's instruction sets are listed in Appendix H, as well as the two Motorola data books mentioned earlier.

AS11 Directives

ORG—SET LOCATION COUNTER TO ORIGIN

This directive can be thought of as the most important, because it, unlike all the rest, is necessary for just about all practical programs. If it is left out, AS11 assumes (usually wrongly) that the program will start at location 0000.

As its description says, the ORG directive is used to set the value of the location counter. The ORG directive instructs the AS11 where to put the next byte it generates after the ORG directive.

One important purpose of this directive is to tell the assembler where the program starts. For instance, in our DAYALERT program the line of code,

```
ORG $B600
```

tells the assembler we want to start our code at $B600. However, telling the assembler where the program starts is just one purpose of the ORG directive. This directive is often used in several other places in a program. Every time it is used, it changes the location counter. It is often used to properly place the INTERRUPT vectors. For instance, the code segment

```
ORG $FFFE
FDB $E000
```

sets up the all-important RESET vector so that the program will automatically start at $E000. The assembly-language source code for this program must have another ORG statement to reset the location counter to $E000. This statement would be

```
ORG $E000
```

After this statement, the actual source code starts.

EQU—EQUATE SYMBOL TO A VALUE

The EQU directive assigns the value of the expression in the operand field to the label, which is in the label field. For example, the line of code

```
LED1        EQU  $7000 ;Green LED1
```

assigns the value $7000 to the label LED1. In effect, the AS11 replaces the 4-character text string LED1 with the number $7000 everyplace it finds it in the source code. The EQU directive's purpose is to help the programmer by replacing abstruse numbers with descriptive labels.

FCB—FORM CONSTANT BYTE

The FCB directive puts a byte in memory for each operand of the directive. With multiple single-byte operands, each operand is separated by commas and is stored in successive memory bytes. For example, the statement

```
TEN         FCB  00,01,02,03,04,05,06,07,08,09
```

will store the 10 different data bytes in 10 consecutive bytes of memory, and it also will tell the AS11 that the label TEN is the symbolic address of the first byte, whose value is 00. Since this is a statement that is part of an unknown program that has an ORG directive somewhere, there is no way of knowing the address where each byte of data is stored. However, we can change this to make it more concrete and thus clearer. For example, the two statements

```
        ORG  $2000
TEN FCB  00,01,02,03,04,05,06,07,08,09
```

will cause the AS11 to store 00 in the memory location $2000, 01 in the memory location $2001,..., and 09 in the memory location $2009.

If FCB seems at all confusing, read the following   Note that some people who are experienced in AS11 assembly language refer to the directives that deal with memory storage, such as FCB, as *data storage directives* rather than *assembler directives*. Giving the two directives different names may be helpful in avoiding confusion in the minds of novice programmers when they work with such directives as FCB.

Where is the possible confusion? Well, the FCB directive places data right in the memory locations. This may seem, to new programmers, as something only the CPU can do with such instructions as STAA. For those who think this, think again. For instance, say you start your program at $3000 and the first instruction is

```
               LDS   #$50
```

What does the assembler do? It stores $8E (the opcode for LDS) at $3000, $00 (the MSB of the operand) at $3001, and $50 (the LSB of the operand) at $3002. The same thing could have been accomplished with the statements

```
               ORG  $3000
               FCB  $8E,$00,$50
```

Both ways of storing data will accomplish the same thing, although the second way is almost never used when the same thing can be accomplished by using an HC11 instruction mnemonic such as LDS. If you doubt any of this, try assembling the preceding program segment as well as the following program segment:

```
               ORG  $3000
               LDS  #$50
```

You will find (hopefully, not surprisingly) that the .S19 files of both are identical.

FCC—FORM CONSTANT CHARACTER

The FCC directive is one of the most helpful and widely used of the data-storage directives. This directive translates the string of printable ASCII characters (e.g., A, z, ., #, 2, 9, etc.) in the arguments of the directive into their ASCII code. The string delimiter is specified by the first nonblank character after the FCC directive. However, I will use the apostrophe (') as the delimiter because the BUFFALO monitor's source code uses it in its FCC statements.

Most FCC directives have a label. The label is assigned to the address of the first byte in the string. The following example of an FCC directive was taken from BUFFALO's source code. It may appear familiar to you by now.

```
   MSG1           FCC   'BUFFALO 3.4 (ext) Bit User Fast Friendly Aid to'
                  FCC   'Logical Operation'
```

FDB—FORM DOUBLE BYTE CONSTANT

The FDB directive is similar to the FCB directive except, as its name indicates, it handles 2 bytes of data at a time, not just 1. The 16-bit values corresponding to each operand, which are separated by commas, are stored in 2 consecutive bytes. This directive is useful when working with 16-bit registers such as the index register or the D accumulator. It also is useful for initializing HC11's vectors.

The following statements were taken from the M11DIAG2.ASM source-code file and modified slightly:

```
               ORG  $FFFE
               FDB  $F800      ;RESET VECTOR
```

These statements insert the data $F8 at $FFFE and $00 at $FFFF. Notice that an FCB directive could have been used, but the FDB directive is more appropriate when working with 16-bit numbers.

BSZ—BLOCK STORAGE OF ZEROS

The BSZ directive causes AS11 to allocate a block of bytes and assign each byte the value of zero. The number of bytes you want to allocate and store zero in is set by the expression in the operand field. For example, the two statements

```
ORG   00
BSZ   40
```

reserve the first 40 bytes of the HC11's memory space, which is in internal RAM, and initially store 00 in each of these memory locations.

The BSZ directive is not used as frequently as FCC, FCB, or FDB. By the way, AS11's ZMB directive is the same as BSZ.

FILL—FILL MEMORY

The FILL directive does just what it says: instructs the AS11 to fill an area of memory with a constant value. Two expressions, separated by commas, placed in the operand field tell the AS11 the value of the byte and the number of bytes to use to fill memory. The first expression is the value of each byte; the second expression the total number of bytes. For example, the statements

```
ORG 00
FILL 00,40
```

do exactly the same thing as the preceding example did using the BSZ directive. Like the BSZ, the FILL statement is only used infrequently.

RMB—RESERVE MEMORY BYTES

This frequently used directive causes the location counter to be advanced by the value of the expression in the operand field. What this action does is reserve a block of memory, usually for scratchpad or a table area. The size of this reserved block is the value of the expression in the operand field.

Because of its importance, we will look at the RMB directive a little deeper by looking at an actual example of its use in the BUFFALO monitor. Here we take a program snippet from near the start of the source code:

```
        ORG $2D
*** Buffalo ram space ***
                RMB 20      user stack area
USTACK          RMB 30      monitor stack area
STACK           RMB 1
REGS            RMB 9       user's pc,y,x,a,b,c
SP              RMB 2       user's sp
```

Next, we run this code snippet through the AS11 and get the following assembly listing file:

```
Freeware assembler ASxx.EXE Ver 1.03.
0001 002d                           ORG $2D
0002          *** Buffalo ram space ***
0003 002d                RMB 20      user stack area
```

```
0004 0041        USTACK    RMB 30      monitor stack area
0005 005f        STACK     RMB 1
0006 0060        REGS      RMB 9       user's pc,y,x,a,b,c
0007 0069        SP        RMB 2       user's spy,x,a,b,c
Number of errors 0
```

Notice a bunch of interesting things about this listing. However, before we discuss the fascinating aspects of this listing, recall that the four-digit decimal numbers at the extreme left (e.g., 0001, 0002,...) are just line numbers put there by AS11—they have no real importance in the program. The next four-digit numbers (e.g., 002d) are in hexadecimal format, and they pertain to the AS11 location counter, which is intimately related to the actual memory location in the HC11.

Line-by-line discussion

Line 0001.   The ORG 2D statement sets the location counter at 2d.

Line 0002.   This is solely a comment line and is ignored by AS11.

Line 0003.   The statement RMB 20 sets aside the next 20 (decimal) memory locations for the user stack. Notice that 20 in decimal is 14 in hexadecimal and that 2d + 14 = 41 in hexadecimal—remember this number, 41.

Line 0004.   Notice that the location counter is now at 0041. It was advanced with the preceding statement, which reserved 20 bytes for the user stack. The label USTACK here corresponds to the memory address $0041. Now, the AS11 recognizes USTACK + 1 as address $0042, USTACK + 2 as address $0043, and so on. Also notice that 30 (the number, in decimal, of consecutive memory bytes AS11 has reserved) is 1E in hexadecimal and that 0041 plus 001E is 005F—again, remember this number!

Line 0005.   Now, the location counter has jumped up to 005F—exactly 20 (decimal) higher than the preceding location of 0041! The label here is STACK and is located at address $005F. One more than $5F is? Right, $60. And it is no surprise that the location counter at the next line is 0060. I leave it to you to continue this line-by-line discussion for lines 0006 and 0007.

Looking at this listing, you may have noticed something interesting—the ORG and RMB directives are really closely related. In fact, the RMB directive really could be completely replaced by the ORG directive. The primary difference between the two is that the ORG directive is absolute, while the RMB directive is relative. (Imagine that RMB stands for "relative memory bytes" instead of "reserve memory bytes.") For instance, the source code that would accomplish the same thing could be

```
                  ORG $2D
USTACK            ORG $41
STACK  ORG        $5F
REGS              ORG $60
SP                ORG $69
```

To show the truth of the preceding statement, take a look at the assembly listing of the preceding source code:

```
Freeware assembler ASxx.EXE Ver 1.03.
0001 002d                 ORG $2D
0002 0041  USTACK         ORG $41
0003 005f  STACK ORG      $5F
0004 0060  REGS           ORG $60
0005 0069  SP             ORG $69
0006
Number of errors 0
```

Notice the basic similarity between this code and the source code using the RMB directive? They both do the same thing!

OPT—ASSEMBLER OUTPUT OPTIONS

I recommend that this assembler directive be used with every nontrivial program. The actual options are placed in the operand field and are separated by comments. These options include

l.   Print the listing from this point on. This option is the one that is most important.

c.   Enable cycle counting in the listing.

cre.   Print a cross-reference table at the end of the source listing.

e.   Print a symbol table at the end of the source listing.

   Often, it is smart to use all of these options. The statement

```
        OPT l,c,e
```

does just this.

   One important point: Unless you redirect output by using the greater than symbol ">" at the end of the command line, all options will be printed on the screen, and the only file created is the .S19 one. It is quicker and smarter to print the listing to a file. Assuming that the source-code file is DAYALERT.ASM, the following command line does just this:

```
    AS11 DAYALERT.ASM >DAYALERT.LST
```

   There is another way to create an assembler listing other than using the OPT directive— add switches directly on the command line. To do this, place a hyphen after the source-code file, and then add the option. For example, the command line

```
    AS11 DAYALERT.ASM l,c,e >DAYALERT
```

will produce a file with the assembler listing even if the OPT directive is not included in the source code.

PAGE—TOP OF PAGE

The PAGE directive tells AS11 to advance the paper to the top of the next page. This directive only applies if an assembler listing is produced. While PAGE was a nice afterthought by the creator(s) of AS11, its function can be better filled by a text writer/word processing program that can format the assembler listing in various ways.

END

While the END directive is not used by AS11, it does not screw things up either. The END directive apparently was "allowed" by the designers of AS11 so that the source code written for other cross-assemblers, which require this directive, would be compatible with AS11.

Read the File AS11.TXT on the CD-ROM

The file AS11.TXT on the CD-ROM provides more details on using the AS11 cross-assembler. Be sure to read this file.

A Sampling of HC11 Instructions

In the rest of this chapter I discuss the more common and useful of HC11's instructions as well as all instructions used in DAYALERT.ASM. Appendix H reproduces a portion of the HC11's data sheet where all the HC11's instructions are listed. For a more complete description, refer to Appendix A in the HC11's bible (i.e., M68HC11 Reference Manual, Motorola ref. no. M68HC11RM/AD). Moreover, before continuing here, it may be instructive to go back to Chap. 3 and review the description of HC11's CPU registers.

Hopefully, a careful reading of this chapter will enlighten the reader not only about HC11's instructions but also about using these instructions to write programs. However, I am not going to go into specific programming techniques. While these are important, I believe that assembly-language programming techniques are best learned by programming and not by reading or even studying about them. However, there is much to be gained from discussing basic HC11 instructions and using them in relatively simple programs. So let's get right into it.

I will start off here by examining loading the index register. This is not the simplest set of instructions, but since we will look at index addressing rather closely, it is nice to know how the data got into the index register in the first place.

Note that while it may appear from the instruction mnemonics that the X and Y index registers are identical in makeup, there is a major difference between the two: *Most instructions using the Y index register take an extra clock cycle to complete compared with the X index register.* Normally, this is bad. Nonetheless, when writing delay loops, this is good. Machine-language programmers, as well as those who are looking to write tight code, also should notice that two opcodes are required when working with the Y index register rather than just one, as with most instructions.

LDX AND LDY

These are 16-bit (2-byte) instructions that load the X and Y registers, respectively, with 2 consecutive memory bytes. The register's MSB (most significant byte) is loaded from the byte of memory specified by the program, while the register's LSB (least significant byte) is loaded from the next byte of memory (1 + address specified by program).

Examples of addressing modes used by LDX and LDY

Immediate addressing. LDX #$1000 loads the X index register with $1000. Since the index registers are 16-bit registers and $1000 is 16 bits, everything should be obvious.

Direct addressing. LDY $40 loads the Y index register's MSB with the data at address $0040 and the Y index register's LSB with the data at address $0041. Remember when dealing with 16-bit registers that you require 2 bytes in memory—this can get confusing, especially if you are just starting out in programming with assembly language.

Extended addressing. LDX $F800 loads the X index register's MSB with data at $F800 and its MSB with data at $F801.

Index addressing. No example will be given because of possible confusion. (For goodness sakes, even saying "loading an index register using index addressing" can get confusing.) For index addressing techniques, see the following discussion of STAA, STAB, LDAA, and LDAB instructions.

LDAA AND LDAB

These 8-bit instructions load the contents of the specified memory location into accumulator A (LDAA) or accumulator B (LDAB). An example of these instructions is given for each addressing mode.

Examples of addressing modes used by LDAA and LDAB

Immediate addressing. LDAA #$55 loads accumulator A immediately with $55.

Direct addressing. LDAB $55 loads accumulator B with the data at address $0055. Notice that data are not actually "moved" from a memory location to the accumulator. Rather, they are *copied,* since the data in the memory location are not changed.

Warning:

Do not get the data in the operand (immediate addressing) confused with the address in the operand (direct addressing). The only way AS11 can tell the two addressing modes apart is with that teeny, weeny number prefix (#). While writing programs, I have left out this itsy, bitsy prefix a number of times and have always wasted more time than I would like to admit (even to myself) in finding the screwup. This is where machine language and the assembler listing come in really handy.

Extended addressing. LDAA $ABCD loads accumulator A with the data at address $ABCD. (Believe it or not, the address $ABCD does exist.)

Quick question What do you think would happen if you wrote the statement LDAB #$ABCD? How about LDAB $ABCD?

Hint Accumulator A is only an 8-bit register.

Index addressing. LDAB 5,X loads accumulator B with the data at the address that is obtained by adding five (the offset value) to the current value stored in the X index register. If the previous instruction was LDX #$2000, then LDAB 5,X would load accumulator B with the data at address $2005. *Note that if you use index register Y here, an extra cycle is required (5 versus 4) for execution of this instruction.*

Index addressing may seem most confusing, but it is quite handy and not prone to as many goofs as some simpler addressing modes (e.g., the immediate mode.)

STAA AND STAB

These instructions store the contents of accumulator A (STAA) or accumulator B (STAB) at the specified memory location. Since the accumulator's contents are unchanged, it is more appropriate to think of these instructions as "COPYA" and "COPYB."

It is important to realize that these load instructions perform write operations, since they cause the R/_W line to go to 0 during the last cycle (memory address is on address bus) of the instruction's execution. Of course, this R/_W line info is only interesting to those who are designing circuits with the HC11 that uses its normal expanded mode.

Examples of addressing modes used by STAA and STAB

Immediate addressing. There is none—doesn't make sense here.

Direct addressing. STAA $2A stores (actually copies) the contents of accumulator A at the address $002A.

Extended addressing example 1. STAA $002A does the same as STAA $2A but takes an extra clock cycle to accomplish the same thing.

Extended addressing example 2. STAA $7000 stores (actually copies) the contents of accumulator A at $7000.

Extended addressing example 3. STAA 1000 stores the contents of accumulator A at the address 1000 (decimal), which is $03E8.

Index addressing example. The code

```
(line 1)  LDX #$7000;
(line 2)  LDAA #$01;
(line 3)  STAA 1,X
```

stores 01 at address $7001.

In Mag-11, these three lines of source code would turn on LED2. How? Well, line 1 loads index register X with $7000, and line 2 loads accumulator A with 01. Line 3 stores (remember, actually copies) 01, the contents of accumulator A, at the address $7000 + 1 = $7001. Remember from our discussion of Mag-11 that addresses $7000 through $7007 correspond to LED1 through LED8, respectively. Also recall that the decoding circuit is so designed that bit 0 of the data bus turns the LED off when 0 (clear) and on when 1 (set).

LDD

This 16-bit instruction loads the double accumulator D with the contents of memory locations M and M + 1. Accumulator D actually consists of accumulator A and accumulator B hooked up together, so watch it when using the LDD instructions. *When LDD is used, you change the contents of both accumulator A and accumulator B*. The data at location M are loaded into accumulator A (MSB), while the data at location M + 1 are loaded into accumulator B (LSB).

Immediate addressing.   LDD #$F800 loads 16-bit double accumulator D with $F800, which is a 16-bit (2-byte) number. This is simple to use and understand because accumulator D is a 16-bit register and $F800 is a 16-bit number. The problem, as mentioned a few sentences back, is what the LDD instruction does to accumulators A and B. *Keep alert when using any LDD instructions!*

Direct addressing.   LDD $23 loads 16-bit double accumulator D with the data at addresses $0023 and $0024. After this instruction is executed, LDD's MSB (i.e., accumulator A) contains the same data that are at $0023 and its LSB (i.e., accumulator B) contains the data that are at $0024.

Extended addressing.   LDD $1A35 loads 16-bit double accumulator D with the data at addresses $1A35 and $1A36. After this instruction is executed, LDD's MSB (i.e., accumulator A) contains the data that are at $1A35 and its LSB (i.e., accumulator B) contains the same data that are at $1A36.

Indexed addressing.   The two statements

```
(line 1) LDX #$1000;
(line 2) LDD 0,X
```

first put $1000 in the X index register and then load double accumulator D with the data at addresses $1000 and $1001. After this instruction is executed, LDD's MSB contains the same data that are at address $1000 and its LSB contains the same data that are at $1001.

STD

This 16-bit instruction stores the contents of double accumulator D at two consecutive locations in memory. The contents of accumulator D's MSB (i.e., accumulator A) is copied to the memory address indicated to by the operand, while the LSB (i.e., accumulator B) is copied to this memory address plus one.

Immediate addressing.   As before, this makes no sense, so it would be a real screwup if I placed an example here.

Direct addressing.   STD $B9 stores (a copy of) 16-bit accumulator D's contents to memory locations $00B9 and $00BA. MSB (accumulator A) goes to $00B9, and LSB (accumulator B) goes to $00BA.

Extended addressing.   STD $3F4E stores accumulator D's MSB at $3F4E and its LSB at $3F4F.

Indexed addressing.   The statements

```
(line 1) LDX #2000;
(line 2) LDD #$1234;
(line 3) STD $A0,X
```

wind up storing $12 at $20A0 and $34 at $20A1. If it doesn't take you long to verify this, you have successfully mastered HC11's indexed addressing mode.

LDS

This 16-bit instruction loads the 16-bit stack pointer with the contents of memory locations M and M + 1. Since this instruction is similar to other load instructions that pertain to 16-bit registers (e.g., LDX, LDD), I will only give one common example—loading the stack pointer using immediate addressing.

Immediate addressing.   LDS #$00FF loads the stack pointer immediately with $00FF. By the way, this example is typical. However, do not load the stack pointer with $00FF when using the BUFFALO monitor. BUFFALO uses RAM addresses greater than $41 for its own purposes. When using the BUFFALO monitor, it seldom is necessary to initialize the stack pointer, since BUFFALO does it for you. If you wish, however, to set up your own stack pointer, load the stack pointer with $2D.

   At this point, I am going to change my description of HC11 instructions. Previously, I gave detailed examples of the different addressing modes. I trust that you have already learned, with the help of these examples, the differences between immediate, direct, extended, and indexed addressing. From now on, I will only provide one brief example of each instruction.

ADDA, ADDB

This instruction adds the contents of a memory location to the contents of accumulator A or accumulator B, respectively. The sum is in the respective accumulator.
Example   Immediate addressing:

```
(line 1) LDAA #$34;
(line 2) ADDA #$23
```

After the HC11 executes both lines of instructions, accumulator A contains $57 ($34 + $23).

SUBA, SUBB

This instruction subtracts the contents of a memory location from the contents of accumulator A or accumulator B, respectively. The difference (the result of the subtraction operation) is in the respective accumulator.

Example Extended addressing:

```
(line 1)  LDAA #120;
(line 2)  STAA $4000;
(line 3)  LDAB #215;
(line 4)  SUBB $4000
```

After the HC11 executes these four lines of instructions, accumulator B contains 95 (215–120). This example demonstrates not only the subtraction instructions but the fact that AS11 can handle decimal numbers. Recall that numbers without a prefix are assumed to be decimal by AS11. What would happen if line 1 were LDAA #288 instead of LDAA #120? Well, AS11 would shoot out an error message. Remember, accumulator A is an 8-bit register, and the largest number you can represent with 8 bits is %11111111 = $FF = 255.

JMP

This straightforward and simple instruction causes a jump to occur to another instruction stored at the effective address. The effective address is obtained according to the rules of extended or indexed addressing.

Example Extended addressing: JMP $E000 causes a jump to the instruction at $E000.

Before continuing the discussion on common HC11 instructions, I will take another look at the HC11's CCR (condition code register). This register is vital to conditional branching instructions. These instructions, more than any other, are responsible for the HC11's "virtual intelligence."

Condition Code Register

As described earlier, the CCR (condition code register) is an internal 8-bit register. A simple pictorial of the CCR is shown in Fig. 8-5. To keep things simplified, we will only look at bits 0, 2, and 3, which are, respectively, the carry flag (C), the zero flag (Z), and the negative flag (N).

Z bit (the zero flag).   This is not only the simplest and easiest to understand of all the flags but also probably the most useful. This flag is set or cleared after most, but not all, operations that involve a CPU register. The instruction TST, which involves only memory locations, also affects the Z bit. Check Appendix H for details on how each instruction affects the Z bit. The zero flag is set (i.e., made a 1) if the operation resulted in all bits in the register of interest being cleared (i.e., made 0). If even one of the register's bits is 1, the zero flag is cleared (i.e., set to 0). For example, the instruction CLRA simply sets all of accumulator A's bits to zero. It also, of course, sets the CCR's Z bit to 1.

Register's Name: Condition Code Register (CCR)					Register's Address: None (Located in CPU)		
BIT#	7	6	5	4	3	2	1	0
BIT NAME	S	X	H	I	N	Z	V	C

S—stop disable flag
X—XIRQ interrupt flag
H—half-carry flag (from bit 3)
I—interrupt flag
N—negative flag
Z—zero flag
V—overflow flag
C—carry/borrow from MSB (most significant bit)

FIGURE 8-5 The condition code register.

C bit (the carry flag).   Whenever a carry is generated as a result of an operation, the C flag will be set to 1. If there is no carry, the C flag will be set to 0.

Example 1

```
LDDA #$F0
ADDA #$10
```

After these two instructions are executed, accumulator A will contain $00, and both the zero and carry flags will be set. While setting of the zero flag should be obvious, a little investigation may help explain why the carry flag is also set. To see this, think of the C bit as the 9 bit of the accumulator. Now $F0 is %11110000 and $10 is %0001000. Add these two binary numbers together and you get %100000000, where the most significant bit is really the C bit (carry flag) and the rest of the bits of the sum (all zero) reside in accumulator A. In other words, since accumulators are 8-bit registers, this sum just will not fit, and you can think of the CCR's C bit as an extension of the accumulator. You also may want to go back to Chap. 1 and look at the carry extension to the circa 1965 hobbyist computer circuit in Fig. 1-8.

Example 2

```
LDAB #$F0
SUBB #$F1
```

After these two instructions are executed, accumulator B will contain $FF. In this case, too, the carry flag was set (C = 1). This time it was set to indicate that a borrow operation occurred and not a carry.

Every time you try to subtract a larger number from a smaller one, the C bit is set to 1. The C bit is also set to 1 when you add two numbers and the result is greater than $FF (255 in decimal). It should be obvious here that the carry flag can have different meanings depending on the operation involved. The fact that the C bit is a 1 can mean either that a carry occurred or that a borrow occurred. The exact meaning of the 1 depends on whether the operation was an addition or subtraction.

Before we look at the CCR's negative flag, let's look at "signed" numbers.

SIGNED NUMBERS

As you may know, the first digital computers were decimal system–based computers that used 10 different voltage levels. However, it quickly became apparent that it was easier and more reliable to design circuits with just two voltage levels. Voila! The binary number system.

After the binary number system was nearly universally regarded as superior, the next controversy erupted—how to represent negative numbers. For a while, there were three different ways of doing this. However, all three ways had one thing in common—negative numbers were differentiated from positive numbers by a 1 in bit 7, the most significant bit in a byte. The HC11, and just about all MPU/MCUs in use today, use two's complement numbers to represent signed numbers. (Signed numbers are the number set that includes both positive and negative numbers.) To find the two's complement of a binary number, just take the one's complement (i.e., replace 1s with 0s and 0s with 1s) and add 1. Keep in mind, however, that numbers up to 01111111 (127 in decimal) are treated just like unsigned numbers. You only take the two's complement when dealing with negative numbers. While the really important thing to know is that if there is a 1 in bit 7 of a number, the number is treated as a negative number, I will still give an example here of two's complement numbers for the heck of it.

Example   Say you want to express -18 (decimal) as an 8-bit two's complement number. First, you express it in plain old unsigned binary, which is 00010010 (i.e., $12). Then you take its complement and get 11101101. Finally, you add 1 to get 11101110. This number is a two's complement number and represents -18. The really important thing to remember here is that the little 1 in bit 7—that tiny little *numero uno*—means that the number is negative—if you think of it as a "signed" number, that is. This qualification brings up an important point concerning signed numbers and two's complement numbers: *The HC11's CPU adds bit patterns as if they are unsigned binary numbers. It is our interpretation of the binary numbers that decides if signed or unsigned numbers are indicated.*

Most, if not all, CPUs use the two's complement number system when they subtract. Here, they just make the subtrahend negative and then add. To see this, look at our preceding example, which was binary 00010010 ($12). Let's try to subtract this number from 01101111 ($6F) by adding. You know the two's complement of 00010010 is 11101110, so you add this to 01101111 to get 101011101. This is nearly the same as if you subtracted directly 00010010 from 01101111. The only difference is that 1 way, way, way over to the left. This little 1 is what sets the carry flag when you subtract a larger number from a smaller one. I referred to this before as a *borrow operation*. Now you can see that I told a "white lie" earlier and it really was a carry after all.

WARNING! WARNING! INFORMATION OVERLOAD!

To tell the truth, the preceding discussion was probably more information than you needed. Keep this in mind, however: When dealing with signed operations, bit 7 of the number indicates whether the number is negative (bit 7 is a 1) or positive (bit 7 is a 0). Finally, let's look at the negative flag, which is bit 3 (N bit) of the CCR.

N bit (the negative flag) The negative flag monitors bit 7 of the accumulators. Immediately after an operation involving an accumulator, the N flag examines bit 7 to check if bit 7 is 1, which means that the number is viewed as negative. (Of course, this assumes that we care if it's negative or not. Usually, we do not give a darn!) If so, the N flag is set to 1. If not, it is set to 0. You see, it wasn't so hard after all!

One other tiny, little point. Operations on the CPU's 16-bit registers, such as the index registers, also affect the N bit. Here, the most significant bit, bit 15, determines whether the negative flag is cleared or set. As before, if this bit is a 1, the CCR's N bit is set, if it is a 0, the N bit is cleared.

Example

```
            ORG $0020
TWOSCOM     FCC %00010010       ;put $12 in $0020
            NEG   TWOSCOM        ;start of program
            WAI                  ;wait here
            END
```

This example introduces the NEG instruction. As its mnemonic indicates, this instruction makes negative (takes the two's complement of) the number referred to in the operand, which here is %00010010. (By the way, this number was put here by the assembler directive FCC %00010010.)

Running this little program snippet will result in the two's complement number of %00010010 ($12), which is %11101110 ($EE), being stored at $0020. Since the result has a 1 at bit 7, the N bit is 1 (the negative flag is set).

It may be very interesting and informative if you actually try to run this little program snippet. Assuming that the program is in an ASCII file called neg.asm, type the following command line:

```
AS11 NEG.ASM -l >NEG.LST
```

The resulting assembler listing file is

```
Freeware assembler ASxx.EXE Ver 1.03.
0001 0020               ORG $0020          ;start of source code
0002 0020 12 TWOSCOM    FCB %00010010      ;put $12 in $0020
0003 0021 70 00 20      NEG   TWOSCOM      ;start of program
0004 0024 3e            WAI                ;wait here
0005                    END
Number of errors 0
```

Now enter the program into the Mag-11 using the BUFFALO monitor's MEMORY MODIFY command along with the machine-language code provided by the assembler listing. Here you enter the following:

```
MM 0020 <CR> 12 <SP> 70 <SP> 00 <SP> 20 <SP> 3E <CR>
```

Now run the program. However, if you enter "Go 0020," you will have problems. Why? The program really does not start until the first instruction—not the first assembler statement—and that is at address $0021! Line 0002 was created by AS11, under the control of the FCB directive we inserted in the source code, *that forced us to enter $12 at the $0020 address.*

After running the program, by entering "Go 0021," press the reset switch and check out the data at $0020 using the MM command. It should be what? Is it? Now we return to the discussion of selected HC11 instructions.

Additional HC11 Instructions

BRA

Similar in action to the JMP instruction, BRA differs from JMP in the addressing mode used. The BRA instruction uses relative addressing, which was described briefly in Chap. 3. This addressing mode differs, in principle, from the other modes mentioned in the descriptions of other instructions. The relative addressing method uses a formula in which the data in the operand field are the relative offset stored as a two's complement number. If the relative offset is a positive number (bit 7 = 0), the branch is forward; if the offset is negative (bit 7 = 1), the branch is backward.

Branching forward is simple. The CPU just adds the positive offset to the program counter so that the next instruction starts at the place pointed to by the new program counter number. However, since the highest positive number can be $7F, *127 is the farthest forward a BRA instruction will take you.*

Just as branching forward is elementary, branching backward can be confusing if you are writing machine-language programs. Luckily, AS11 eliminates this confusion. With AS11 (and all other reputable HC11 cross-assemblers), the programmer never puts down the often confusing relative offset in the operand field. *Rather, the programmer specifies the destination of BRA (and all other branch instructions) by its absolute address, either as a numerical value or more commonly as a label or expression.* The AS11 then converts this absolute address into a relative offset. This relative offset shows up nice and plainly in the machine-language section of the assembler listing.

Note that branch instructions are limited as to how far in the program they can go—127 forward or 128 backward. If you attempt to go further, you will get an error message something like, "Branch out of range." Don't worry about it until you get this message, and then correct it.

Example

```
LOOP        STAA #$7000
            INCA
            BRA  LOOP
```

This short, trivial, and probably useless program will loop until the music dies. If you try this program snippet out on Mag-11, it will flash LED1 on and off at a rate of more than 100,000 times a second—which just means it will appear dim to your eyes!

BEQ

This conditional branching instruction causes a branch instruction if the CCR's Z = 1 (the zero flag is set). It is usually used immediately after an instruction that affects the Z bit.

It is interesting to note here that if a BEQ instruction immediately follows a CMP instruction, one can think of them as logically working together. Why do I say this? Well, CMP instructions compare the contents of an accumulator with the contents of memory specified in the operand field. It does this with a virtual subtraction operation. However, while the CCR register's flags are set according to the subtraction's result, neither accumulator nor memory contents are affected. This is the reason for using the adjective *virtual* when referring to subtraction. If the contents of the accumulator equal the contents of the memory location, then the zero flag is set, and the BEQ instruction will cause a branch. Here, BEQ truly can be thought of as an acronym for "branch if equal." Very logical! Very useful! Very easy to use!

Note:

The instruction's official name is Branch if Equal to Zero.

Example

```
LOOP1       LDAA    $7000
            CMPA    #$11
            BEQ     LGHTLED7
            BRA     LOOP1
LGHTLED7    STAA    $7007
            WAI
```

Assuming that this program is used on Mag-11, if the contents of the memory location $7000 equal $11, then the program branches and lights LED1 and then waits and waits and waits. Because of the decoding circuitry, dipswitch S1 is located at memory location $7000. If switches 1 and 5 are on and dipswitches 2, 3, 4, 6, 7, and 8 are off, then the first instruction (LDAA $7000) will stick $11 in accumulator A—which is the data at $7000.

BNE

This is BEQ's complementary instruction. BNE causes a branch when the Z bit is *not equal to zero*—just the opposite of the BEQ instruction. Again, when used immediately after a CMP instruction, the two form a neat pair. This time you can look at BNE as standing for "branch if not equal." Also very logical, very simple, and very easy to use.

Note:

The instruction's official name is Branch if Not Equal to Zero.

Example

```
LOOP2       LDAA $7000
            CMPA #$11
            BNE  LOOP2
            STAA $7007
            WAI
```

This short program does the same as the example for BEQ. Notice that there is one less line. It seems that often BNE is a better choice in many programs than BEQ.

BPL

The BPL (branch if plus) instruction tests the state of the N bit of the CCR and causes a branch if N is clear. As the discussion on signed numbers indicated, the N bit is clear if the most significant bit of a register (or memory location) is a 0. This 0 indicates that the number (looking at it as a "signed" number) is positive (plus). A simpler way of looking at the BPL instruction is that it looks directly at the most significant bit and causes a branch if this bit is 0. This is a surprisingly easy-to-use and efficient instruction.

Example

```
*Get data from A/D and wait until data is done
LOOP2           LDAB ADCTL
                BPL  LOOP2
*Light data are now in accum B
```

This program snippet was taken from our DAYALERT.ASM file. Recall that bit 7 of the ADCTL (A/D control register) is set when the A/D result registers contain valid conversion results. The BPL monitors this bit 7, and if bit 7 is not set, it branches back to LOOP2, where accumulator B is loaded again with the ADCTL's data. The program keeps looping until bit 7 of the ADCTL is set, and then the program continues. The comment indicates that the light data are now in accumulator B.

BMI

The BMI (branch if minus) instruction is the complement of BPL. A simple way to look at this instruction is that if you use it after an instruction that affects the N bit (e.g., most accumulator-related instructions), it causes a branch if the MSB in the register (or memory location) is set to 1.

Example

```
LOOP4    CLR     $7000      ;Make sure LED1 is OFF
LOOP3    LDAA    $7000      ;accum A now has S1 data
         BMI     LOOP3      ;if bit 7=1 GOTO LOOP3
         STAA    $7000      ;light LED1
         BRA     LOOP4      ;go back to start
```

Question: Will LED1 light when S1(8) (dipswitch S1's position 8) is on (bit 7 = 1) or off (bit 7 = 0)? If you have doubts, you can try it out.

Hint:

Add OPT I and ORG $B600 statements before the first line and then use AS11 file.asm .file.lst to obtain an assembler listing.

BLO

The BLO (branch if lower) instruction is identical to the BCS (branch if carry set), although it has a more logically appealing name. If used immediately after a compare or subtract instruction, a branch will take place if the number in the accumulator was lower than the number in the memory location referenced to in the operand field. This is not useful with instructions that do not affect the C bit. It is sometimes frustrating to realize (sometimes too late) that some simple instructions such as INC (adds 1 to the register or number at a memory location) or DEC (subtracts 1 from the register or memory location) do not affect the C bit, and thus the BLO instruction is futile here.

Example

```
*If it's not dark is it daytime?
NOTDARK     CMPB #DAY
            BLO   NOTDAY
            STAA LED4          ;if it's daytime
            STAA LED5          ;light these LEDs
            STAA LED6
            STAA LED7
            STAA LED8
            BRA   LOOP2 ;get new light data
********************************
*If neither dark nor day then
*light red LED3 and make it blink
NOTDAY      STAA LED3
               . . .
```

This program snippet was taken from our DAYALERT.ASM file. The memory location indicated by the label DAY contains information concerning brightness, and accumulator B contains data indicating the current light intensity. The BLO NOTDAY instruction will cause a branch to NOTDAY if the number in accumulator B is lower than the number in DAY. If not, LED4 through LED8 will be lighted, and the program will jump back to the memory location indicated by LOOP2 to get new light data.

BHS

The BHS (branch if higher or same) instruction is the complement of the BLO instruction just discussed *and* is identical to the BCC (branch if carry clear) instruction. If used immediately after a compare or subtract instruction, a branch will take place if the number in the accumulator was higher or the same as the number in the memory location referenced to in the operand field. Restrictions on this instruction are identical to those for the BLO instruction.

Example

```
LOOP6       CLR    $7001      ;LED2 is off
LOOP5       LDAA   $7000      ;Load A with SW data
            CMPA   #$1F       ;If Sw data is >=$1F
            BHS    LOOP5      ;branch to LOOP5
            STAA   $7001      ;If not lt LED2
            BRA    LOOP6
```

This is a short program, but a tricky program. Nonetheless, you should be able to figure it out by now, at least with the help of several hints. It may be fun to actually try this program out. *Questions* What positions does S1 have to be in order to light LED2? Will only one set of positions work, or will a number of sets do the job? Will LED2 light if all switches are on?

Remember that you are using a BHS instruction *and* remember that bit 0 of the data at $7000 turns on the respective LED. Also keep in mind that turning on one of S1's switches grounds the respective bit. By the way, the answer may be simpler than you think!

BHI

The BHI (branch if higher) instruction seems like it should be the complement of the BLO instruction, but it is not. In fact, it is the most complex conditional branching instruction that I have discussed so far. Why do I say this? All other conditional branching instructions made their decision on whether to branch or not on a single CCR flag. BHI, however, uses two flags—C and Z. Both these flags must be zero (C + Z = 0).

Common-sense interpretation If the BHI instruction is used immediately after a compare or subtract instruction, a branch will take place if the number in the accumulator was higher than the number in the memory location referenced to in the operand field. Restrictions on this instruction are identical to those for the BLO and BHS instructions.

Example

```
*Light data is now in accum B
        BSR   CLRLEDS    ;shut off all LEDs
        CMPB  #DARK
        BHI   NOTDARK    ;If not dark goto NOTDARK
*If it is dark light green LED1
*and then get next byte of light data from AN1
        STAA  LED1
        BRA   LOOP2
```

This program snippet again came from our DAYALERT.ASM source-code file. As this program snippet commences, the light data are already in accumulator B. The first actual instruction here (BSR) is an instruction that will be discussed shortly. For now, just think of the statement BSR CLRLEDS as a simple instruction that clears all eight of Mag-11's LEDs. The next two instructions work together: If the light data in accumulator B are higher than the data in the memory register that has the DARK label associated with it, the program will branch to the instruction with label NOTDARK. What it does from this point was detailed in the discussion of the BLO instruction. If the light data in B are not higher (lower than or same), the next instruction lights LED1, and the program branches to LOOP2 to get new light data.

BLS

This instruction is complementary to BHI. BLS (branch if lower or same) branches under the exact opposite conditions from BHI (i.e., C + Z = 1).

Common-sense interpretation If the BLS instruction is used immediately after a compare or subtract instruction, a branch will take place if the number in the accumulator was lower or the same as the number in the memory location referenced in the operand field. Restrictions on this instruction are identical to those for the BLO, BHS, and BHI instructions.

Question Refer to the example in the discussion on the BLO instruction. If BLS replaced BLO in the source code, how would the execution of the program change?

Hint: Remember the equate statement DAY EQU 150?

Answer   If the original source code was used with the BLO instruction, LED4 through LED8 would light when accumulator B registered 150 (decimal) or over. With the BLS instruction substituted, the LEDs would not light until accumulator B registered more than 150. Here, there was not much difference. However, one must be cautious in using this conditional branching instruction.

WARNING! WARNING! BUG WATCH!

A relatively common and hard-to-find bug that occurs with conditional branch instructions is one in which a particular value is ignored completely. For instance, say you are programming an EPROM that is suppose to control the amount of gasoline to inject depending on, say, a temperature of 20°F. Say your program uses the BLO instruction to provide, say, 0.0005 gal/s of gasoline when the temperature is below 20°F. Similarly, your program uses the BHI instruction to provide 0.00045 gal/s of gasoline when the temperature is above 20°F. What happens when the temperature is right at 20°F? Who knows for sure? In some systems the engine may actually stall. There is a definite bug in such a program. Bugs similar to this have occurred. I know! Keep alert!

This discussion of conditional branching instructions has left out BGT, BLE, BGE, BLT, BVS, and BVC. These instructions are more complex than the ones discussed, and new programmers can get in trouble using them. For more information, see the HC11 bible and the HC11 data booklet.

I now continue my discussion of HC11 instructions with an instruction that can manipulate a single bite. It is one of a set of bit manipulation instructions that are new to the 680x series of MPU/MCUs.

BSET

The BSET (bit set) instruction does just as it says—it sets bit(s) in the data of a memory location. The bits to be set are specified by ones in the mask byte. The *mask byte* is the last byte in the operand field. *Important: All other bits in the data are unaffected.* HC11 only allows direct or indexed addressing with this instruction. *Beware! It is possible to confuse the two because of format similarities!*

Example

```
*Turn on the A/D system and enable RC osc
          BSET OPTION,X #$CO
          BSR  DELAY       ;gives A/D time to stabl
```

This program snippet was lifted from our DAYALERT program. Here, the first line uses the BSET instruction, along with indexed addressing, to set bit 7 and bit 6 of the OPTION register. You will recall that these two bits turn on the A/D system and enable the internal RC oscillator.

COM, COMA, COMB

The COM (complement) instruction also was used in the DAYALERT program. All it does is switch the bits in the accumulator or memory location. Where there were zeros before the instruction was executed, there are 1s after, and vice versa. Data that were $00 (%00000000) become $FF (%11111111), and vice versa.

Example

```
        LDAB ADR2,X
*Light data is now in accum B
*We now inverse it to be more logical
        COMB
        BSR   CLRLEDS      ;shut off all LEDs
        CMPB #DARK
        BHI  NOTDARK       ;If not dark goto NOTDARK
*If it is dark light green LED1
```

Again, this example came from a code snippet from the DAYALERT.ASM source code. Here, COMB reverses the data, resulting in a more logical program. Before the COMB instruction was inserted, low values in accumulator B corresponded to high light levels, and vice versa. By using the COMB instruction, high levels of light correspond to high values in accumulator B. This simplifies the logical construction of the program and helps avoid those tension headaches.

Circuit Note

If you look at the schematic in Fig. 5-16, notice the 10-kV resistor connected to +5 V (remember this resistor is built into Mag-11). When light hits the phototransistor, it draws more current, and thus it acts as if its resistance drops. This causes the voltage at pin 11 (the input to the A/D converter) to drop. Thus it is obvious that the more light there is, the lower is the voltage. If there was no light at all, the voltage at pin 11 would approach +5 V.

Subroutines and Subroutine-Related Instructions

I will now introduce subroutines, as well as the instructions, BSR, JSR, PSH, PUL, and RTS, that really make assembly-language subroutines practical. Of course, I will also take another look at a CPU register that makes all of this possible.

HC11 assembly-language subroutines are very similar to BASIC subroutines. In most variants of BASIC, you will recall, the GOSUB "label" statement is used to call a subroutine. In the subroutine itself, the last statement is a RETURN instruction. The HC11 instructions BSR (branch to subroutine) and JSR (jump to subroutine) are very similar in principle to the GOSUB statement. Also, the HC11 instruction RTS (return from subroutine) is very similar to BASIC's RETURN statement. However, there are differences. Unlike BASIC, HC11 assembly language deals directly with the "personal" registers and other inner workings of the chip. BASIC hides much of this from the programmer. If you treat BSR and RTS just like their BASIC counterparts, you can wind up in trouble—especially if you use the same registers in your subroutine that you use in your main program. To avoid these problems, there are eight related instructions that can help you out here. Let's look at them before I talk more about BSR, JSR, and RTS.

PSHA, PSHB, PSHX, PSHY

These instructions store the contents of the respective register on the stack (e.g., PSHA stores accumulator's A contents on the stack). Two memory locations are required to store 16-bit registers (e.g., the X index register), whereas only one memory location is needed to store 8-bit registers. If you want to save accumulator's D contents, use PSHA and PSHB instructions. When 16-bit registers are loaded, two is subtracted from the stack pointer (SP), but only 1 is subtracted from the SP when 8-bit registers (accumulator's A or B) are pushed onto the stack. *Question.* "Where is the stack? I know by skimming over Chap. 3 that the HC11 has a 16-bit stack pointer register, but where is the stack itself where all these data are being stored?" *Answer.* It is in RAM. Often a portion of the HC11's internal 256 bytes of RAM (addresses 0–$FF) is used for the stack. Some MPUs have their own separate stack, which avoids trying to keep track of the stack pointer. However, most of these built-in stacks are small and must be used sparingly and with great care. *Author's wish list.* An HC11 with a huge built-in stack that does not require a stack pointer. *Hint.* Go back to Chap. 3 and review the discussion on HC11's stack and stack pointer. This time, do not skim, but study intently.

Example

```
*Simple subroutine creates about
*a 1/2 sec delay with 4-MHz crystal
DELAY      PSHY
           LDY  #$FFFF
LOOP3 DEY
           BNE  LOOP3
           PULY
           RTS
```

This subroutine was lifted from our DAYALERT program. Notice that it starts with a PSHY instruction. This instruction stores the Y index register in the stack in order to preserve its contents. Here, this instruction was superfluous because the Y register was not used in the main program. However, what if we lifted this subroutine and used it in another program that used this register? In this case it could mean the difference between a potentially monumental invention and the collapse of the designer's ego.

PULA, PULB, PULX, PULY

These instructions, which should always be used *with* the PUSH instructions just mentioned, load the respective register from the stack. When 16-bit registers are loaded, two is added to the SP. However, only 1 is added to the SP when 8-bit registers (accumulator's A or B) are loaded from the stack.

Example

```
*Simple subroutine creates about
*a 1/2 sec delay with 4 MHz crystal
DELAY      PSHY
           LDY  #$FFFF
LOOP3 DEY
           BNE  LOOP3
           PULY
           RTS
```

Notice from these same program snippets as discussed in the preceding section that the instruction PULY loads the Y index register with the same data it had before the subroutine was called.

> Make sure you only pull if you push and then only pull once. Also make sure that you pull every time you push.

BSR

This is a branch to subroutine instruction that causes a branch to occur to the location specified by the branch offset. Of course, with AS11, one usually uses a label (or address) instead of a branch offset. By convention, this label is the name of the subroutine. On the surface, BSR is very similar to the simple BRA instruction. However, beneath the surface, there is much more going on with this instruction. While both BSR and BRA cause the program counter to be incremented by two, unlike BRA, BSR pushes the contents of the program counter onto the stack. The program counter needs this "number" to return to the next instruction in the main program.

JSR

This is identical to BSR except for the addressing modes. JSR can use direct, extended, or indexed addressing, whereas BSR uses only relative addressing. JSR, like BSR, pushes the contents of the program counter onto the stack.

RTS

This return from subroutine instruction takes the program back to the main program, where the instruction located immediately after BSR or JSR is executed. Before returning to the main program, the RTS instruction pulls the contents of the program counter from the stack. The original contents of the program counter are what allow an orderly return to the main program. What if, just what if, no one initialized the stick pointer and it pointed to a location where there was no usable RAM memory? Obviously, there would be a program crash after the RTS instruction was executed. Now you know why I have stressed loading the stack pointer early in the program.

Example

```
*If neither dark nor day then
*light red LED3 and make it blink
NOTDAY      STAA LED3
            BSR  DELAY
            CLR  LED3
            BSR  DELAY
            BRA  LOOP2
*******************************
*Simple subroutine creates about
*a 1/2 sec delay with 4 MHz crystal
DELAY       PSHY
            LDY  #$FFFF
```

```
LOOP3   DEY
        BNE   LOOP3
        PULY
        RTS
```

Again, I lift a section of the source code from the DAYALERT program. The first BSR DELAY instruction causes a branch to the subroutine called DELAY. The DELAY subroutine's last instruction, RTS, causes the program counter to load the contents so that the program starts at the next instruction, which is CLR LED3 the first time DELAY is called and BRA LOOP2 the second time it is called.

Note that this DELAY subroutine is quite useful. The time delay here has to do with the fact that each instruction takes a certain number of clock cycles to execute. With a 4-MHz crystal, each clock cycle is exactly 1 μs long. The PSHY instruction takes five cycles, and the PULY instruction takes six cycles. The LDY #$FFFF instruction takes four cycles. However, the DEY and BNE instructions are the instructions that cause *real* program delay here, since they are the ones that are executed repeatedly—actually, they are executed 65,535 times. Since DEY takes four cycles and BNE takes three cycles, it becomes obvious that this program causes a delay of just over $65,535 \times 7 = 458,745$ μs, or just under $^1/_2$ s. Note that the Y index register was used here because DEY takes four cycles to execute, whereas DEX takes only three.

USE SUBROUTINES UNTIL YOU GET SICK OF THEM!

Thomas Alva Edison once said: "Time is the most precious thing in the universe. A billion dollars can't buy you one minute!" This is something to think about when you start to program. If you can save programming time, you are saving something that is priceless. Now, subroutines used properly can save you lots and lots of time—this is obviously true in long programs. Perhaps surprisingly, they are even more useful in small programs that are designed to fit in the 512 bytes of the HC11's EEPROM. Here, not only do they save precious programming time, but they also save precious EEPROM space as well. There is more. *Extensive use of subroutines makes the program easier to understand later if you need to modify it.*

MAKE UP A FILE OF YOUR FAVORITE HC11 SUBROUTINES

One of the really neat things to do is to make your own compilation of subroutines. If you have them all on a computer disk in ASCII format, all you have to do to write some programs is use your text editor/word processor's Copy/Paste tools to write your program. One example of a subroutine that could be useful is a SQRTD_A subroutine that takes a number in the double accumulator D and returns the square root of the number in accumulator A. Such a subroutine will save priceless minutes and probably even hours. Of course, the DELAY subroutine should be on every HC11 devotees's must list.

Motorola's HC11 Instruction Set Details

We have just looked at a relatively few HC11 instructions. The "HC11 bible" has them all in its Appendix A. For an example of how to use this appendix, let's look at the LSR (logic shift right) instruction on page A-70 of the "HC11 bible." A copy of this sheet is shown here in Fig. 8-6.

OPERATION SECTION

This section gives a picture of what is happening in the register. Note that it shows basically every bit in the register being shifted right. It also indicates, pictorially, that bit 7 (the most significant bit) is set to 0 during the execution of this instruction and that the CCR's C (carry) bit is loaded from bit 0, which is the least significant bit.

DESCRIPTION SECTION

This section is often easiest to understand from the viewpoint of those who are in the HC11 learning stage—which probably is everyone, including me. Here there is one thing new we

LSR   **Logical Shift Right**   # LSR

Operation: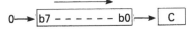

Description: Shifts all bits of ACCX or M one place to the right. Bit 7 is loaded with zero. The C bit is loaded from the least significant bit of ACCX or M.

Condition Codes and Boolean Formulae:

S	X	H	I	N	Z	V	C
—	—	—	—	0	↕	↕	↕

N   0
   Cleared.

Z $\overline{R7} \cdot \overline{R6} \cdot \overline{R5} \cdot \overline{R4} \cdot \overline{R3} \cdot \overline{R2} \cdot \overline{R1} \cdot \overline{R0}$
   Set if result is $00; cleared otherwise.

V $N \oplus C = [N \cdot \overline{C}] + [\overline{N} \cdot C]$ (for N and C after the shift)
   Since $N = 0$, this simplifies to C (after the shift).

C   M0
   Set if, before the shift, the LSB of ACCX or M was set; cleared otherwise.

Source Forms: LSRA; LSRB; LSR (opr)

Addressing Modes, Machine Code, and Cycle-by-Cycle Execution:

Cycle	LSRA (INH)			LSRB (INH)			LSR (EXT)			LSR (IND, X)			LSR (IND, Y)		
	Addr	Data	R/W̄	Addr	Data	R/W̄	Addr	Data	R/W̄	Addr	Data	R/W̄	Addr	Data	R/W̄
1	OP	44	1	OP	54	1	OP	74	1	OP	64	1	OP	18	1
2	OP+1	—	1	OP+1	—	1	OP+1	hh	1	OP+1	ff	1	OP+1	64	1
3							OP+2	ll	1	FFFF	—	1	OP+2	ff	1
4							hhll	(hhll)	1	X+ff	(X+ff)	1	FFFF	—	1
5							FFFF	—	1	FFFF	—	1	Y+ff	(Y+ff)	1
6							hhll	result	0	X+ff	result	0	FFFF	—	1
7													Y+ff	result	0

FIGURE 8-6 The LSR (logic shift right) instruction.

have learned about the LSR instruction: It applies to accumulators A and B as well as locations in memory. In other words, this is one of the few data-manipulation instructions that applies to memory locations and not just CPU registers.

CONDITION CODES AND BOOLEAN FORMULAS SECTION

This section includes a pictorial representation of the CCR showing which flags are affected as well as a bit-by-bit Boolean formula showing under what conditions each respective flag is affected by the instruction.

In the pictorial representation of the CCR, dashes in the rectangles that represent the S, X, H, and I flags mean that these flags are not affected by the LSR instruction. Since bit 7 always has a 0 after an LSR instruction is executed, the N flag is always cleared (N = 0). The two-headed arrows in the boxes labeled Z, V, and C indicate that these respective flags are either cleared or set, depending on the result of the shift.

The Boolean formula for the Z flag is _R7*_R6*_R5*_R4*_R3*_R2*_R1*_R0, where _R7···_R0 represent the inverse of the respective bit of the result and * is the Boolean AND symbol. Thus we can see that if all bits are 0, their inverse is 1, and the Boolean formula, in this instance, can be viewed simply as 1*1*1*1*1*1*1*1=1, which means simply that if all bits of the result are 0, the Z flag is set.

The Boolean formula for the V flag is

$$N+C=[N+C]+[N+C]$$

for N and C after the shift. Here + is the Boolean symbol for EXCLUSIVE OR and simply means N or C but not both N and C. (+, of course, is the symbol for regular OR and means what it says.) Since N = 0, it is obvious that this reduces to simply V = C after the shift. In other words, if the carry flag is set, so is the overflow flag; if the carry flag is not set, neither is the overflow flag.

The Boolean formula for the C flag is M0 (bit 0 of memory location), which simply means that C after the shift is exactly the same as bit 0 of the accumulator or memory location was before the shift.

SOURCE FORMS SECTION

This section gives the exact mnemonic forms recognized by most assemblers and cross-assemblers. Here, LSRA pertains to accumulator A and LSRB pertains to accumulator B. The (opr) in LSR refers to the operand. More information on this operand is given in the next section.

ADDRESSING MODES, MACHINE CODE, AND CYCLE-BY-CYCLE EXECUTION SECTION

This section is set up as a table. The number of headings depends on the number of variations of the instruction. With the LSR instruction, there are five table headings because there are five variations. To simplify things, we will look at only one heading—LSR (EXT).

This heading [LSR (EXT)] details use of the extended addressing mode. Notice that the table assumes that the user already knows quite a bit about the HC11 in general and the

LSR in particular in order for the table to be used effectively. In other words, it is *not* meant for HC11 tenderfeet. For instance, the creators of the table assume that the user knows all about shifting data in a memory location. While this is an assumption that is necessary to make, it is not spelled out in *big block letters.*

The subheadings under LSR (EXT) are Addr, Data, and R/_W. These subheadings indicate what is on the address bus, data bus, and R/_W lines, respectively. Also notice the column at the far left. This heading is "Cycle." Beneath this heading are the numbers 1 through 7. Each number corresponds to a CPU clock (E) cycle.

During cycle 1, the address bus contains the address of the opcode (OP), the data bus has the respective machine-language code (which is 74), and the R/_W line is high, indicating a read operation. During the next cycle (cycle 2), the address bus is simply 1 higher than the last cycle, and the data bus now has the most significant byte (hh) of the extended address. During the third cycle, the address bus is now 1 higher than the preceding cycle, and the data bus now has the least significant byte (ll) of the extended address. (The R/_W line is still high, just as it was in the first two cycles.) During the fourth cycle, the address bus now has the full extended address (hhll) of the memory location, while the data bus has the data at the memory location. R/_W is still high. During the fifth cycle, the CPU is actually manipulating the data in the memory location, and *all* bits on the address bus are 1 (FFFF) and the address bus has irrelevant data. R/_W is still high. Finally, in the sixth and final cycle, the address of the memory location is on the address bus, and the result of the operation is on the data bus. Now, however, the R/_W line is 0, indicating a write operation. It is during this operation that the actual memory location is changed to reflect execution of the instruction.

Appendix A of the "HC11 bible" provides an enormous amount of information. However, much of it can be confusing to new users because of the nomenclature used. Pages A1 through A3 in the "HC11 bible" provide a semiconfusing guide to this nomenclature.

Talking Frankly About the Location of Internal RAM and Registers

So far we have assumed that the HC11's 256 bytes of RAM are located between 00 and $FF in the HC11's memory. This is why we loaded the stack pointer with $FF or $2D or sometimes some number in between. I also have stated that the 64-byte block of internal registers is located between $1000 and $103F. Thus, in order to light Mag-11's CPU check LED, LED10, we stored $08 at $1000, which is port A. These memory locations are "normal" and will not be changed in any experiment or software in this book. However, because of HC11's versatility, these locations are not etched in silicon. *They are only chalk marked on silicon!* Both internal registers and internal RAM can be remapped to the beginning of any 4-kB page in the HC11's memory map. This is accomplished in the INIT register, which is "initially" located at $103D (see Fig. 8-7).

RAM3–RAM0—RAM MAP POSITION

These four bits, which specify the most significant nibble of the RAM address, control the position of the RAM in the memory MAP. After reset, these bits are zeros, and the RAM is located "normally" (i.e., #0000–$00FF). If RAM0 is a 1, all others off (i.e., most significant

REGISTER'S NAME: INIT					REGISTER'S ADDRESS: $103D			
BIT#	7	6	5	4	3	2	1	0
BIT NAME	RAM3	RAM2	RAM1	RAM0	REG3	REG2	REG1	REG0
BIT'S STATUS AFTER RESET	0	0	0	0	0	0	0	1

FIGURE 8-7 The INIT register.

nibble a 1), then the RAM's location is changed to $1000 to $10FF. (Details on possible conflicts with the internal registers are found in the fine print section.) If RAM1 is a 1, all others off (i.e., most significant nibble a $2), then the RAM's location is changed to $2000 to $20FF. If RAM3 = RAM2 = RAM1 = RAM0 = 1 (i.e., most significant nibble an $F), then the RAM is located at the start of the last 4-kB page of memory or at $F000 to $F0FF.

REG3–REG0—64-BYTE REGISTER BLOCK POSITION

The discussion here is nearly identical to the RAM map position. However, at reset, REG0 = 1, while all others are off. Thus the registers are located initially between $1000 and $103F.

Fine print

(1) Whereas, if a specified user attempts to change the INIT register *after* 64 clock cycles have taken place, *after* reset, and before the attempt to modify has been made, the attempt is null and void and the location of the RAM and registers will remain unchanged.

Exception to subparagraph (1): If the attempt to modify the INIT register is conducted, knowingly and willingly, while the HC11 microcontroller is in a "special" mode, subparagraph (1) should be ignored, since the INIT register will so be changed whether or not 64 clock cycles have passed from the time of reset.

(2) Whereas, if a specified user, whether out of full cognition of fact or out of total disregard of manufacturer's communiqué, locates RAM and registers at the start of the same 4-kB page of memory, the specified register will have complete and absolute priority over RAM. In other words, the first 64 bytes of RAM will be useless in this special circumstance.

Large print

Hey, now you know why I left these details out until now.

Experiment a Bit More with DAYALERT

Before leaving this chapter, run the DAYALERT program again. With dim room light, the red LED should flash. If you shine a flashlight on the phototransistor, five LEDs, LED4 through LED8, should light. Put your hand over the phototransistor to simulate darkness, and only the green LED1 should light.

Now try replacing the COMB instruction with the NOP instruction. To do this, replace the 53 (code for COMB) in line 2 of Table 8-1 with 01 (code for NOP.) Just as its mnemonic indicates, the NOP instruction does "no operation"—in other words it does nothing, although it takes one clock cycle for doing this nothing. NOP's primary purpose is during the debug phase of program development, where NOP instructions are sometimes used to temporarily replace other machine-code instructions. In other words, NOP instructions are to machine language as first-column asterisks are to assembly language.

What do you think will happen if you now run the program with the NOP instruction replacing the COMB instruction? Try it to see if you are right.

Now try something a little different. First, change the NOP instruction back to COMB, and then look at the DAYALERT's source code. Notice the equate directives:

```
DARK        EQU  20       ;Set darkness level
DAY         EQU  150      ;Set day level
```

These directives set the DARK and DAY levels. Why not try changing these levels? Recall that the equates statements used decimal and that the machine code uses hexadecimal. Twenty in decimal is 14 in hexadecimal and 150 in decimal is 96 in hexadecimal. The DARK level byte is located at address $B618 and the DAY level at $B623. If you would like to see what will happen by changing these levels, use BUFFALO's memory modify (MM) command to change the data at these two addresses ($B618 and $B623.)

The next chapter starts the second part of the book, which concentrates on practical applications of the theory you have learned in the first eight chapters.

9

INTRODUCING A SIMPLER,
CHEAPER WAY

So far we have looked at two different HC11 systems. You may have already forgotten about the first one. Remember way, way back in Chap. 2 where you (hopefully) wired up a basic HC11 system right on a breadboard? The other HC11 system, of course, is Mag-11, which I like to refer to as the "HC11's experimenter's dream board."

Both Mag-11 and the breadboarded HC11 system have one thing in common: They use the HC11's expanded multiplex mode and thus require an external EPROM and address decoding. However, the very concept of microcontrollers, including the HC11, is based on a minimal chip system. *Minimal chip* here means a system without external firmware and the resulting need for address-decoding circuits. The original, inexpensive version of the HC11 was designed so that the instructions that control its operations normally would be stored in its own 8-kB ROM (read-only memory). HC11 is designed to operate under the instructions contained in ROM, and these instructions are etched into the silicon during the manufacturing process. This is great if you are General Motors or Ford and plan on using the HC11 in a production run of a car or truck. The cost of producing the ROM mask can be spread over hundreds of thousands of units. However, for the rest of us, whose grand plans may include a maximum production run of, say, a thousand units, ROMs just are not practical. Even if you have hopes some day of producing a million units, there is little doubt that between your initial five prototypes and that million you will need to change your program many, many times—and each change will be terribly costly if you are using ROM. One solution is to use the HC711 series, which replaces the ROM with EPROMs that you program yourself. The problem with the HC711 series is that they are roughly twice as expensive as your garden variety HC11 and they can only be programmed once—which, again, means that it is a rather costly alternative.

Is there a practical solution for those of us who want to use the HC11 in specialized limited-production products such as an experimental robot or a computerized weather station and do not want to see the mass of our wallets slip below the critical point? Smile. There is a solution. And it resides deep within the HC11 itself—it is HC11's own little EEPROM. If the 512-byte EEPROM in the garden variety HC11 is not enough to store the instructions for your ingenious thingamajig, then how about the 2048-byte EEPROM in the HC811 version? And yes, the HC811 is significantly cheaper than the HC711! And the program, burnt into the EEPROM, can be revised over and over and over—actually over 10,000 times. Of course, this solution has costs and limitations. For instance, practically speaking, using the HC11's EEPROM for program storage means that *you must program using HC11 assembly language.* BASIC and even C are out—there just is not enough room in 512 bytes. But is this really so bad? Often, it is easier to program an HC11 with assembly language than BASIC for the simple reason that you are frequently dealing intimately with the HC11's registers. Besides, learning about the HC11's assembly language is not a waste because the HC11, along with its grandparents, is a lively series of MPUs/MCUs that should be around for a long, long time.

Think here of the fellows who in the mid 1970s delved into 8080/Z80 assembly language. Many 8080 assembly-language programs can be used on 80386, 80486, and even Pentium machines. Was it a waste of these guys' time and brains to learn about the 8080 chip? Ask that Bill what's-his-name in Seattle.

I will show in the next chapter a single-board computer with only one other chip in addition to the HC11 itself. Before I do this, let's look at the HC11's internal EEPROM and how to use it to store the necessary instructions that tell the HC11 what to do and when to do it. However, even before we do this, we will again play a bit with Mag-11, and then I will show you how to build a simple and cheap EEPROM programmer that also doubles as a basic program development board.

Running Programs from the HC11's EEPROM

There really is nothing new here in using the HC11's internal EEPROM to store our program. In fact, most of the experiments using the Mag-11 do just that. Remember, we used the BUFFALO line assembler to write our programs that were stored somewhere in the 512-byte block of memory between $B600 and $B7FF. However, to run such programs, we used the BUFFALO's Go command. Obviously, if we want a simple design, this is a dumb way of doing things, since you would need to include a BUFFALO EPROM along with an address decoder with every design.

The alert reader will recall that Version 3.4 of the BUFFALO monitor has another way of starting programs that are located at the start of EEPROM ($B600.) If it detects that bit 0 of port E is 1, then the program jumps to $B600. But still, even here, you need a BUFFALO monitor in the system.

Apparently, the original purposes of the EEPROM were

1. A more economical method of selecting product options
2. The storage of setpoint and calibration data

3. During the product development cycle for the storage of data and the testing of program routines

4. For logging daily, weekly, or monthly data

5. The implementation of self-adjusting/self-adapting systems

Despite the original purposes, one or more designers foresaw the desire of some users of the HC11 to employ the internal EEPROM to store complete programs, eliminating the need for ROM or an external EPROM. These designers looked for a practical way to force the MCU to jump directly into EEPROM after a reset. Nonetheless, these designers did not want to add another special mode, which likely would increase the cost of the chip and certainly would increase the steepness of the HC11's already steep learning curve. As a compromise, they provided a nearly hidden way to accomplish this. First, the HC11 must be placed in the special Bootstrap mode. Then the SCI system must receive a break character on the RxD line. After this, the HC11 jumps directly to the start of EEPROM ($B600.) This break character can be initiated automatically simply by tying the RxD and TxD pins together and then to a pull-up resistor and then reset in special bootstrap mode. For those who are concerned about not being able to use the SCI because the RxD and TxD lines are tied together, do not fear. Instead of tying TxD to the RxD, all you need do is have your software send the break character to the RxD line. Details on this will be supplied later. All you need know for now is that a break character simply can be looked at as a longer than normal NUL (00) character. (By the way, a simple NUL character usually will do the job. However, there are baud-rate restrictions. More on this later.)

You may be asking the following question: "Well, it all sounds great except for one little thing. I don't trust EEPROMs. When I go to the trouble of designing a product, I want the thing to last and last and last. Well, at least a month longer than the 1-year warranty my company puts on its products. EEPROMs sound so ah, ah, well, nonsubstantial. You know, all you need is one tiny bit to go haywire, and the whole product becomes somebody's landfill problem."

If you are unsure of EEPROM's reliability, let me attempt to allay your fears a bit with a short story about the EEPROM's close relative, the EPROM. Each memory cell bit in an EPROM consists of a storage capacitor isolated and buried deep in the silicon so that it is difficult for this capacitor to gain or lose a charge. The charge on this isolated capacitor is sensed by an MOS transistor that gives out a 1 if the capacitor is uncharged and a 0 if it is charged. The actual process of programming involves supplying a relatively high voltage (sometimes as much as 25 V) to the memory cell to be programmed that charges the associated capacitor through the process descriptively called *avalanche injection.*

EPROMs are not the only kid in town that can be programmed by the end user. There are also fusible-link PROMs (programmable read-only memory). Unlike EPROMs, there is something really tangible taking place in PROMs—tiny little fuses are burnt out by the programmer. In fact, while they are extremely small, it is possible to see these burned out fuses with a good microscope. Unlike EPROMs, there is nothing hypothetical here. This is a good, simple, concrete principle. PROM theory reeks of reliability, while the theory behind EPROMs is vague and abstract. When General Motors first started to use MPU/MCUs in its vehicles, it used PROMs—the company was very suspicious of EPROMs. The company doubted the long-term reliability of EPROMs despite the fact that the manufacturers guaranteed their data retention for 10 years. When several years went by, General Motors decided that EPROMs seemed reliable after all.

While I did not hear anything directly from General Motors, through personal experience and by word of mouth I discovered a fascinating thing—the so-called reliable technology of PROMs did not seem so reliable after all (rumor has it that the fuses grew back several years after programming), while EPROMS seemed to be like the Energizer bunny—they keep going, and going, and going. To be honest, EPROMs seem more reliable than they should, theoretically anyway. Facts are facts, though, and who is thee to deny thy apparent truth!

Now for EEPROMs. They are based on the same basic theory as EPROMs, so it would seem that they should last just as long. However, EEPROM memory is more complex than EPROM memory. For instance, each byte in EEPROM requires 17 transistors, while a similar byte in EPROM requires only 8. Of course, the really neat thing about an EEPROM is that you can erase it using only electrical energy, whereas with a UV-erasable EPROM, you need a good-quality UV eraser. (Many EPROMs today no longer have windows and cannot be erased. These types are called *OTP EPROMs,* an abbreviation for *one-time programmable EPROMs,* despite the fact that there is an apparent conflict in terms.)

Don't be a worry wart about EEPROMs, fearing that they'll be worn out before you wear out. While the HC11's internal EEPROM is expected to degrade, it is guaranteed to last at least 10,000 erase/write cycles. Apparently, this figure of 10,000 is conservative and is the estimated lifetime at extremely high temperatures. At more normal temperatures, this internal EEPROM should last closer to 100,000 cycles. If one erases and writes once a day, this 10,000-cycle lifetime would last until past A.D. 2020; a 100,000-cycle lifetime would end in the year 2275. Since EEPROMs are only guaranteed to hold their data for 10 years, I would not be overly concerned about wearing them out by programming them too often. By the way, this 10-year lifetime is a conservative figure. Under normal temperature conditions, they will likely last longer than you will. However, both EPROMs and EEPROMs are naturally sensitive to radiation, so it is not recommended that you use either near a nuclear bomb blast.

A Little Magic Trick

I will next show an apparent magic trick. I will show how Mag-11 will still operate properly even when its EPROM is pulled mercilessly from its socket, thrown on the floor, and stepped on. Actually, it is no trick at all. All I will do is demonstrate what I have just discussed—placing a program's instructions inside the HC11's internal EEPROM and running the program by starting it in special bootstrap mode with the RxD and TxD lines tied together. By the way, please do not accuse me of EPROM abuse. My references to "thrown on the floor and stepped on" are meant solely in the metaphorical sense. Honestly, you have to truly believe that I would never, ever abuse an EPROM—especially one I had to replace by depleting my own net worth.

Tale of the "Invisible" Address Bus

So far, to be able to run a program automatically in EEPROM, we have seen that you must use either the BUFFALO monitor, or similar software, to "point to" the EEPROM. (In other words, somehow you have to get the program started at address $B600.) However, as mentioned a few paragraphs ago, the designers of the HC11 foresaw that some users might like to jump directly to internal EEPROM after reset rather than to the reset vector located at $FFFE and $FFFF. They provided for this feature in the internal bootloader firmware that every HC11 model A contains, whether it has internal ROM or not.

Now let's write a simple program that will make the green LED1 blink. LED1, you will recall, is controlled with bit 0 at the address $7000. The listing in Fig. 9-1 should make LED1 blink.

Use BUFFALO's line assembler to enter this program, and then run it with BUFFALO's Go command. LED1 should blink. Now we will try something more mysterious. Run this program without using the BUFFALO monitor—in fact, without using an EPROM at all.

In order to implement this feature, you must first place the chip in the special bootstrap mode (with Mag-11, this is accomplished by installing a shorting jumper at JP11 and placing a shorting jumper between pins 2 and 3 of JP1). Then you must tie the RxD and TxD pins together and to a pull-up resistor (with Mag-11, installing a shorting jumper between pins 1 and 2 of JP12 does this). After reset, the chip will jump directly to $B600, and the program should run. Right? Well let's try it!

Turn the power off, and remove the BUFFALO monitor EPROM from Mag-11. Turn on power and reset the system. What happens? Nothing! Right? After reset, LED1 does not blink, as you might expect. Now use a logic monitor or an oscilloscope to examine the HC11's address and data lines. There is nothing. The HC11 appears dead. What has happened? If you tried this without reading about it first here, you might have panicked, fearing that the HC11 went to that infinite silicon sand pile in the sky. Well, be assured that nothing that drastic has happened. Remember, the chip is now in special bootstrap mode, which means that its external data and address buses are disconnected from the internal workings. While the HC11 is most likely executing all the instructions stored in its EEPROM perfectly, it is operating invisibly, since LED1 requires external data and address lines to operate. If you are thinking what good is a perfectly working MCU if it does not interact with the outside world, keep in mind that the HC11 is loaded with

---

Notes: 1. Line numbers at extreme left are meant for discussion purposes only. 2. BUFFALO's ASM line assembler assumes that all numbers are already in hexadecimal. *Do not add "$" prefix.* 3. Numbers in the second column are addresses. *Do not type these numbers in.* However, these addresses are displayed on the BUFFALO screen. 4. To enter this listing, start BUFFALO and then type "ASM B600 <CR>" and then start entering the text in *third column only.* After you are done typing the line, hit the ENTER key (i.e., carriage return). 5. After entering last line, hit the decimal point key to leave the ASM line assembler.

```
Line#
1    B600    LDS    #2D
2    B603    STAA   7000
3    B606    LDY    #FFF
4    B60A    DEY
5    B60C    BNE    B60A
6    B60E    COMA
7    B60F    BRA    B603
```

FIGURE 9-1 Program listing, in BUFFALO's ASM line assembler format, that blinks Mag-11's LED1.

input-output (I/O) ports that are still active. To demonstrate this, try a variation of the program (see Fig. 9-2). With this new listing, LED10 will blink; it is connected to bit 3 of port A.

In order to reprogram the HC11's EEPROM, again you must reinstall the BUFFALO EPROM and place JP12's shorting jumper between pins 2 and 3. Also, HC11 should be in expanded multiplex mode. Remove the jumper at JP11, and install a jumper between pins 1 and 2 of JP1. Remember that the starting address is $B600, and enter the program in Fig. 9-2 using BUFFALO's ASM line assembler.

Now try this program out by setting the jumpers for special bootstrap operation. Recall that for the bootstrap mode you must place a jumper between pins 2 and 3 of JP1 and install a jumper at JP11. Also, again move a shorting jumper at JP12 from pins 2 and 3 to pins 1 and 2. If you want to be really brave, also remove the BUFFALO EPROM from socket U7 Now press Mag-11's RESET switch. LED 10 should now blink on and off.

Notice that the system is extremely simple. Basically, only U1, U2, Q1, and voltage regulator U4 are operating. It is obvious that you could build a complete MCU project with only two ICs: a 68HC11 and a microprocessor supervisory IC such as MAX690 or Motorola's MC34064. Projects such as these can be simple in physical design and sophisticated in function, but best of all, they can be "el cheapo" to build!

Notice that we have been operating in the special bootstrap mode, which limits our visions of grandeur. However, even though we must start out in the special bootstrap mode to enable the jump directly to EEPROM, no one says we have to stay there. We can use software to jump into the normal expanded multiplex mode. The procedure to get back to the expanded multiplexed mode is straightforward. All that has to be done is to modify the HPRIO register (located at $103C) using software. Examine the HPRIO in Fig. 9-3. Here, "x" stands for the bits we are not interested in right now.

Table 9-1 shows that the condition of bits SMOD and MDA after reset depends on the operating mode at reset. Notice with jumpers set for the special bootstrap mode that bit SMOD is 1 and MDA is 0, just the opposite from the expanded multiplex mode. There is one requirement: MDA can only be changed by software when SMOD is set. Thus it is obvious that MDA must be changed before SMOD is cleared.

As can be seen, in order to switch to expanded multiplex mode, we want the SMOD bit cleared and MDA set. In practice, you first write a 1 to bit MDA (causing the MCU to temporarily go into the special test mode). The next set of instructions should clear SMOD, which causes the MCU to go into the expanded multiplex mode.

Note: Start entering program listing by typing "ASM B600 <CR>" at BUFFALO's prompt.

```
Line #
1    B600    LDS    #2D
2    B603    STAA   1000
3    B606    LDY    #FFFF
4    B60A    DEY
5    B60C    BNE    B60A
6    B60E    COMA
7    B60F    BRA    B603
```

FIGURE 9-2 Modified program listing, in BUFFALO's ASM line assembler format, that blinks Mag-11's LED1.

REGISTER'S NAME: HPRIO						REGISTER'S ADDRESS: $103C	
BIT#	7	6	5	4	3	2	1	0
BIT NAME	RBOOT	SMOD	MDA	IRV	PSEL3	PSEL2	PSEL1	PSEL0
BIT'S STATUS AFTER RESET	*	*	*	*	0	1	0	1

*Refer to Table 9-1 for the condition of these bits immediately after reset.

FIGURE 9-3 The HPRIO register.

TABLE 9-1   SETTING AND CHANGING HC11'S MODES

INPUTS		MODE	CONTROL BITS IN HPRIO (LATCHED AT RESET)			
MODB	MODA	DESCRIPTION	RBOOT	SMOD	MDA	IRV
1	0	Normal single chip	0	0	0	0
1	1	Normal expanded	0	0	1	0
0	0	Special bootstrap	1	1	0	1
0	1	Special test	0	1	1	1

	SMOD	MDA
Single chip	0	0
Expanded multiplexed	0	1
Special bootstrap	1	0
Special test	1	1

The program listing in Fig. 9-4 is a modification of the program given in Fig. 9-2, which is designed to have LED1 blink. Notice that the first and second lines set the MDA bit and the third and fourth lines clear the SMOD bit. The fifth line sets the stack pointer. To program the EEPROM, you will need to configure the jumpers for expanded multiplex operation and install the BUFFALO EPROM in socket U7. Again, place a shorting jumper between pins 2 and 3 of JP12. (Remember to start the program at address $B600. With BUFFALO's ASM, you would type the following after receiving the prompt ">": "ASM B600.")

To try out the program, again shut off power and set the jumpers for special bootstrap mode. (The BUFFALO EPROM can be removed if desired.) Again, place a shorting jumper between pins 1 and 2 of JP12. Turn on the power, and press the RESET switch. LED1 should now blink, since we have switched back in the program to the expanded multiplex mode that activated both address and data buses.

Note: Start entering program listing by typing "ASM B600 <CR>" at BUFFALO's prompt.

```
Line #
1    B600    LDAA  #75
2    B602    STAA  103C
3    B605    LDAA  #35
4    B607    STAA  103C
5    B60A    LDS   #2D
6    B60D    STAA  7000
7    B610    LDY   #FFFF
8    B614    DEY
9    B616    BNE   B614
10   B618    COMA
11   B619    BRA   B60D
```

FIGURE 9-4 **Program listing, in BUFFALO's ASM line assembler format, that does have Mag-11's LED10 blink.**

Before leaving Mag-11 and introducing MagPro-11, let us look at using a MOT S19 file to program an EEPROM. Recall that .S19 files are created automatically by AS11 (and most other HC11 assemblers and cross-assemblers). Once you get the knack of using the .S19 file to program HC11s, all you need do is write an assembly-language program and then let your computer and Mag-11 or MagPro-11 do all the actual programming for you. While the use of BUFFALO's line assembler is great for programs of under, say, 20 lines, long programs become difficult to write and understand, as well as time-consuming to enter, using merely the line assembler.

Note:

> You may want to skip over the next section, which details programming a 40-pin DIP HC11 (e.g., MC68HC11A1P) with Mag-11, and go directly into MagPro-11, which details programming a 52-pin PLCC HC11 (e.g., MC68HC11A1FN). Much of the discussion here will be repeated.

Programming MC68HC11A1P EEPROMs Using Mag-11 and .S19 Files

Programming of the HC11's EEPROM will use routines from BUFFALO. For instance, the LOAD T command uses a downloaded object file, in MOT S19, to load the file into an HC11 system's memory. However, because of certain programming restrictions with Version 3.4 of the BUFFALO monitor, this command cannot be used to load files directly into EEPROM. However, there is a way to get around this little problem, since BUFFALO's MOVE command does program the EEPROM. To use this command, though, the data to be moved must already be some place in the HC11's memory space. Luckily, Mag-11 uses a 6264 8-kB RAM chip that provides more than enough space for even the HC811, which has 2 kB of EEPROM.

If it is not done already, carefully install a 6264 8-kB RAM chip in the socket at U9. Before installation, make sure power is removed from Mag-11.

When writing the source code for the file, use the ORG $B600 directive, since we want the program to start (have its ORiGin) at $B600. Use AS11 to create the .S19 object file, and keep track of its name and its path.

For the next step, since the file starts at $B600 and you want to download it to RAM starting at $0100, you must use the OFFSET command. (Remember, addresses $0100 through $0FFF and $1040 through $1100 now point to U9, the 6264 RAM chip.) To use the OFFSET command, enter

```
OFFSET -B500 <CR>
```

Notes:

(1) $B600 – $B500 = $0100. (2) $100 through $0FFF are located in the U9 RAM chip.

Do not add a prefix of "$" because BUFFALO only understands hexadecimal. Next, enter

```
LOAD T <CR>
```

and then start downloading the .S19 file. After a successful download, "Done" will be displayed on the screen.

Now we have the program in RAM. However, we still have to *move* it to EEPROM. To do this, enter

```
MOVE 0100 02FF B600 <CR>
```

This command moves data from addresses $0100 through $02FF to $B600 through $B7FF. This procedure takes a bit longer in time because the data are being burnt into EEPROM.

To show how all this is actually done, let's take an actual example and go step by step. See the source code, in AS11 format, in Fig. 9-5. This program simply flashes LED1. However, it flashes it at a much faster rate than the program in Fig. 9-4 so that we can tell that we actually did something.

Assuming that you named the source code file BLNKLED1.ASM and you have the file on the same directory as AS11.EXE, then simply type

```
AS11 BLNKLED1.ASM l,c,e >BLNKLED1.LST
```

(Note that the extension to the line ">BLNKLED1.LST" has no effect on creation of the MOT S19 object code and can be eliminated.) The resulting BLNKLED1.S19 file is given in Fig. 9-6. Keep track of the file's name and directory because you will need this information when it is time to download this file to your Mag-11 board.

Next, set up Mag-11 as before with the BUFFALO EPROM installed at U7, and then access your computer's communication program and operate Mag-11 under BUFFALO's control as before.

```
*BLNKLED1.ASM program
*Will flash LED1 at a fast rate
*start program at beginning of EEPROM
              ORG     $B600
*the next two instruction sets:
*RBOOT=0, which disables Bootstrap ROM
*SMOD=1 so MDA can be written to
*MDA=1 for expanded mode (normal or test)
              LDAA #$75
              STAA $103C
*the next two instruction sets:
*SMOD=0 so normal exp mode is in effect
              LDAA #$35
              STAA $103C
              LDS  $#2D   ;setup stack pointer
AGAIN         STAA $7000  ;send A's bit 0 to LED1
*the next three instructions cause a delay
*so LED1 doesn't blink so fast we can't
*see it, the delay is 5 times as fast
*as it would be if LDY $FFFF was used
LOOP          LDY  #$3000
              DEY
              BNE  LOOP
*the next instruction simply switches
*bit 0 of accum A so if LED1 was lit
*it will turn off and vice versa
              COMA          ;if bit0=1
              BRA  AGAIN
              END
```

FIGURE 9-5 Source code, in AS11 format, for BLNKLED1, which is a program that flashes Mag-11's LED1 at a much faster rate.

```
S11DB6008675B7103C8635B7103C9E00B7700018CE3000180926F84320F201
S9030000FC
```

FIGURE 9-6 The object code, in .S19 format, for BLNKLED1.

The first step is to make arrangements so that LOAD T will place the program at the desired addresses. Since we want to start loading it at $0100, we use the following OFFSET command (recall that $B600 − $B500 = $0100):

```
OFFSET -B500 <CR>
```

Then, at the BUFFALO's prompt (>), type "LOAD T <cr>." Now, Mag-11 is waiting for the .S19 file to be downloaded. Set up your communication program so that it will send the file BLNKLED1.S19. Often this just takes a few clicks of your mouse. (You first select the file, and then click SEND.) However, every program is different, so consult your documentation before attempting this step.

The downloading takes just a second or two to complete, and "Done" should appear on the BUFFALO screen when it is complete. Now, the instructions for our BLNKLED1 should be in RAM starting at $0100. Next, we have to *move* these instructions to EEP-ROM, which starts at $B600. This is done with the following BUFFALO command line:

```
MOVE 0100 02FF B600 <CR>
```

Since it takes much more time to program a byte of EEPROM (here the write time is measured in milliseconds and not microseconds), this operation takes longer than preceding ones.

Now use BUFFALO's line assembler mode to make sure that the EEPROM has been programmed correctly with the new program. Note that if you entered the other program, the only change might be on line B60F, which should show "18 ce 30 00."

Now, to actually try running the program, first remove power to Mag-11, and then set the jumpers for special bootstrap mode. To do this, place a jumper at JP1 between pins 2 and 3 and also install a jumper at JP11. Finally, place a jumper at JP12 between pins 1 and 2; this ties lines RxD and TxD together, which forces a jump at reset to the start of EEPROM.

Now run the program by applying power. Nearly immediately, LED1 should flash, now at a rate of five times faster than before. If everything is working OK, try making things more interesting by shutting off the power, removing U7 (the BUFFALO EPROM), and again applying power. It should not surprise you that LED1 flashes the same as it did with U7 installed.

Now that you learned how to use Mag-11 to program HC11's EEPROM, I will show how to construct MagPro-11, whose primary purpose is to program 512-byte HC11 EEPROMs and 2048-byte HC811 EEPROMs. Whereas Mag-11 was only able to program 40-pin DIP-packaged HC11s, MagPro-11 programs the more popular and less expensive 52-pin PLCC. MagPro-11 is also smaller and quite a bit cheaper to build than Mag-11. Since MagPro-11 uses the BUFFALO monitor and has a built-in MCU check LED, it also has use as an inexpensive HC11 development system.

The Origins of MagPro-11

Our purpose here is to design an inexpensive HCx11-based product. The most economical way to accomplish this is to use the HCx11's internal EEPROM and initially place the MCU in the single-chip bootstrap mode. In this mode, it is possible to automatically jump to the EEPROM after reset. The primary limitation here is the EEPROM's relatively small size: $1/2$ kB for the economical garden variety HC11 version or 2 kB for the higher-priced HC811. For many controller purposes, $1/2$ kB is sufficient even though its small size often forces the designer to write tight code. Nonetheless, there is nothing intrinsically wrong with tight code. Writing tight code is somewhat challenging, which just makes things more fun.

There are several ways to go about the design of an EEPROM programmer for the HCx11. The design that was chosen here makes use of the supremely excellent (if it depleted one's net worth, it would only be rated excellent) BUFFALO monitor. While Version 3.4 of BUFFALO does not apparently support the direct downloading of .S19 files (a common type of object code) into EEPROM, it is possible to use an indirect method that I will describe shortly. This indirect method requires external RAM. Voila! MagPro-11's 6264 8-kB RAM.

While the original purpose of MagPro-11 was as a simple programmer, it became obvious that it could easily be turned into an inexpensive HCx11 mini-development system/trainer. The reason for this rests with the BUFFALO monitor's abundant capabilities. Keep in mind that MagPro-11 is far from a complete development or trainer system. While one can test or learn about most of the HCx11's software capabilities, MagPro-11 has very limited hardware capability. For those who are mainly interested in hardware development, take a look at Mag-11 (described earlier). One of the primary advantages of MagPro-11 as a development system is that it is cheap and has great documentation.

A Look at MagPro-11

Refer here to the simplified schematic in Fig. 9-7 as well as the detailed schematic that can be printed out from the SCHEMAT program on the CDROM. The HCx11 MCU, whose EEPROM we plan on programming, is inserted into the PLCC socket at U1. This is decidedly different from EPROM programmers, which usually use a dedicated MPU or MCU. Here, the intelligence is accomplished by the chip to be programmed. The result is a simple and relatively inexpensive programmer that also functions as a mini-development system.

While HC11s can operate at 2.1 MHz, MagPro-11 only pushes it to 1 MHz. There is little advantage to speed here. The mode for U1 is set by jumper blocks JP2, JP3, and JP4. For normal EEPROM programmer operation (normal expanded mode), jumpers at JP1 and JP2 are not used. For normal serial communication with a computer or terminal, a jumper is placed between pins 2 and 3 of JP4.

U6, an MAX709L MPU supervisory IC, keeps the HCx11 from having problems by pulling the MCU's reset line down to ground whenever the +5-V power line drops below 4.65 V. U7, an MAX232, provides driving power and decoding for the asynchronous serial interface.

LED1, the MCU check LED, is connected to bit 4 of the HC11's port A. *When bit 4 of port A is cleared, LED1 is lighted; when it is set, LED1 is off.* This LED is used to verify that the circuit is performing correctly. It is not used directly by the BUFFALO monitor but has uses if MagPro-11 is employed for development or educational purposes. Since the HC11 has a multiplexed data/low-order address bus, we must separate the low-order address bus from this multiplexed bus. This task is accomplished by U3, a 74HC373 octal D-type transparent latch.

The BUFFALO monitor is contained in the 32-kB 27C256 EPROM U2. U2 responds to hex addresses from 8000H to $FFFF. However, the MCU disables the external address and data buses when addresses between $B600 and $B700 are selected. Recall that these addresses are used by the internal 512-byte EEPROM.

U5, the 8-kB 6264 RAM, is partially address decoded by U4:A and U4:B. U5 potentially will respond to addresses between 0000$ and $1FFF, as well as between $2000H and $3FFF. Of course, addresses between 0000 and 00FF and $1000 and $103F are used internally by the HC11, and thus external RAM is useless at these addresses. Also, because U5 is only partially decoded, a conflict with internal RAM exists between addresses $2000 and $20FF. Because of this, it is wise to think of U5's memory space as only existing between $2100 and $3FFF.

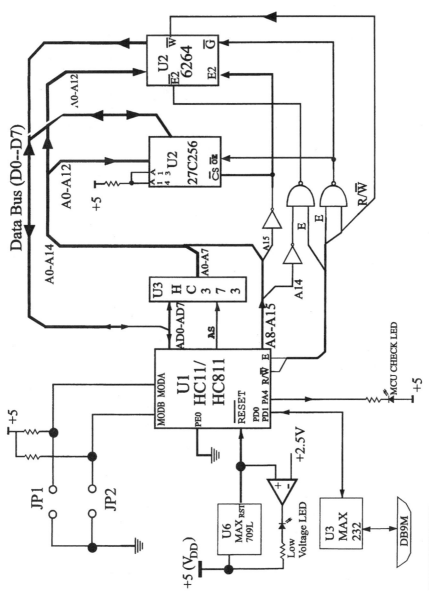

FIGURE 9-7 (A) MagPro-11's simplified schematic.

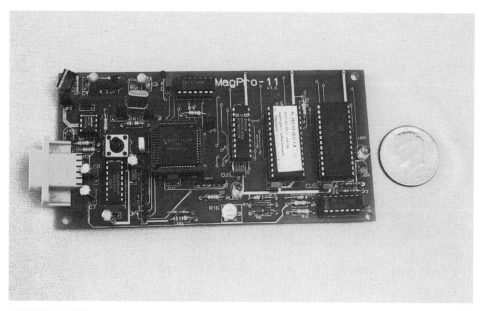

FIGURE 9-7 (*Continued*) (B) Photo of MagPro-11.

MagPro-11 can be powered by either a regulated +5-V volt source or an unregulated +7- to +13-V dc source. As shown in the schematic in Fig. 9-11, a jumper is placed at JP3 only if the source is not regulated +5 V. If the power source is +5 V, connect the positive voltage source to pin 3 of J1; if it is not regulated, connect the positive source to pin 1 of J1. In either case, pin 4 of J1 is grounded. When power is supplied correctly, green LED1 lights. MagPro-11 has a built-in unique voltage monitor circuit. This circuit consists of U8:A, U8:B, D1 (3.6-V zener), LED3, LED4, and associated resistors. It also makes use of U6's voltage-detecting capabilities. The overvoltage circuit, which controls red LED3, is basically a voltage comparator with D1 providing the reference voltage. When the power supply voltage rises above about 5.5 V, U8:A's noninverting input exceeds 3.6 V, which causes its output to switch high, and red LED3 lights. The reason for pot R16 is to compensate for D1's lack of precision. The low-voltage circuit operates slightly differently. Here, the decision making is basically done by U6, whose main purpose is to function as an MPU supervisor. Whenever the supply voltage drops below 4.65 V, U6's pin 7, its reset pin, is pulled down to near-ground potential. In addition to being connected, through a 4.7-kΩ resistor, to U1's reset input, pin 7 of U1 is also connected to the noninverting input of U8:B. U8:B's inverting input is connected to 1/2 VCC, and thus, whenever the voltage drops below 4.65 V, the output of U8:B drops low, which lights yellow LED4. (The anode of LED4 is connected to +5 V.) This voltage monitor circuit is optional, and its primary purpose is as an inexpensive HC11 insurance policy.

CONSTRUCTION

Except for the power supply, MagPro-11 is a complete self-contained design. All parts are mounted on a 3 × 6 in double-sided circuit board. Figure 9-8 shows the solder-side foil pattern, and Fig. 9-9 shows the component-side foil pattern. Figure 9-10 is MagPro-11's component mounting guide. If you want to use a double-sided board, it probably would be most economical to purchase a bare board from Magicland. However, as with Mag-11, it is possible to use a single-sided board with jumper wires. See Appendix E for tips on making and wiring your own handmade boards.

Before installing J1, use long-nose wires to pull out its pin 2. This is necessary for polarizing (insurance) purposes. Obviously, a socket is necessary at location U1. Also, it is wise to have a socket at U2. IC sockets at other locations are optional. The installation of U8, the LM324 quad op-amp, D1, and LED3 and LED4 and associated resistors is optional. Do not install ICs in their sockets until initial voltage tests have been completed.

INITIAL TESTS

MagPro-11 can be powered by a variety of dc power sources—anywhere from a regulated +5 V to an unregulated +13 V. Because of its low current requirements (about 50 mA), it can be even powered for a short period of time by a 9-V transistor battery. A blowup of the power supply schematic section is shown in Fig. 9-11.

While it is possible to wire power supply leads directly to the board, it is recommended that you make use here of a four-circuit connector. Notice from Fig. 9-11 that only two pins are used: pins 3 and 4 if a regulated +5 V is used, 1 and 4 if an unregulated but filtered dc supply is used between +7 and +13 V. For both supplies, pin 4 is ground. Also, for safety's sake, polarize the plug by inserting a polarizing plug in hole 2 of the plug. If an unregulated supply is used, be sure to place a jumper at JP3; failure to do so will not hurt a thing, but nothing will work either.

Apply power. Green LED2 should light. Connect the negative (usually black) lead of a dc voltmeter, capable of measuring 5 V, to a circuit ground. Using the positive (usually red) lead of the voltmeter, check the +5-V pin on all IC sockets. For instance, make sure pin 26 of U1 measures between 4.65 and 5.5 V. Similarly, check potentials at the power pins at locations' U2, U3, U4, U5, U6, U7, and U8.

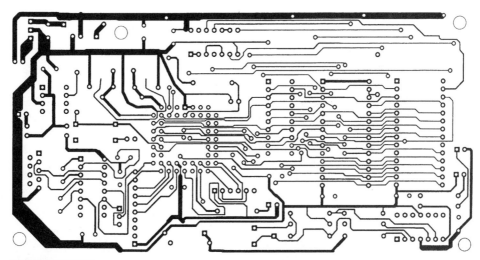

FIGURE 9-8 MagPro-11's solder-side foil pattern.

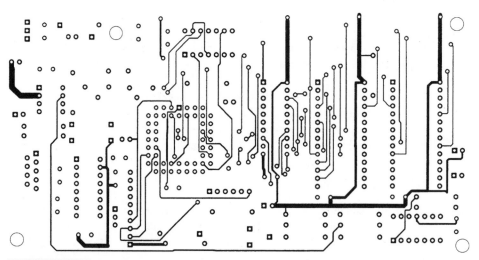

FIGURE 9-9 MagPro-11's component-side foil pattern.

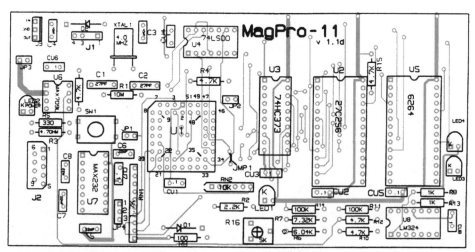

FIGURE 9-10 MagPro-11's component mounting guide.

Next, disconnect power and then install the LM324 at U8 (if used) and the MAX709L at U6. Again apply power. Yellow LED4 should light briefly immediately after power is applied. If it stays lighted, the voltage is below 4.65 V, and the HCx11 will not function properly. Check out your power supply. Now carefully adjust R16 so that LED3 just lights; then back off a bit on R16's adjustment so that LED3 goes out.

TESTING MAGPRO-11

If everything so far appears correct, turn off all power, and then install all remaining ICs. Remember to install a 27C256 EPROM programmed with the BUFFALO monitor firmware at U2. In order to test MagPro-11, you also will need to install an MC68HC11A1FN at U1.

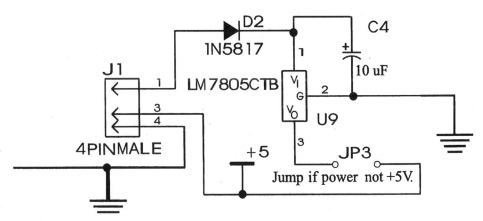

FIGURE 9-11 Enlarged view of MagPro-11's power supply circuit. *Note:* If power source is +5V, connect +5 to pin 3 of J1. If power source is +7 to +13 V, connect +V to pin 1 of J1, connect GROUND to pin 4 of J1.

Use a cable to connect the 9-pin serial interface connector J2 to the serial interface of a computer running a suitable communication program. Notice that pin 3 of J2 is an output, pin 2 is an input, and pin 5 is ground. Make sure the communication program is set for 4800 baud, 8 data bits, 1 stop bit, and full duplex. Also set the jumper blocks as follows:

JP1—No jumper

JP2—No jumper

JP3—Depends on power supply: no jump if +5 V is used

JP4—Place a jumper between pins 2 and 3

   After setting up the communication program, power up MagPro-11. The screen should show "BUFFALO 3.4 (ext)—Bit User Fast Friendly Aid to Logical Operation." If it does not, press MagPro-11's RESET switch. If still no luck, connect a logic probe or a lead of an oscilloscope to pin 11 of U7. Again press the RESET switch. If MagPro-11 is operating properly, you should detect pulses for a fraction of a second at pin 11 every time the RESET switch is pressed. If you do detect pulses here, most likely your problem has to do with either your communication program or your interface cable. If pulses are not detected, there is a problem with your MagPro-11 board. Most likely it is a solder connection if you used a two-sided board with plated through holes or a misplaced jumper connection if you used a single-sided board.

Programming HC11s

Since it is a bit simpler to program HC11s (MC68HC11A1FN), which have a 512-byte EEPROM, than HC811s (MC68HC811E2FN), which have 2 kB of EEPROM, we will first look at programming HC11s. The first step here is to insert the MCU into U1, which is a standard PLCC socket. Be sure to have power disconnected from the circuit when either installing or removing any IC. Also, take standard precautions from static such as touching a ground before handling any chip. While it is simple to "push" this chip in, removal is a bit harder. A neat, special-purpose tool is available that makes this job a snap. However, it costs nearly $50, and it only works with 52-pin PLCC packages. Another general-purpose tool is available for a double sawbuck—and you get change back. However, it takes a bit of skill to use effectively. Alternatively, one can ease the chip out of its socket with the blade of a tiny screwdriver. This works, and I have done this a number of times. Just be careful and proceed gently and at a snail's pace.

The code/data you want to program the EEPROM with must be 512 bytes or less. A 513-byte program just will not do. A 123-byte program is fine, however. (Of course, I am assuming that you are programming an HC11 and not an HC811. See the next section for additional tips on programming HC811s.) Also, the file must be in standard MOT S19 format (.S19) and should start at $B600. For most cross-assemblers, such as the AS11 on the CD-ROM, this is no problem.

Connect MagPro-11 to the serial interface of your computer, and power it up. After the BUFFALO prompt is seen, you can start. As mentioned in the discussion on programming EEPROMs with Mag-11, it is not possible, with Version 3.4 of BUFFALO, to download an .S19 file directly into EEPROM. The reason for this is the manner in which an EEPROM is programmed. So what you want to do is download the file into external RAM located between $2100 and $2FFF and then use BUFFALO's MOVE command to move the data from RAM into EEPROM. (MagPro-11's RAM addresses actually extend from $2000 to $3FFF, but because U4 is not completely address decoded, you can experience address conflicts between $2000 and $20FF and between $3000 and $303F.) Since our file starts at $B600 and we want to download it starting at $2100, we use the OFFSET command. Here we simply enter:

```
OFFSET -9500 <CR>
```

Do not add a suffix of "H" or a prefix of "$" because BUFFALO only understands hexadecimal. Next, enter

```
LOAD T <CR>
```

and then start downloading the .S19 file. After a successful download, "Done" will be displayed on the screen. Next, enter

```
MOVE 2100 22FF B600 <CR>
```

This command moves data from addresses 2100 through 22FF to B600 through B7FF. This procedure takes a bit longer because the EEPROM is being programmed.

Programming HC811s

As mentioned earlier, the HC811 is basically an HC11 with 2048 bytes of EEPROM instead of just 512 bytes. However, there are more differences between the two than just this. For instance, the all-important CONFIG register has been changed somewhat. See the pictorial representation of the HC811's CONFIG register in Fig. 9-12.

Notice from Fig. 9-12 that bits 1 and 3 (B1 and B3) are not shown. You may recall that in Fig. 4-11, bit 3 was called the NOSEC bit and bit 1 the ROMON bit. These bits do not apply to the HC811—at least at present. Motorola ships HC811 with these bits set, but they do not seem to mean anything.

BIT DEFINITIONS (REFER TO FIG. 9-12)

EE0–EE3. These bits set the location of the EEPROM in HC11's memory map. See Table 9-2.

Note:

These bits have no meaning in the single-chip mode of operation, since the 2-kB EEPROM is forced on at locations $F800 through $FFFF. Don't doze off here! This is important! This means that no matter where we originally stick the EEPROM (say, at $800–$FFF so that it is out of the way of the BUFFALO monitor), we can still use the "normal" interrupt vectors located between $FFC0 and $FFFF. This makes things really neat.

NOCOP. A 1 in this bit disables COP, whereas a 0 enables it.

EEON. A 1 in this bit enables on-chip EEPROM that is mapped at the address pointed to by the four bits EE0 to EE3. Again, see Table 9-2.

While it is obvious I have a crush on the HC11 chip, I have to admit when it comes to the HC811 that it appears that Motorola is not sure what it is doing. For instance, in revision 3 of the HC11's 1-in-thick reference manual, there is a note concerning the all-important CONFIG register. It states the HC811's CONFIG register is shipped programmed with $FF (all bits set), but a change was being considered that would make this value $0F—a "fantabulous" idea. Well, I purchased HC811s produced in 1995, and they still contain $FF! What's the big deal here?

REGISTER'S NAME: CONFIG (HC811)						REGISTER'S ADDRESS: $103F		
BIT#	7	6	5	4	3	2	1	0
BIT NAME	EE3	EE2	EE1	EE0		NOCOP		EEON
BIT'S STATUS AFTER RESET	*1*	*1*	*1*	*1*	*?*	*1*	*?*	*1*

Note: From tests by the author, bit 3 and bit 1 seem to be programmed with 1 by manufacturer. However, the data sheets do not fully support this. Whatever, these bits do not seem to mean anything for the HC811E2 model.

FIGURE 9-12 HC811's CONFIG register.

TABLE 9-2 M68HC811'S 2-KB EEPROM MEMORY MAP POSITIONS

EE3	EE2	EE1	EE0	LOCATION
0	0	0	0	$0800–$0FFF
0	0	0	1	$1800–$1FFF
0	0	1	0	$2800–$2FFF
0	0	1	1	$3800–$3FFF
0	1	0	0	$4800–$4FFF
0	1	0	1	$5800–$5FFF
0	1	1	0	$6800–$6FFF
0	1	1	1	$7800–$7FFF
1	0	0	0	$8800–$8FFF
1	0	0	1	$9800–$9FFF
1	0	1	0	$A800–$AFFF
1	0	1	1	$B800–$BFFF
1	1	0	0	$C800–$CFFF
1	1	0	1	$D800–$DFFF
1	1	1	0	$E800–$EFFF
1	1	1	1	$F800–$FFFF

Well, as just mentioned, the four most significant bits of the CONFIG register determine the memory location of the 2-kB EEPROM. With all ones in the four most significant bits (i.e., an "F" in the most significant nibble), the EEPROM is located between $F800 and $FFFF. The problem here is that this memory location conflicts with the BUFFALO monitor in MagPro-11 (located between $E000 and $FFFF). In other words, this presents a problem. Do not fear; it is an easily solved problem. (Note that if the indicated change were made, the EEPROM would be located between $0800H and $0FFF. This would allow direct use of the BUFFALO monitor.)

Assuming that the HC811 is shipped with $FF in its CONFIG register, it seems obvious, because of conflicts with the BUFFALO monitor, that we cannot use the HC811 exactly as it comes from the factory. Or can we? The answer is no and yes. The trick here is to install a jumper at JP2 so that the special test mode is selected after reset. When in the special test mode, the CONFIG register's EEON bit is cleared, which means the 2-kB EEPROM is removed from the memory map. Voila! No more conflict. None. However, when in the test mode, the all-important interrupt and reset vectors are located at $BFC0 through $BFFF instead of $FFC0 through $FFFF as with the normal expanded mode. Now what? No problem with MagPro-11. It was designed purposely (actually redesigned) with redundant address decoding so that the 27C256 EPROM responds to both ranges of memory addresses.

One other little point that may soon confuse you. In the last paragraph I mentioned that HC811s are shipped with $FF in their CONFIG register. This is true. However, I just explained that when in special test mode the EEON bit in the CONFIG register (i.e., bit 0) is forced to *zip*. This means that if you attempt to determine the contents of the CONFIG register while in test mode, you will find that it contains an apparent $FE and not $FF. Now, does it make sense?

To program HC811s, which have 2048 bytes of EEPROM instead of just 512, a jumper must be installed at JP2. Reset MagPro-11. Make sure the .S19 file starts at $F800 and includes the appropriate reset vector at locations $FFFE and $FFFF. Also make sure the file ends at $FFFF. Next, type the following:

```
OFFSET -D700   <CR>
LOAD T         <CR>
```

and then download the .S19 file.

The following line of instructions sets the EEPROM address range from $F800 to $FFFF:

```
EEMOD F800     <CR>
```

Next, use BUFFALO's memory modify (MM) command to change the CONFIG register located at 103F from FE to FF. This is done by first typing

```
MM 103F        <CR>
```

and then typing

```
FF   <CR>.
```

Finally, use the MOVE command to program the EEPROM. Type

```
MOVE 2100 28FF    F800 <CR>
```

and wait a minute or so for the programming to be completed. When programming is completed, BUFFALO's prompt (>) appears.

Before removing the MCU from its socket, remove the jumper at JP2, and then reset MagPro-11. After programming is completed, the EEPROM resides in locations $F800 through $FFFF. Thus it can be used easily in a normal single-chip mode, with your garden variety interrupt vectors located right in your (actually HC11's) backyard—from $FFC0 through $FFFF.

Now that you've learned how to program both HC11s and HC811s, you are prepared to design real honest to goodness gizmos. Chapter 10 will introduce a simple HC11 board that will help in your quest.

CONSTRUCTION AND USE OF
MAGTROLL-11

A Simple and Inexpensive HCx11 Controller Board

In the preceding chapter I described the construction and use of MagPro-11. MagPro-11's primary purpose is as an EEPROM programmer, although it doubles as a mini-development system. In this chapter I will use an HCx11's programmed EEPROM to create a simple and cheap minimum chip, multipurpose controller board. The program instructions for MagTroll-11 will be placed in the HCx11's EEPROM. Useful tips on employing a program in EEPROM will be divulged. In the next chapter I will look at a spectacularly unique practical demonstration project that uses MagTroll-11—a solid-state wind direction indicator.

About MagTroll-11

As you can see from Fig.10-1, MagTroll-11 is basically a single-chip computer board, although the board contains three other integrated circuits. MagTroll-11 was designed with simplicity and cheapness in mind. Refer here to the simplified schematic in Fig.10-2. In addition, it is recommended that you print out the detailed schematic using the SCHEMAT program on the enclosed CD-ROM.

U4, which can be either an LM7805CTB or an LM2931T +5-V voltage regulator, is optional if the power source for the board is to be a regulated +5 V. For details on the features of the LM2931, refer to Chap. 1.

FIGURE 10-1   Close-up of MagTroll-11.

U3, an IC +5-V reference voltage IC, is also optional. If U3 is not used, a jumper is placed between pins 2 and 3 of JP4. Leaving out these options results in a two-chip board.

U2's purpose is threefold. First, it provides the protective supervisory function of keeping the HCx11's reset input low whenever U1's VCC pin is below 4.65 V. This protects the CONFIG register and the EEPROM. Second, it provides automatic RAM backup control when the power fails. Third, it provides an interrupt signal whenever it detects that the power is failing. This interrupt signal is applied to U1's XIRQ input, which can force the program to instruct the HCx11 to enter its ultra-low power STOP state. This procedure is another way of saving the HCx11's internal RAM contents. More on this later.

Note that some real cheapos will want to leave out U2 and connect U1's reset pin directly to +5 V through a 4.7-kΩ resistor. Before you attempt this, however, remember that "fools rush in where angels fear to tread." This old saying definitely applies here, since there is a chance, albeit slim, that if you omit U2, you will screw up the CONFIG register.

Jumper blocks' JP3 and JP5 are used to select HC11's operating mode, while JP2 is used to select a direct jump to EEPROM for MC68HC11A1FN chips when special bootstrap mode is selected. Also see Table 10-1.

Like MagPro-11, a low-power LED, LED1, is connected to bit 4 of U1's port A. LED1 functions primarily as an MCU check LED. It is turned on when bit 4 of port A is cleared; otherwise, it is off.

Male headers J-PORTA through J-PORTE and J-MISCL are used to connect to external circuits. Refer to the section "Connecting Up to MagTroll-11." Notice that while

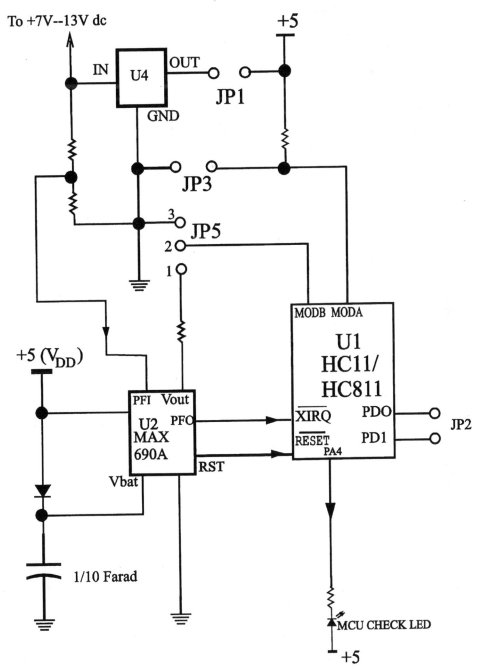

FIGURE 10-2 Simplified schematic of MagTroll-11.

TABLE 10-1	MAGTROLL-11'S JUMPER POSITION GUIDE
JP1	Jump if power is not a regulated +5 V. If jumper is placed here, power source should be between +7 and +13 V and positive source connected to pin 1 of J1:1. Ground is always connected to pin 4 of J1:1.
JP2	Place jumper here only if you are using the special bootstrap mode *and* you want an immediate jump, after RESET, to EEPROM starting at B600H. Only use when you have installed an MC68HC11A1FN at U1.
JP3	If a single-chip or special bootstrap mode is desired, place jumper here. For normal expanded mode or special test mode, do not install jumper. Normally, jumper should be installed here.
JP4	Jumper location here determines source of reference voltage for the MCU's internal A/D converter. If U3 is not installed and/or a +5-V regulated power source is used, choose pins 2 and 3. Otherwise, choose pins 1 and 2. For optimum accuracy here, place a jumper between pins 1 and 2, but note the restrictions.
JP5	Place jumper between pins 1 and 2 if you want the normal single-chip or normal expanded mode and between pins 2 and 3 for the special bootstrap or special test mode. Generally, if an HC11 is installed at U1, place between pins 2 and 3; if an HC811 is used, choose pins 1 and 2.

MagTroll-11 was designed primarily to be used in the single-chip mode (either normal or special bootstrap), it can be used in expanded modes (either normal or special test). However, to be used in expanded mode, port C must be demultiplexed in order to extract the low-order address bus from the data bus. This can be accomplished with a 74HC373 octal transparent latch on an external board. This is explained later in this chapter.

CONSTRUCTION

Despite its simple nature, a printed-circuit board should be used to construct MagTroll-11 to increase reliability. The solder-side foil pattern is given in Fig. 10-3, and the component-side foil pattern is given in Fig. 10-4. Figure 10-5 shows the component mounting guide. A double-sided board with plated through holes is available from Magicland. Refer to Appendix C. If you wish to make your own single-sided board, also refer to Appendix E, which provides several useful tips.

Notice that all 10-pin male header connection blocks have pin 9 removed for polarizing purposes. (Save two of these pins for test points TP1 and TP2.) This polarization of connectors reduces the chance of connection errors. Be sure to place a keying/polarizing plug at pin location 9 of all cable sockets that will mate with MagTroll-11's connectors. Also, J1 has pin 2 removed, so insert a keying/polarizing plug at pin location 2 of its respective socket.

A 52-pin PLCC socket at location U1 increases MagTroll-11's versatility. Chapter 9 gives tips on PLCC removal. A socket for U2 is optional, though still recommended.

C1 is a 1000-μF axial lead electrolytic capacitor. To conserve precious PC board real estate, we mount it vertically instead of horizontally. Use care when installing this capacitor.

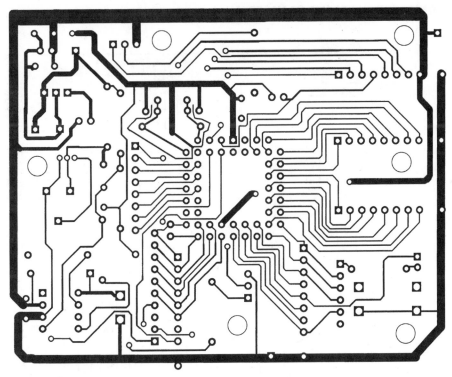

FIGURE 10-3 Solder-side foil pattern for MagTroll-11.

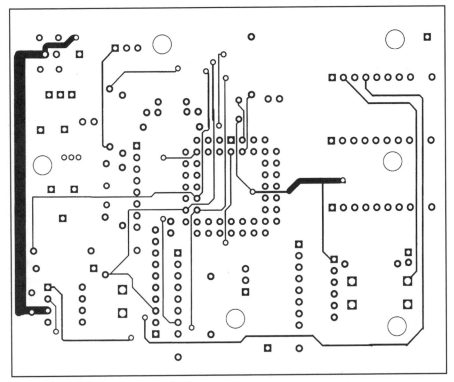

FIGURE 10-4 Component-side foil pattern for MagTroll-11.

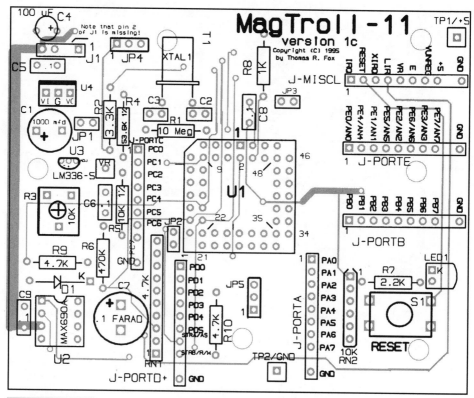

FIGURE 10-5 Component mounting guide for MagTroll-11.

You may wish to use an insulating sleeving (spaghetti) on the long bare lead. Alternatively, you can use a 330-μF radial lead capacitor here with only minor performance restrictions.

If C7 is omitted, there will no longer be "normal" RAM backup. For longer RAM backup, you may want to increase C7's size to 1.0 F. However, you may have to jerry-rig up something to accommodate the larger physical size of these capacitors in the circuit board described here.

PRELIMINARY TESTS

Do not install ICs at U1 and U2 until requested. Connect a suitable dc power source to MagTroll-11 via J1: 1. If the source is regulated +5 V, connect the positive source to pin 3, and do not connect a jumper at JP1; if it is unregulated and between +7 and +13 V, connect the positive source to pin 1, and place a jumper at JP1. In either case, connect the ground to pin 4 of J1: 1. If the power source exceeds 8 V, place a jumper between pins 1 and 2 of JP4; if not, place a jumper between pins 2 and 3.

Apply power. Connect a voltmeter's negative lead (usually black) to the test point JP2/GND. Using the positive (usually red) lead, measure the voltage at TP1/+5, pin 26 and 52 of U1, and pin 2 of U2. The voltmeter should show between 4.75 and 5.25 V. If the power source exceeds 8 V, adjust R3 so that pin 52 of U1 is at 5.00 V.

To simplify design calculations, you may want to readjust R3 later for a 5.12-V reading at pin 52 of U1. With this reference voltage, each analog-to-digital (A/D) input bit will correspond to .02 V.

Now turn off power and install U2. Turn on power and check pin 7 of U2. Under steady-state conditions, this voltage should be close to +5 V. However, when power is removed from the circuit, this voltage should drop quite fast; when power is applied, there should be a slight delay before the voltage suddenly jumps up to 5 V.

SETTING UP JUMPERS

Jumper JP1 is used to set up the type of power source; JP4 is used to select the type of reference voltage (standard +5 V or, more accurately, that produced by U3); JP3 and JP5 select the MCU's operating mode; JP2 is used to cause an immediate jump to $B600 when the special bootstrap mode is selected. Refer to Table 10-1 for appropriate jumper installations.

WORKING TESTS

Unless you have an MC68C11A1FN or MC68HC811E2FN that has its EEPROM programmed with a suitable test program available, you cannot go any further with tests. However, if you constructed MagPro-11, described in the last chapter, or have a similar board available that uses the BUFFALO monitor, you may want to try the following test program to check MagTroll-11. The listing in Fig. 10-6 was written for the HC11 and is printed in standard BUFFALO convention and can be entered exactly as shown using BUFFALO's line assembler (ASM). See Chap. 5 for more details. Note that I have left out the line numbers here. I have done this because you should be able to understand this simple test program by now. Remember, the line numbers were added for discussion purposes only.

After the HC11's EEPROM is programmed with the preceding program, install it at U1. Set up MagTroll-11 for special bootstrap mode by placing a jumper between pins 2 and 3 of JP5 and installing a jumper at JP3. Installing a jumper at JP2 instructs the special bootstrap ROM to jump immediately to $B600 (also see Table 10-1).

After power is applied, the LED should flash at a $^1/_2$-s rate with a 4-MHz crystal installed.

If you wish to test MagTroll-11 with an HC811, use the listing in Fig. 10-7. Note that programming an HC811 using the BUFFALO monitor Version 3.4 is slightly more complicated than programming an HC11. See Chap. 9 for details. Briefly, you must place the programming board (e.g., MagPro-11) into the special test mode and then change the data at location $103F (the CONFIG register) to $FF. This is done with the memory modify (MM) command. You also must use Buffalo's EEMOD command to ensure that the BUFFALO monitor "knows" that the EEPROM is located between $F800 and $FFFF and *not* between $B600 and $B7FF. (BUFFALO is not smart enough to know you put an HC811E2FN in the socket until you do this.) The following command line does this:

```
EEMOD F800
```

Notes: (1) BUFFALO's ASM line assembler assumes that all numbers are already in hexadecimal. *Do not add "$ prefix.* (2) Numbers in the first column are addresses. *Do not type these numbers in.* However, these addresses are displayed on the BUFFALO screen. (3) To enter this listing, start BUFFALO and then type "ASM B600 <CR>" and start entering the text in *second column only.* After you are done typing the line, hit the ENTER key (i.e., carriage return).

```
B600    LDS  #00B0
B603    LDAA #10
B605    STAA 1000
B608    JSR  B622
B60B    CLRA
B60C    STAA 1000
B60F    JSR  B622
B612    BRA  B603
. . . . . . . . . . . . .
. . . . . . . . . . . . .
B622    LDY  #FFFF
B626    DEY
B628    INY
B62A    DEY
B62C    BNE  B626
B62E    RTS
. . . . . . . . . . . . .
. . . . . . . . . . . . .
```

FIGURE 10-6 Program listing that tests MagTroll-11. The program is embedded in an HC11's EEPROM.

Overall, while the MC68HC811E2FN is easier to use and has a much larger EEPROM than the MC68HC11A1FN, it is harder to program, at least using the BUFFALO monitor.

After the HC811 EEPROM is programmed, install it at U1. Set up the MagTroll-11 for single-chip operation by installing a jumper at JP3 and one between pins 1 and 2 of JP5. There is no need to install a jumper at JP2. If everything is working properly, the LED should flash at a 1-s rate with a 4-MHz crystal.

CONNECTING UP TO MAGTROLL-11

While the next chapter will describe using MagTroll-11, it is appropriate here to mention at least one way of making connections to MagTroll-11. In order to connect to MagTroll-11 you will require 10-contact, single-row connectors with 0.1-in spacing. For polarization purposes, a keying plug is recommended at pin 9. While several manufacturers make suitable connectors, I have used AMP MT connectors. Sources for these connectors include Digi-Key Corporation and Mouser Electronics. Here, you will require, for each connector, a 10-position single-row MODU housing (AMP 1-87499-7, Digi-Key A3019-ND, Mouser 571-1874997), nine contacts (AMP 87667-3, Digi-Key A3000-ND, Mouser 571-876673), a keying plug (AMP 86286-1, Digi-Key A3077-ND, Mouser 538-15-04-9209), and a suitable cable. While special, expensive tools are available to crimp the wire to the contacts, with some care, long-nosed pliers and a soldering iron can be used with great effectiveness.

```
F800       LDS   #00B0
F803       LDAA  #10
F805       STAA  1000
F808       JSR   F822
F80B       CLRA
F80C       STAA  1000
F80F       JSR   F822
F812       BRA   F803
. . . . . . . . . . . . . . .
. . . . . . . . . . . . . . .
F822       LDX   #E700
F825       NOP
F826       NOP
F827       NOP
F828       NOP
F829       NOP
F82A       NOP
F82B       DEX
F82C       BNE   F825
F82E       RTS
. . . . . . . . . . . . . . .
. . . . . . . . . . . . . . .
For the following two bytes use BUFFALO's Memory
Modify command (MM) and not the ASM:
FFFE       F8
FFFF       00
```

FIGURE 10-7 Program listing that tests MagTroll-11. The program is embedded in an HC811's EEPROM.

Note:

When connecting to J-PORTE, make special note of its rather odd pinout arrangements. Refer to the schematic and component mounting guide.

Using the HC11's Internal EEPROM

As mentioned a number of times earlier, Motorola designed the original MC68HC11s primarily for use in single-chip circuits. The program that controls the application was to be stored in the chip's internal 8-kB ROM. The 512-byte EEPROM was meant primarily to be used to store information that needed only to be changed once or at most a few times, although it can be changed more than 10,000 times. One example is a unique serial number for each finished product. I already mentioned other examples. Nonetheless, the designers of the 68HC11 anticipated the need in certain circumstances to have a way to force the MCU to jump directly into EEPROM after reset. This feature would enable a simple single-chip system to use the 512-byte EEPROM for program storage. Since the EEPROM's starting address is at location $B600, the MCU has to jump directly to this address after reset. Normally, the HC11 jumps to the address pointed to by the vector located at address $FFFE and $FFFF. For instance, if the address $FFFE contained $B6 and $FFFF contained $00, under normal operation, the MCU would jump to the address $B600. However, in a simple, single-chip system using the standard MC68HC11A1FN, there is no external memory at the addresses $FFFE and $FFFF. The solution to this problem was to cause the jump to the start of EEPROM under two conditions automatically that are easily fulfilled by the designer. While I have mentioned all this before, its importance makes it worth mentioning again and again and again.

First, the chip has to be in the special bootstrap mode. This is accomplished by placing a ground potential at pins 2 (MODB) and 3 (MODA) of the MC68HC11A1FN at the time of reset. (Note that if you simply ground these pins, it is not possible to use the battery-backup feature of the chip's internal RAM because pin 2 also functions as a battery-back-up voltage source.) Second, the SCI's RxD line must receive a break character. This can be accomplished easily by tying the TxD and RxD lines together (shorting pins 20 and 21 of the 68HC11A1FN). However, if this is done, it is not possible to use the asynchronous ser-ial interface. Before you shake your head in disgust, there is a way to get around this prob-lem so that the HC11's SCI can still be used. In brief, it amounts to having the software send a NUL or a break character instead of tying TxD and RxD lines together. More on this near the end of this chapter.

With MagTroll-11, an automatic jump to EEPROM at $B600 occurs at reset when we install JP2 and JP3 and place a jumper between 2 and 3 of JP5.

Using the HC811's Internal EEPROM

Unlike the original 68HC11, the MC68HC811E2FN was designed for program code to be contained within its 2-kB EEPROM. Because of this, it comes shipped with its 2-kB EEP-ROM located between $F800 and $FFFF, and its data sheet does not even show a working 8-kB ROM like the MC68HC11A1.

The HC811 does not present the problems associated with the original HC11. Here, one can place the chip in single-chip mode and place the starting address of the program (typ-ically $F800) at its reset vector, which is located at the "normal" addresses of $FFFE and $FFFF. As of this writing, the primary problem with this chip is not technical; rather, it is cost and supply. The HC811 is rather hard to obtain, and it also costs roughly 50 percent more than the garden variety HC11.

If you are using an HC811, do not install jumpers at JP2 and JP3. However, install a jumper between pins 1 and 2 of JP5. Also make sure that the starting address of your pro-gram is stored at the reset vector's address of $FFFE and $FFFF. Notice that the memory map for the HC811's EEPROM is changeable. This is accomplished by modifying the chip's CONFIG register located at $103F. Refer to Chap. 9 for more details concerning this little-used feature.

Another Look at HCx11's All-Important Interrupt Vectors—Where It All Starts

If you only recently discovered that not all decent MPUs and MCUs have an Intel stamp on them, you may be shocked to learn that not all microcontrollers start at $0000 after reset like Intel's 8051 series. The HCx11, like the original 6800, uses a reset vector to "point to" the starting address. This reset vector is contained in the last 2 bytes of the memory space, i.e., $FFFE and $FFFF. Also, unlike the 8051 series, this vector is a true vector—the data stored at these addresses correspond to the starting address. You do not use a JUMP or similar instruction—the HCx11 automatically causes a jump to this location. While it may be a matter of personal preference, I believe that Motorola's way of handling interrupts is superior to Intel's. Since Motorola purposely made the 8051 somewhat downward software compatible with the original practical MPU, the 8080, and Motorola's 6800 came out after the 8080, it seems reasonable to believe that Motorola learned a bit from the perceived deficiencies of the 8080. Apparently, one thing the company did not like was Intel's way of handling interrupts.

Table 6-1 gives the all-important list of vector assignments. I have omitted the register and local masks for simplicity. Consult Motorola's MC68HC11A8/D technical data book for this and other pertinent information.

A glance at Table 6-1 makes it obvious, to even an Intel devotee, that the HC11 is more sophisticated than the 8051 series. If you have any doubts, compare this table with a similar one in Intel's microcontroller handbook concerning the 8051 series, which shows only six interrupt sources. Take caution here. Do not assume that the phrase "more sophisticated" is synonymous with "better." Often, the converse is true.

Descriptions of all interrupt sources are way beyond the scope of this chapter. However, recall that we looked at the illegal opcode trap, COP failure, and COP clock monitor for two reasons: They are interesting, and (more important), their appropriate use adds to a system's reliability. Keep in mind the meaning of these interrupt vectors. For example, say the MCU received a valid IRQ on pin 19 and say $D0 was stored at $FFF2 and $00 at $FFF3. The MCU would automatically jump to address $D000 when it received the IRQ. Other vectors are handled similarly. (Of course, certain other conditions, such as the clearing of the condition code register's I bit, must be met.)

An observant reader may notice a little problem if all we use is an HC11's 512-byte EEPROM for program storage—no usable memory at the addresses listed in the first column in Table 6-1. A reader with a good memory will recall that I brought up the existence of these pseudo-vectors way back in Chap. 6. Because of their importance, I will discuss them again. This time I will go into more detail.

Motorola engineers foresaw the problem created by restricting the program storage to the 512-byte EEPROM between $B600 and $B7FF. To solve this problem, they provided bootstrap mode pseudo-vectors that are in internal RAM. These RAM locations are referred to as *pseudo-vectors* because they can be used like vectors to direct control to interrupt service routines. See the right column in Table 6-1.

Realizing the difference between "normal" and "pseudo" vectors is vitally important. With normal vectors, the CPU already "knows" to jump to the address stored at the vectors. This is not true with pseudo-vectors. Because of this, you, the programmer, must add a JMP ($7E) instruction. This is why each pseudo-vector is allowed 3 rather than only 2 bytes of memory. These 3 bytes are needed because, unlike true vectors, you must add the JMP opcode to cause the desired jump. Again refer to the right column in Table 6-1 for a listing of these pseudo-vectors.

Notice in Table 6-1 that the address for reset stands out prominently from the rest. Obviously, unlike the other addresses, $BF40 is not in internal RAM (0–$FF). Rather, it is located in the bootloader ROM (normally located between $BF40 and $BFFF), which is contained in all

HC11 chips. This short bootloader firmware, which is enabled whenever you start the HC11 in the special bootstrap mode, has a number of features that are beyond the scope of this book. One feature, which has already been mentioned, concerns us here. If the bootloader determines that PD0/RxD and PD1/TxD are connected together (pins 20 and 21 in the PLCC package), it automatically causes a jump to the start of EEPROM at location $B600.

As an example of how to use these pseudo-vectors, let's assume you have an interrupt service routine that starts at $B700 that you wish accessed when the chip receives a valid IRQ interrupt signal at pin 19. The pseudo-vector of interest starts at RAM location $00EE and ends at $00F0. Refer again to Table 6-1. (Do not refer to "Motorola's bible" because it lists the address as $00EE to $00FD, which is an obvious error.) We want the JMP opcode, $7E, stored at $00EE, the address $B7 stored at $00EF, and $00 stored at $00F0. How do we do this? Unless you are absolutely confident in your RAM backup power source, do not rely on programming RAM after construction and then forgetting it. You will most likely have problems. You do not need more problems. Trust me. The best solution is to write a short subroutine and stick it at the beginning of your code in EEPROM. For instance, the following program source code, written in AS11 format, will accomplish this:

```
ORG   $B600
LDS   #$00B0
LDAA  #$7E
STAA  $00EE
LDD   #$B700
STD   $00EF
*
*
```

Life after Death (At Least after a Dead Power Source)

The title of this section is apropos if you view life and the retention of memory as closely related. MagTroll-11 uses two schemes for internal RAM backup. First, when the voltage drops below about 4.65 V, the 0.1-F (or larger) supercap C7 takes over as the RAM's energy source. This capacitor provides backup power for quite a bit of time—usually much longer in practical terms than calculations would assume. (One possible explanation here is that these supercapacitors are grossly undervalued.) Notice that with MagTroll-11 it is not possible to use this means of RAM backup if the special bootstrap or special test modes are used.

MagTroll-11's other memory backup system is more sophisticated, although it is a bit macabre because it occurs when the power supply is dying either from a power failure or a turned-off power switch. It uses the fact that the HC11's STOP instruction causes all clocks to stop; reducing consumption to a few microamps. Besides reset and backup source switchover capability, the MAX690A has a power-fail detector. Along with resistors R4, R5, and R6 and capacitor C6, it triggers an interrupt at PF0 (pin 5) when pin 4 (PFI) detects that the raw power supply (before voltage regulator U3) is starting to fail. In MagTroll-11 this triggers an XIRQ, which is a pseudo-nonmaskable interrupt. What happens then depends on the firmware. Generally, the XIRQ vector points to a service routine that includes the STOP opcode. A sample code segment might be

```
            *
*REM: XIRQ RECEIVED; STOP ALL CLOCKS
            NOP
            STOP
            RTI
            *
```

If you wish to learn about the intricacies of the XIRQ interrupts, consult the "HC11 bible" (M68HC11 reference manual). However, I will discuss a few practical aspects of it here.

As its name suggests, the XIRQ is not a true nonmaskable interrupt. It can be masked with the X bit in the CCR (condition code register). Also, after reset, the X bit is set; this means that the XIRQ input is not enabled immediately after reset. To enable it, along with the STOP instruction, clear the bits in the CCR shortly after initializing the stack pointer. The following program segment will accomplish this:

```
    *
        *
            CLRA
            TAP
            *
```

If you wish to experiment a bit here with the XIRQ and the STOP instruction using the MC68HC11A1FN, try out the listing in Fig. 10-8. The listing again assumes that you will be using the BUFFALO monitor's ASM line assembler.

What this short program in Fig. 10-8 does is initialize the XIRQ pseudo-vector in RAM so that the program jumps to the XIRQ service routine that starts at $B700. It then flashes the LED on and off to verify that everything is working OK. Next, it checks out RAM location $10. If $6F is stored at $10, the LED lights, and the program then pauses (WAI instruction) and waits for an interrupt. If other data (anything but $6F) are stored at $10, then the LED is off, and the program waits for an interrupt. If an XIRQ is detected, which means the power source is failing, the XIRQ service routine stores $6F at $10 and then executes the STOP instruction.

Another way of explaining the code, in even more down-to-earth terms, is that when power fails, a test code is stored in RAM. If this code remains the same after power returns, the LED will stay lighted after the initializing routine (flash on and off); otherwise, it will stay off after this initializing routine. Thus one can easily test RAM backup capabilities here. Tests suggest that RAM backup, using this method, is limited to 3 to 12 seconds, depending on the way the power fails (disconnected power source or power failure). If C1 is increased, longer backup times can be expected. Keep in mind that this limited RAM backup is primarily due to the fact that in the special bootstrap mode MagTroll-11 cannot use the supercap (C7) RAM backup. If an HC811 is used, backup time is increased tremendously because it can be operated in the normal single-chip mode and thus makes use of C7 storage capabilities.

COP WATCHDOG TIMER RESET

Since an enabled COP watchdog system can cause problems unless the software is designed specifically to use it, Motorola ships HC11s with the COP disabled. The COP system is enabled by clearing bit 2 (the NOCOP bit) of the CONFIG register—it is shipped with this bit set, which disables the COP watchdog system.

```
B600        LDS    #B0
B603        CLRA
B604        TAP
B605        STAA   F3
B607        STAA   1000
B60A        LDAA   #7E
B60C        STAA   F1
B60E        LDAA   #B7
B610        STAA   F2
B612        JSR    B650
B615        LDAA   #10
B617        STAA   1000
B61A        JSR    B650
B61D        CLRA
B61E        STAA   1000
B621        JSR    B650
B624        LDAA   #10
B626        STAA   1000
B629        LDAA   10
B62B        CMPA   #6F
B62D        BNE    B633
B62F        CLRA
B630        STAA   1000
B633        WAI
....
....
B650        LDX    #FFFE
B653        NOP
B654        NOP
B655        NOP
B656        DEX
B657        BNE    B653
B659        RTS
....
....
B700        LDAA   #6F
B702        STAA   10
B704        NOP
B705        STOP
B706        RTI
```

FIGURE 10-8 Program listing to demonstrate MagTroll-11's XIRQ/STOP memory backup capability.

Caution:

Do not modify the CONFIG register ($103F) unless you know exactly what you are doing. Modification of this register can cause problems that might not be solvable by those still on the HC11's steep learning curve.

If you use MagPro-11, as described in Chap. 9, or another system that makes use of the BUFFALO monitor, modifying the CONFIG register is trivial. After you obtain BUFFALO's prompt (>), simply type the following:

```
MM 103F  <CR>
```

If you are using an MC68HC11A1FN, the response will likely be

```
        0D
```

Now simply type

```
    09   <CR>
```

and then reset MagPro-11.

However, before you do all this, make sure you understand how to use the COP. Chapter 6 describes the COP system in moderate detail. I will repeat the very basics here. For more details, reread Chap. 6 and consult the "HC11's bible."

Once the COP system detects a problem, a system reset is made automatically, and the MCU jumps to the start of a program pointed to by the vector located at $FFFA and $FFFB. Keep in mind that although the MCU can differentiate between different reset causes, it is most common to direct all reset vectors to the same initialization software. When the bootloader program is used to jump directly to the start of EEPROM, place the following data in the pseudo-vector for the "COP fails interrupt": at $00FA, place $7E; at $00FB, place $B6; and at $00FC, place $00. These data will be interpreted by the CPU of an HC11 as instructions that will cause it to jump to the start of EEPROM ($B600) when the COP system has detected a software failure. This will cause a system reset.

How does the COP system detect the problem? In theory, very simply! Basically, it keeps checking to make sure the software is executed in the proper sequence. Although this was explained in Chap. 6, I will repeat its basic operation here.

This software sequence requires two steps. First, a write of $55 is made to the COPRST ($103A) register. Second, a write of $AA is made to the same register. Any number of instructions can be performed between these two steps as long as both steps are performed in the correct sequence before the timer times out. This sequence of instructions is referred to as *servicing the COP timer*. What about this timeout period? How long is it? Go back to Chap. 6 and find out.

Tips on Setting Bits in a Register or Other Memory Location

The HCx11 has a number of registers whose bits control various aspects of its characteristics. This is one feature that accounts for the HCx11's versatility. Of course, to set or clear a single bit at a memory location, all you need do is store the appropriate number at the register. For instance, say you want to turn on the clock monitor function. Here, you must set bit 3 at the OPTION register, which is located at $1039. The following code segment will accomplish this:

```
        *
        *
   LDAA     #%00001000
   STAA     $1039
        *
        *
```

If you are observant, you may have noticed a problem here. If you did not, think about the other bits in the OPTION register for a second. Most have an important function, and the program segment just cleared them all. There are two solutions here. First, we will look at an example that only uses 6800-compatible instructions to set bit 3 at $1039.

```
        *
        *
LDAA    $1039
ORAA    #%00001000
STAA    $1039
        *
        *
```

And to clear bit 3:

```
        *
        *
LDAA    $1039
ANDA    #%11110111
STAA    $1039
        *
        *
```

If you wish to use the extended opcodes provided to HC11 users, the following code segment will set bit 3 of $1039:

```
        *
        *
LDX     #$1039
BSET    0,X,%00001000
        *
        *
```

and the following code segment will clear this same bit:

```
        *
        *
LDX     #$1039
BCLR    0,X,%00001000
        *
        *
```

However, be careful here because the code is no longer downward-compatible. BSET and BCLR are new opcodes added to the HC11 and the HC05; they are not legal for 6801/6803, 6802/6808, or earlier MCUs.

Tips When Using the MC68HC811E2FN in MagTroll-11

If you plan to install an HC811 in MagTroll-11, you will likely want to operate the MCU in the normal single-chip mode. This is accomplished by placing a jumper at JP3 and installing a jumper between pins 1 and 2 of JP5.

The software design, when the MCU is in normal, single-chip mode, is rather straight-forward. First, the all-important reset vector is located at $FFFE and $FFFF, just like the 6803, 6808, 6802, and even the original 6800. Thus the address of the start of your pro-gram should be placed here. For instance, typically your program will start at $F800, so place $F8 at location $FFFE and $00 at $FFFF. Also, this same address should be placed at the illegal opcode trap vector at $FFF8 and $FFF9. Absolutely make sure that one of the first instructions is to initialize the stack pointer. Typically, this can be accomplished with the following single instruction:

```
LDS     #$00FF.
```

Unless you are experienced, do not use the COP watchdog system or the clock monitor system.

Tips When Using the Standard MC68HC11A1FN in MagTroll-11

While programing an HC11's EEPROM is simpler than programming an HC811's EEP-ROM, it is more complicated to use an HC11 in MagTroll-11 than an HC811. This, of course, assumes that you are going to use the MCU's internal EEPROM for the system's sole firmware source. The primary attributes of the HC11 that make it slightly more com-plicated to use in MagTroll-11 are

1. The EEPROM's weird location ($B600–$B7FF) in memory
2. The fact that the EEPROM cannot be moved in memory
3. Its small size (512 bytes)

It is also possible to use MagTroll-11 in an expanded system. Here, refer to the next sec-tion, "Using MagTroll-11 in an Expanded Multiplex System."

In order to use a MagTroll-11 with an HC11 in a single-chip system, the MCU first must be placed in special bootstrap mode. This is accomplished by placing a jumper at JP3 and placing a jumper between pins 2 and 3 of JP5. In order to cause an immediate jump to the beginning of the EEPROM located at $B600, also place a jumper at JP2.

Notice from Table 6-1, as well as from my previous discussion, that the normal reset vector, located at $FFFE and $FFFF, is no longer valid here. Rather, the reset vector is located at $BF40, which is the start of the bootloader program. In one sense, this simpli-fies things. However, if you are accustomed to designing programs for the 6800 or its off-spring, it can become confusing. As the preceding discussion indicates, if jumper JP2 is installed, the bootloader program causes an immediate jump to the start of the program at $B600. Here one also should initialize the stack pointer, only this time have the top of the stack start at $00B0 instead of $00FF. The other thing to keep in mind here is to initialize the illegal opcode trap pseudo-vector at locations $00F7, $00F8, and $00F9. The follow-ing cross-assembler code segment accomplishes all this:

```
        ORG $B600
    *REM: First we load the stack pointer with $00B0
```

```
        LDS             #$00B0
*REM: We then store the opcode for the JMP statement at $F7
        LDAA            #$7E
        STAA            $F7
*REM: Finally we setup our pseudo-vectors
        LDD             #$B600
        STD             $F8
                *
                *
*REM: place rest of program here
```

Using MagTroll-11 in an Expanded Multiplex System

While MagTroll-11 was designed to be used in single-chip systems, it is adaptable for use in expanded multiplex systems with firmware located in an external EPROM. Refer to the simplified schematic in Fig. 10-9. Also refer to the schematic in Fig. 2-1. Notice that a 74HC373 octal transparent latch is required for demultiplexing. This is used to remove address lines A0 through A7 from the multiplexed address/data port, port C. Notice that all you need do is connect the HC11's AS line to pin 11 of a 74HC373, port C to the chip's inputs, and then *bingo!* The address lines magically appear at the chip's outputs. What you do from here is up to you. Typically, however, a system is set up similar to Fig. 2-1.

To set up the MCU for normal expanded multiplex operation, install a jumper between pins 1 and 2 of JP5 but do not install a jumper at JP3. Make sure that you have initialized the reset vector at $FFFE and $FFFF (i.e., place the starting address of your program in these memory locations).

Use the EEPROM to Store the MagTroll-11's Program and Still Communicate with the SCI

Figure 10-10 shows a schematic that will allow you to communicate with a MagTroll-11 through the serial interface of a computer. However, you may be muttering underneath your breath something like, "Why bother! I plan to use the HC11's EEPROM to store the program, so I must place a jumper at JP2, which makes it impossible to use the HC11's SCI."

There is truth buried deep within these mutterings. It is true that once the RxD and TxD lines are tied together, it is impossible to communicate normally with the HC11 through its SCI. Nevertheless, before you jump to the conclusion that *it is impossible to use the SCI and the EEPROM simultaneously,* try to remember why we tied these two lines together. This was simply done so that the SCI's RxD receives a break signal when the bootloader's program is looking for it.

Actually, if you examine the bootloader's software (the source code of which is kindly provided in the "HC11 bible"), you find that the bootloader first sets up the stack and then

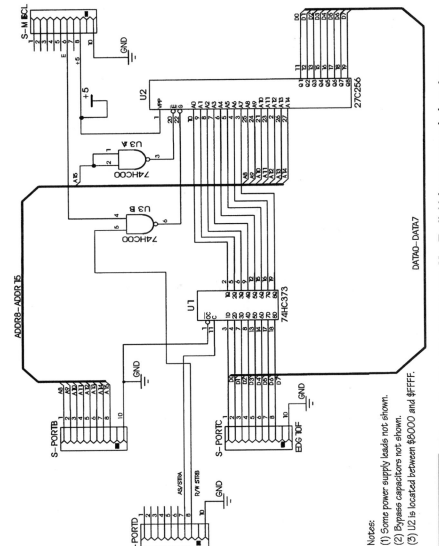

FIGURE 10-9 Schematic showing how to use MagTroll-11 in an expanded system.

Notes:

(1) Some power supply leads not shown.

(2) Bypass capacitors not shown.

(3) U2 is located between $8000 and $FFFF.

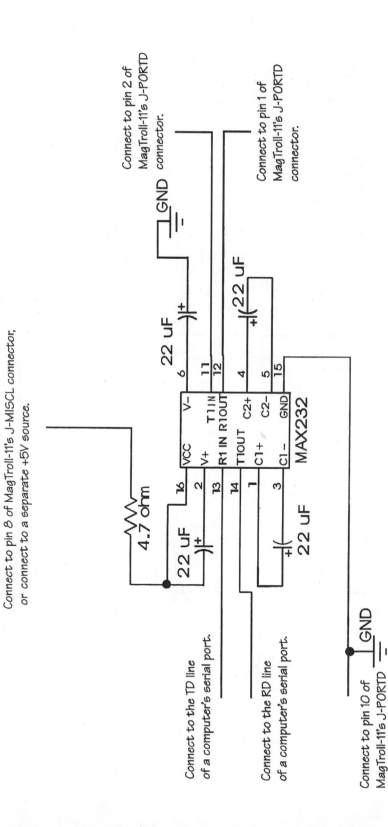

Connect to pin 8 of MagTroll-11's J-MISCL connector, or connect to a separate +5V source.

Connect to pin 2 of MagTroll-11's J-PORTD connector.

Connect to pin 1 of MagTroll-11's J-PORTD connector.

Connect to the TD line of a computer's serial port.

Connect to the RD line of a computer's serial port.

Connect to pin 10 of MagTroll-11's J-PORTD connector as well as to the ground connection of the computer's serial port.

FIGURE 10-10 Schematic showing one way of providing communications between MagTroll-11 and a computer.

initializes the SCI for 7812 baud with an 8-MHz crystal. It then sends a break signal that it clears when it receives a start bit. It then waits to receive the first character. Accumulator A is loaded with this character if it is $00; then the program jumps to the start of EEPROM at $B600. If not, well, let's not go into that.

Nowhere in the Motorola literature is it mentioned that it is possible to use the bootloader program's "jump to EEPROM" feature and the SCI *at the same time*—but it *can* be done! I have done it and even have designed a rather complex HC11 project using it. The only trick is that "normally" you must choose a slower baud rate, such as 4800 with an 8-MHz crystal or 2400 with a 4-MHz crystal. However, even this limitation can be circumvented. There are sophisticated software techniques available that allow you to send a break so that you do not have to choose a slow baud rate after all. You can run the SCI as fast as it can go. While I will not go into details here, these sophisticated techniques are available through special custom controls that are included with professional versions of software development tools such as Visual Basic.

Jumping to EEPROM by Sending a NUL

While you can use MagTroll-11 in this experiment, it is a lot easier to use MagPro-11. *Because of this, it is assumed MagPro-11 will be used.*

Enter the program listing in Fig. 10-11, which is in Buffalo's ASM line assembler format, into MagPro-11. Remember, to program an HC11 using MagPro-11, you must use the expanded multiplex mode, which means that you should not install jumpers at JP1 or JP2. Also make sure there is a jumper at JP4 between pins 2 and 3.

Now you can try out this program by typing "Go B600." If LED1 blinks, this means you have entered the program correctly. Now for the fun!

Question.　How do you get the program in Fig. 10-11 to operate automatically without using the BUFFALO monitor, i.e., without typing "Go B600"?

Answer.　You set MagPro-11 up for special bootstrap mode by installing jumpers at both JP1 and JP2 and then sending a NUL from a computer to MagPro-11's serial interface (J2). This NUL should trigger the bootloader's program so that it will jump directly to $B600, and LED1 should start blinking.

Trying it out.　Set up the communication program for 2400 baud (assuming a 4-MHz crystal), and then, once communication is made with MagPro-11, hit the @ key. This will send a NUL to MagPro-11, and LED1 should start blinking.

Of course, if you really want to communicate with the HC11, you must include sending out this NUL at the very start of your computer program. The following short QBasic program will send a NUL at 2400 baud out the computer's COM1 port:

```
REM A short Qbasic program that sends a NUL out COM1 at 2400 baud
NUL$ = CHR$(0)
OPEN "COM1:2400,N,8,1,CD0,DS0,CS0,RS" FOR RANDOM AS #1
PRINT #1, NUL$
END
```

```
B600        LDS   #20
B603        STAA  1000
B606        LDY   #FFFF
B60A        DEY
B60C        BNE   B60A
B60E        BRA   B603
```

FIGURE 10-11 Program listing that flashes either MagTroll-11's or MagPro-11's LED1. The purpose here is to demonstrate that it is possible to have software send a NUL or break so that the bootloader will cause a jump directly to the start of EEPROM.

Running this little program does the same as hitting the @ key while running a communication program. Try it out on the program listing in Fig. 10-11. See if you can get the LED to flash.

Coming Attractions

Now that you know the basics of MagTroll-11, let's do something practical with it. To make things more interesting, let's look at a unique, at least as far as I can determine, project that is a remote-reading wind direction indicator that does not require any moving parts. It uses all the intelligence that can be packed into a 512-byte program. In fact, it will use the entire 512 bytes in a useful manner. This project, along with a few general hints, should provide enough down-to-earth information for you to construct hundreds, perhaps thousands of different products that make use of MagTroll-11.

AN MCU-BASED SOLID-STATE
WIND VANE

In Chap. 9 we looked at MagPro-11, an EEPROM programmer that can double as a basic development system. In Chap. 10 we looked at MagTroll-11, a simple HCx11-based controller board. This chapter looks at a really down-to-earth, practical project. Not only is it practical, but it's unique. It's different. It also concretely shows the fantastic usefulness of microcontrollers in general and the HC11 in particular. It also is just simply fun.

The Solid-State Wind Vane

I have built remote-reading wind direction indicators in the past. One design consisted of a basic wind vane connected to a precision potentiometer that had 360° rotation capability. The display was a simple analog meter. This electric remote-reading wind vane worked for a while—until a Michigan winter set in and froze the pot in place. The wind direction indicator's meter showed that the wind was blowing steadily out of the northwest for a whole month—until a late January thaw invaded the frozen waste. It then started working again—for a little while, that is. Since the unit was mechanical, and not perfectly waterproof, corrosion caused the unit to completely freeze up after a few years. Warm weather does not solve this type of freezeup, although a bit of warm WD-40 sometimes helps.

After that hapless experience, I resolved that the next wind vane I designed would not have a single moving part. However, for a couple of years, a solid-state wind vane seemed as real as the *Starship Enterprise*'s warp drive.

One cold and windy morning, I had a sudden insight concerning a solid-state wind vane. Why not use temperature sensors? Constructing a solid-state wind vane after all just might be feasible. A simple thought experiment suddenly echoed through my consciousness. Why not, I reasoned, place eight temperature sensors in a circle? One sensor would correspond to each direction (you know, north, northeast, east, etc.). In the middle of this circle of sensors, place a heater. Wind would move the warm air from the heater to the sensor downwind from it and warm it.

This all sounded sensible. But would it work? I quickly jerry-rigged a little circuit with two sensors right next to, but not touching, a hot power resistor. A fan was set up to blow air at the sensors, and a digital voltmeter was used to take voltage readings. I quickly, though sadly, discovered that there wasn't a sufficient difference in sensor temperatures to use as the basis for a practical wind vane. So long warp drive, back to the reality of impulse drive.

Not giving up my basic goal, I conducted another thought experiment. In this experiment, which was based on many years of personal experience in surviving Chicago winters, I pictured myself being outside on a cold windy day. Although I sought warmth, the nearest building was locked. However, I found shelter from the winter winds by standing next to the building. The side of the building that faced the wind was not much warmer than being out in the open. However, the other side—the leeward side—was a lot warmer. A lot! In my thought experiment, it seemed almost comfortable. Swiftly, my thought experiments degenerated into a daydream. Colors suddenly became vivid, and the senses seemed to come alive. I looked around and saw that I stood on the leeward side of an immense building that was warmed by a brilliant sun, low in the sky. Although a huge thermometer on a bank across the street showed 19°F, it still felt oh, so good. Abruptly, the daydream disappeared, and I found myself staring again at a computer screen. The warmth of the sun was replaced by the cold glow from the monitor.

The thought experiment that transmuted into a daydream wasn't only psychologically pleasant but rewarding intellectually as well. Why not, I thought moments after the daydream subsided, place sensors on the leeward side of small windbreaks and then use this information to determine the wind direction? Of course, the sensors had to be heated internally. Nevertheless, this could be accomplished simply by forcing as much current through them as was listed under maximum conditions in the data sheet and then adding a few more electrons per second for good measure.

Initial tests seemed to suggest that just such a device could work. To verify this fact, a simple sensor that used four inexpensive LM335 sensors—one each for north, south, east, and west—was constructed. The output of these sensors was connected to an opamp circuit that provided signal conditioning. The output of this circuit was connected to the A/D input of a MC68HC11A1P that formed the heart of Mag-11. After a bit of revising of Mag-11's application firmware (contained in EPROM, not EEPROM) and playing with a small electric fan, a solid-state wind vane was apparently practical after all.

The Solid-State Wind Vane—More Details

The solid-state wind vane is an excellent HC11 demonstration project for designs that use the HC11's internal EEPROM. What it basically does is it looks at eight different sensors and picks the one with the highest temperature. (As you will soon see, it is more sophisticated than this.)

As in nearly all modern electronic designs, one must choose between either a hardwired or a MCU-based design. One hard-wired approach would be to arrange an array of voltage comparators and associated circuits to pick out the highest voltage and light the proper LED. However, such a scheme not only is unwieldy and expensive, but it also lacks simple modification capability. This hard-wired approach also assumes that all sensors are identical. While it is possible to purchase nearly identical sensors (e.g., the LM135AH), the price of one such premium sensor exceeds that of the MC68HC11A1FN itself, and remember, eight are needed.

A neat way of saving money on sensor purchases is to take advantage of the HCx11's intelligence and use statistical analysis. (An extreme example of this will be given in the next chapter.) While each individual sensor may differ significantly from the others, the average characteristics of three sensors are nearly identical to the average characteristics of any three other sensors. This statistical analysis allows the use of the less expensive LM235H, LM335AH, LM335AZ, or even the cheap LM335Z. What you can do here is compute the average of three adjacent sensors, find the highest of these averages, and then assume that the true direction corresponds to the sensor in the middle. The simple fact is it works—even with LM335Z sensors. However, better-quality sensors, such as the LM235H or LM135H, are still recommended here.

Sensor Assembly's Physical Description

The sensor assembly consists of four simple circuit boards, with one windbreak and two sensors on each board. The assembly is sandwiched with four boards mounted vertically with a separation of roughly $1^1/_4$ in (see drawing in Fig. 11-1 and photo in Fig.11-2).

In the following description, the "referenced" sensor is on the leeward side of the windbreak. For instance, the North sensor is on the "south" side of the windbreak. Thus with a north wind, the North sensor (on the south side) should be warmest. (Remember that it is heated internally.) Of course, the plane of the windbreak of the North-South board is at right angles to the windbreak of the East-West board. Similarly, the plane of the windbreak of the NE-SW board is at right angles to the NW-SE board. Also, the plane of the windbreak of the North-South board is at 45° to the plane of the windbreak of the NE-SW board.

CIRCUIT DESCRIPTIONS

Sensor assembly.   This assembly consists of the heart of the solid-state wind vane. It collects the data necessary to determine the wind direction. If you connect the eight separate outputs of these connectors to eight accurate voltmeters, a skilled observer could determine which way the wind was blowing. In the solid-state wind vane described here, a programmed HCx11 observer replaces the skilled human observer.

The sensor assembly consists of four, electrically simple, printed circuit boards. Any of National Semiconductor's LM135/235/335 series of temperature sensors can be used here. While this temperature sensor is an integrated circuit consisting of more than 15 transistors, it functions as a zener diode with a breakdown voltage directly proportional to the absolute

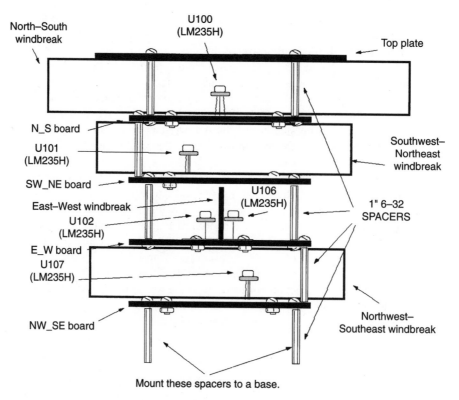

FIGURE 11-1 Drawing of wind vane sensor.

temperature at +10 mV/K. (Remember degrees Kelvin is roughly equal to 273° + degrees Celsius, so 25°C = 298 K.) Thus, at room temperature, 298 K, the breakdown voltage of an LM135 should be at 2.98 V. For the premium LM135AH, the company guarantees that it is between 2.97 and 2.99 V. However, for the garden variety LM335Z, National only guarantees that it is somewhere between 2.92 and 3.04 V. The specs for the other grades lie between these extremes. Although the opamp board uses pots to calibrate this temperature error out, the higher the quality of the sensor, the more reliable is the wind vane.

When measuring the absolute air temperature, internal heating caused by the reverse current of the sensor is not appreciated. Thus the LM135 generally is operated at or below 1 mA. However, here we "love" internal heating, so we operate the sensor slightly above its maximum rating of 15 mA.

Four windbreaks are used, each at 45° angles with each other. Two sensors are used per windbreak, one on each side. Because of internal heating, the sensor on the leeward side of the windbreak is warmest. Since each windbreak is at 45° angles, one sensor should be

FIGURE 11-2 Photo of wind vane sensor.

warmest and the two adjacent sensors only slightly cooler. The other sensors should be substantially cooler when there is a wind blowing. Note that the sensor assembly should have a roof to protect it from the sun as well as rain and snow. One other interesting point: Even a moderately large roof will not protect the sensor from wind-driven rain or snow. However, this does not interfere with accuracy! In fact, rain and/or snow actually may increase accuracy because wind will cause "wet" sensors to cool more than dry sensors. Watch the sun, though! Direct sunshine on a sensor can throw off the solid-state wind vane. Thus a good roof is recommended.

Opamp board. This board provides signal conditioning so that we can make full use of the HCx11's A/D converter; it also provides a means of compensating for variances in sensors. The simplified schematic for this board is shown in Fig. 11-3. Keep in mind here that there are eight separate sensors and that we need a circuit for each sensor. If you wish to examine the detailed schematic, you can print it out by using the SCHEMAT program on the CD-ROM.

U5, an LM336 2.5-V reference IC, provides the stable reference voltage for all eight circuits. Seven of the circuits are identical and have an adjustable output. One circuit, centering about U1:A and U1:B, is slightly different because there is no pot, and thus its output is not adjustable. This circuit is treated, while calibrations are being done, as the standard, since the other circuits are adjusted so that their outputs are identical to this one when the air is calm. Let us look briefly at this circuit. Refer here to Fig. 11-4. By the way, the operation of the seven other opamp circuits is quite similar except that they include an adjustable potentiometer.

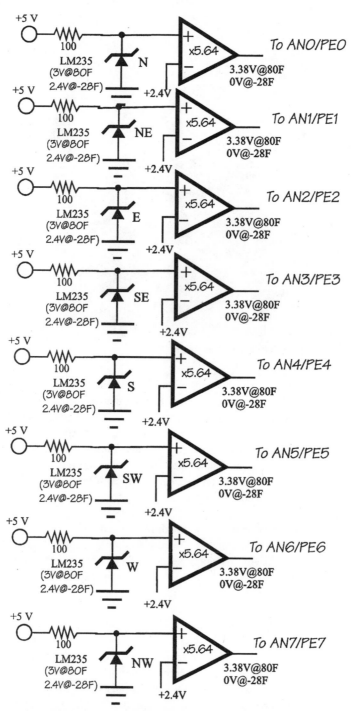

FIGURE 11-3 Simplified schematic for opamp board for solid-state wind vane. *Note:* **LM235 sensors are located on the sensor board.**

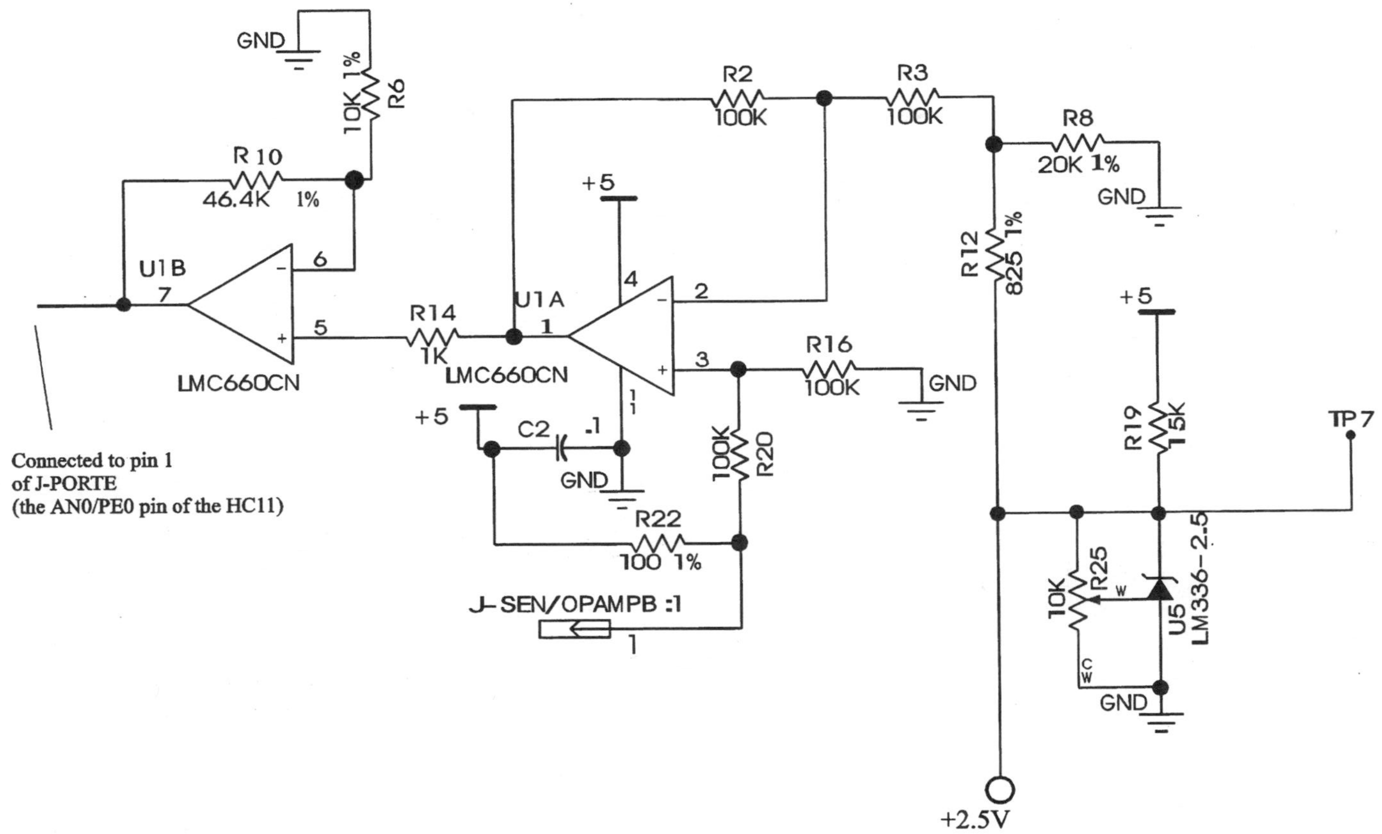

FIGURE 11-4 Detailed schematic of one of the opamp circuits.

For an illustrative look at this circuit, assume a North sensor temperature of about 80°F (about 300 K). This is a good assumption when the air temperature is in the sixties and there is a moderate north wind. (Remember, the North sensor is located on the south side of the windbreak.)

The output of this North sensor is connected to pin 1 of JSEN/OPAMP. We now can assume an input voltage to the noninverting input of the differential amplifier centered about U1:A of 3.00 V. From simple math [(20/20.825) × 2.5 = 2.4] we can see that the voltage at the differential amplifier's inverting input is 2.4 V. This voltage appears at the input of the noninverting amplifier centered about U1:B. Since this amplifier has a gain of 5.64, its output voltage is 5.64 × 0.6 = 3.384 V. This voltage is transmitted to pin 1 of port E (the A/D converter input) of the HCx11 located on the MagTroll-11 controller board. When the A/D converter converts this channel (AN0), it stores the hexadecimal number $AD (173 decimal) at the ADDR1 register [173 = (3.384/5) × 256]. As mentioned earlier, the seven other circuits are similar except that each has a separate pot adjustment.

An astute observer will notice limitations here. For instance, once the sensor hits about 130°F, the circuit becomes saturated, and the wind vane will not function properly. However, the wind vane should function up to an air temperature of 120°F if there is at least a moderate wind.

It also becomes apparent that at a sensor temperature below about −20°F (−29°C), there is another problem. However, keep in mind that the sensor is heated internally, so unless there is a strong wind, the wind vane probably will function down to an air temperature of around −35°F. However, it has not been tested at these extremes, so the limits here are theoretical. (It has been tested at "normal" temperatures from about 20 to 90°F.) Try to design a mechanical wind vane that will work at these temperature extremes. While it isn't impossible, it isn't easy either, unless cost is no problem.

The opamp board uses National Semiconductor's LMC660CN quad CMOS opamp IC. The primary purpose for the choice here is that this IC provides "true" single-supply operation. The inputs and outputs of this chip truly operate down to ground and up to the supply voltage. The popular (and cheap) LM324 promises this with a cursory reading of its data sheet. Nonetheless, it never quite lives up to its promise, since it always stays roughly 20 mV above ground with a single supply. Still, you can pull an LMC660CN out of its socket and slip in an LM324 and get the wind vane to work, at least at "normal" temperatures.

Display board. This simple board consists of 9 LEDs and a 74LS42 BCD-to-decimal decoder. Refer to the schematic that can be printed out with the SCHEMAT program on the CD-ROM. Eight of the LEDs are arranged in a circle to indicate the appropriate direction (e.g., North, Northeast, East, Southeast, etc.). The ninth LED is located in the center and shows no wind, extreme high/low temperature, or another error condition. The input of the 74LS42 is connected (through connector J-PORTB) to the low-order nibble (PB0–PB3) of the HCx11's port B. To light a LED, simply store a number at port B. For instance, to light the North LED, store 07 at port B. To light the West LED, store 01 at port B.

Controller board (MagTroll-11). The MagTroll-11 controller board is about as simple as one can get. If you look at the board, all that you really notice are the HCx11 and a bunch of connectors. The firmware on this board is contained in the HCx11's internal EEP-ROM—512 bytes for the HC11 and 2048 bytes for the HC811. Of course, it is also possible to use MagTroll-11 using a HC11's internal 8-kB ROM, although the problem here is high unit cost when the production run is small. For more details on MagTroll-11, refer to Chap. 10.

Note that other HCx11 controller boards can be used here if they contain an eight-channel (for the full-blown version) A/D converter and have connectors to port E and port B. The 48-pin DIP version of the HCx11 (MC68HC11A1P) only contains a four-channel A/D converter and thus is only suitable for use with the low-cost ("voodoo designed") version of the solid-state wind vane. I will describe a "voodoo designed" version of the solid-state wind vane in the next chapter.

A LOOK AT THE SOLID-STATE WIND VANE'S SOFTWARE

I assume here that the firmware will reside in the HCx11's EEPROM. This restricts us to either 512 or 2048 bytes, depending on the type of HCx11. This size restriction limits us to assembly language. The source code, assembler listing, and MOT S19 code are included on the CD-ROM. Let's look at a few highlights here, but before we do this, let us look at the simplified flowchart given in Fig. 11-5. The following discussion is more detailed than the flowchart, which is meant only as a general guide for the program.

The first step is to set up the stack by loading the stack pointer at an appropriate place in RAM, set up NUMREAD by storing a number here (say, 30), and light the center LED for $1/2$ s. After this, we input all eight temperatures from the sensors. Next, the temperatures from each set of three adjacent sensors are added together and the sum divided by three. The "average" temperature of this set of three sensors is stored at eight different word-sized (2-byte) variables called AN0MOD, AN1MOD, AN2MOD,..., AN7MOD. The next routine determines if all ANxMOD temperatures are equal. If they are, the display board's center LED lights for at least $1/2$ s. This can happen, for instance, if there is no wind or if there is a combination of excessively high temperature (over 105°F) and only a very light wind.

If ANxMOD temperatures are not all equal, the next step is to determine which is highest. Once this is determined, the MCU, through port B, lights the LED that corresponds to the center sensor that makes up the trisensor combination. For instance, if the program determines that the average temperature of the North, Northeast, and East sensors is higher than any other three sensor average, the MCU sends a Northeast signal (actually 06) via its port B to the display board. The display board will then light the Northeast LED. The program then branches back to the start where it starts to take new sensor readings for an update.

A BRIEF LOOK AT A SMALL PORTION OF SOURCE CODE

"Once begun is half done" may be true when washing dinner dishes, but it seldom applies to writing nontrivial software. Despite this, or perhaps because of this, let's look at the beginning program segment of the source code for the solid-state wind vane. Refer here to the source code Fig. 11-6. For convenience, instructions and assembler directives are preceded by a line number; comments are not. Of course, because of these line numbers, you may be deluged with error messages if you attempt to assemble this program by invoking the AS11 cross-assembler. The complete source code is on the CD-ROM in the CHAP11

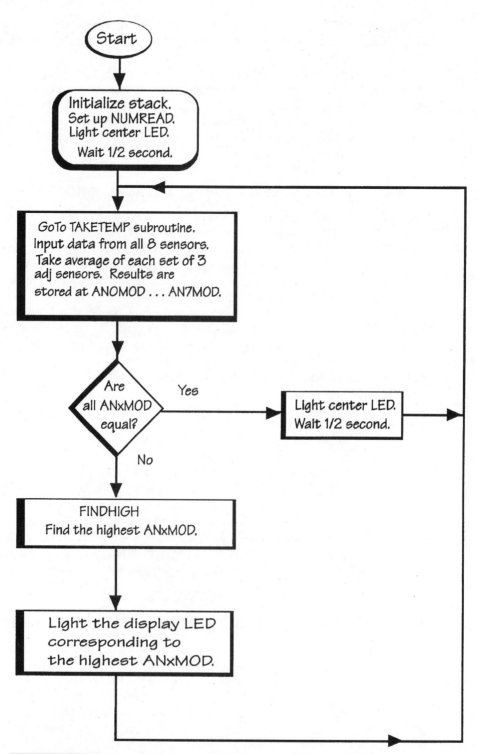

FIGURE 11-5 Simplified flowchart for the wind vane software.

```
*start of program is at $B600
1                   ORG        $0B600
*load stack pointer—very important
2                   LDS        #STACK
*setup the illegal opcode trap pseudo-vector
*which is located in RAM since we are using
*the bootstrap mode
*now if something goes awry the program
*will (hopefully) automatically start again
*at the very beginning
3                   LDAA       #$7E
4                   STAA       $0F7
5                   LDD        #$0B600
6                   STD        $0F8
*set up to take 30 sensor readings each
*time we go to the subroutine TAKETEMP
7 START      LDAA        #30
8                   STAA       NUMREAD
*first-light center LED so we know
*HCx11 is performing properly
9                   BSR        CENTRLED
*leave it on for 1/2 second—500 ms
10                  JSR        DLY500
*now check if ANxMOD temperatures
*are all equal
*if any are not equal branch to FINDHIGH
*the program segment FINDHIGH will locate
*the highest modified temperature
11    NEWCYCLE       JSR        TAKETEMP
12                  LDAA       AN0MOD+1
13                  CMPA       AN1MOD+1
14                  BNE        FINDHIGH
15                  CMPA       AN2MOD+1
16                  BNE        FINDHIGH
17                  CMPA       AN3MOD+1
18                  BNE        FINDHIGH
19                  CMPA       AN4MOD+1
20                  BNE        FINDHIGH
21                  CMPA       AN5MOD+1
22                  BNE        FINDHIGH
23                  CMPA       AN6MOD+1
24                  BNE        FINDHIGH
25                  CMPA       AN7MOD+1
26                  BNE        FINDHIGH
```

FIGURE 11-6 Beginning program segment for the MCU-based solid-state wind vane. This partial source code listing is in AS11 cross-assembler format. However, because of the line numbers in the first column, it can't be assembled without modification. Also refer to the complete source code on the CD-ROM: CHAP11<\WINDVANE.ASM.

```
*if all ANxMOD temperatures are equal
*light center LED for 1/2 second
*then start over and take new measurements
27                        BSR        CENTRLED
28                        JSR        DLY500
29                        BRA        NEWCYCLE
*******************************************
*subroutine CENTRLED lights the
*display board's center LED
30           CENTRLED     PSHB
31                        LDAB       #CENT_LED
32                        STAB       PORTB
33                        PULB
34                        RTS
*******************************************
```

FIGURE 11-6 (*Continued*) **Beginning program segment for the MCU-based solid-state wind vane. This partial source code listing is in AS11 cross-assembler format. However, because of the line numbers in the first column, it can't be assembled without modification. Also refer to the complete source code on the CD-ROM: CHAP11 \WINDVANE.ASM.**

directory/folder. Its name is WINDVANE.ASM. Also in the same directory/folder is the assembler listing (.LST) and the MOT S object code (.S19).

Referring to Fig. 11-6:

Line 1 sets up the starting address for the program with the assembler's ORG directive. We assume here that we will be using the 512-byte EEPROM in the MC68HC11A1FN for our firmware. If we were using the 2-kB EEPROM in the MC68HC811E2FN, this first line would likely be ORG $0F800. If a separate EPROM is used, this starting address could be just about anything between 0 and $FF00.

Line 2 sets up the stack. [The variable STACK was declared previously (not shown) in an RMB (reserve memory byte) directive.]

Lines 3 through 6 set up the pseudo-vector for the illegal opcode trap. Now, if the program goes wild and the CPU detects an illegal opcode, things should calm down because the program will start over again at the beginning—just as if you pressed the RESET switch.

Lines 7 and 8 store 30 at NUMREAD.

Lines 9 and 10 light the center LED for $1/2$ s (assuming a 4-MHz crystal and a 1-MHz clock) to let us know the MCU is operating correctly.

Line 11 calls the all-important subroutine TAKETEMP that does a whole bunch more than you might think. This subroutine (which is too long and complicated to include here) initializes the A/D converter and takes the average of 30 readings (NUMREAD) of each sensor. Next, it takes the average reading of each set of three adjacent sensors and stores the result in ANxMOD. For instance, the first sensor (North) is initially stored at location AN0. This routine takes the average of AN7 (Northwest), AN0, and AN1 (Northeast) and stores it at AN0MOD. Do not get confused here with names. In the software, AN0 is the temporary storage location for the data at HCx11's AN0 pin. While the names are identical, they actually refer to different entities—a storage location and a pin label. These separate entities have a one-to-one correspondence, however. (Perhaps different names here would help; perhaps not. While the names refer to separate entities, they actually refer to the same basic element—the storage location that stores the data that are at the pin with the same name.)

Lines 12 through 26 merely check if all sensor (actually modified sensor) readings are the same. If they are not, the program branches to FINDHIGH, which does what you would think—finds the highest sensor reading. Notice in lines 12 through 26 that we do not use ANxMOD variables but rather ANxMOD + 1 variables. The reason here, though not obvious, is simple. ANxMOD variables are actually 2 bytes wide, and ANxMOD points to the high-order byte, while we are actually looking for the low-order byte, which is one address location higher, so we add "1" to ANxMOD.

If all modified sensor readings are the same, then lines 27 through 29 are executed, which again cause the center LED to light for $^1/_2$ s, and then we start over.

For interest sake, the subroutine that lights the center LED is given in lines 30 through 34. This subroutine should be obvious to anyone with some understanding of the 6800 series of MPU/MCUs.

As you can tell from the S19 file in the CD-ROM, this program fills a 512-byte EEPROM. However, if you use an HC811 with its 2048-byte EEPROM, you can do more sophisticated things such as provide automatic sensor adjustment at power up and routines where only an adjacent LED will be allowed to light, which will eliminate possible confusing flashing due to swirling wind. In addition, with a 2-kB EEPROM (HC811), it is possible to connect another display to MagTroll-11 and measure the wind chill. The next chapter will look at wind chill as well as wind speed and temperature. While hints and tips will be generously provided, the concrete software to accomplish this will be left to you.

Construction

This project requires seven separate circuit boards, plus a power supply. Nonetheless, it is not as complex a project as one might think. One circuit board, MagTroll-11, is described in the last chapter. Four ultrasimple circuit boards are used for the sensor. Also required is a single-sided display board and a double-sided opamp board.

SENSOR ASSEMBLY

Not only is this the solid-state wind vane's heart, it is also the hardest part to build and, unless cheap LM335Z sensors are used, the most expensive. Unlike the opamp board, there is more to constructing this assembly than stuffing a PC board full of electronic parts. Since this assembly is three-dimensional, it is a bit hard to illustrate on a two-dimensional medium such as the pages of this book. However, Figs. 11-1, 11-2, and 11-3 should help. Magicland is making available a semikit for the sensor assembly: the four circuit boards, spacers, screws, nuts, color-coded wires, and connector. The sensors are available separately in sets of four. Refer to Appendix C. Not included in this semikit are the windbreaks. You must custom construct these yourself. The patterns for the windbreaks are given in Figs. 11-7 through 11-10. The prototype used 28-gauge sheet aluminum, which is available at some building supply stores. The actual cost of the windbreaks is almost nil. Each windbreak takes about 10 minutes to make if you work slowly.

Notice the dotted lines on the patterns for the windbreaks. Bend the sheet aluminum together along the dotted lines. Bend the four small tabs outward—they are used to mount

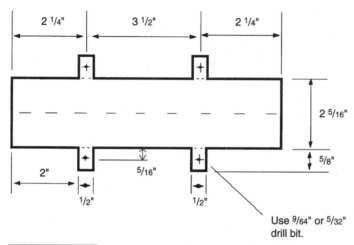

FIGURE 11-7 Construction diagram for North-South windbreak.

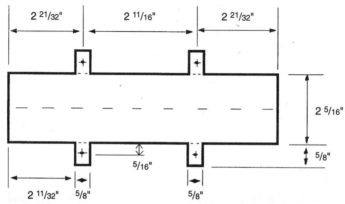

FIGURE 11-8 Construction diagram for SW-NE windbreak.

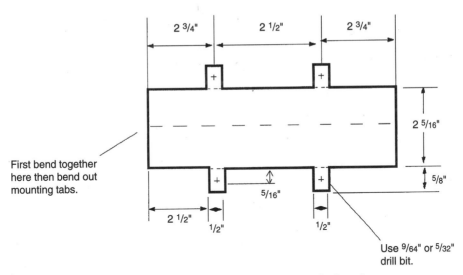

First bend together here then bend out mounting tabs.

Use $^9/_{64}$" or $^5/_{32}$" drill bit.

FIGURE 11-9 Construction diagram for East-West windbreak.

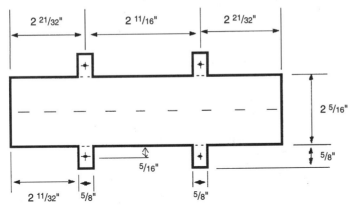

FIGURE 11-10 Construction diagram for NW-SE windbreak.

the windbreak to the circuit boards. A $^9/_{64}$-in hole is drilled in each tab as shown. Then $^1/_4$-in 6-32 hardware is used to mount the windbreak to the circuit boards. Each windbreak is slightly different, so mount the correct windbreak to each board. The exact length of the windbreak is not important so long as they are at least as long as shown in the patterns.

The simple foil patterns for the four sensor circuit boards are given in Fig. 11-11. Notice that this one foil pattern is for all four boards. After etching has been done, cut where shown. You will then have four separate boards. The component mounting guide is given in Fig. 11-12. The foil patterns have four holes for the sensor, despite the fact that they have only three leads (actually, only two leads are used electrically). This allows the easy insertion of two basic types of sensors—the cheap plastic TO-92-cased sensor (part number suffix "Z," e.g., LM335Z) and the metal-cased TO-56 (part number suffix "H," e.g., LM235H). I have discovered that the metal-cased part ("H" suffix) does better in this project. When soldering, be

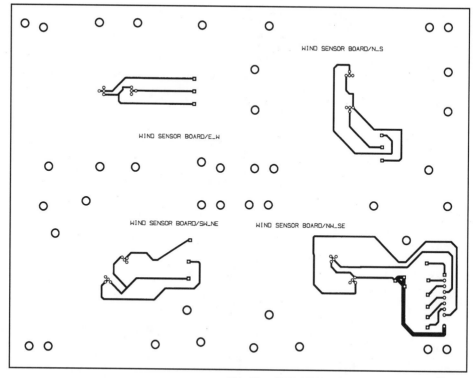

WIND SENSOR BOARD/N_S

WIND SENSOR BOARD/E_W

WIND SENSOR BOARD/SW_NE

WIND SENSOR BOARD/NW_SE

Note: Use a 9/64" drill bit to drill holes marked with a ⭕

FIGURE 11-11 Solder-side foil patterns for all four sensor boards.

sure to use a heat sink, such as an alligator clip, between the soldering iron and the case. Excessive heat can damage the sensors, so work quickly. Soldering should be done in less than 5 s per lead. Wait at least 1 minute between each soldering attempt per sensor.

Before mechanically attaching the boards together, they should be wired together. Here, use colored wire to keep things straight. For simplicity, you can use the colors recommended in the component mounting guides.

The sensor circuit boards are stacked vertically with the Northwest-Southeast board on the bottom and the North-South board on top. Refer here to Fig. 11-1 (as well as Fig. 11-2), which gives a detailed side view of the sensor assembly. Then 6-32 spacers are used, along with screws, to attach the boards to each other, as well as the base and the top plate. (Note that the top plate functions merely as a secondary roof—it is not meant as the primary roof.) Four spacers/screws mount the top plate to the North-South board, the middle two boards, and the bottom board (Northwest-Southeast board) to the large bottom plate. The other two boards use only two spacers and screws. If necessary, and if clearances permit, one or two additional spacers/screws can be added to increase strength. Also, if necessary, a nut or two can be added to a spacer to increase clearances.

Figure 11-13 shows one possible way of mounting a roof over the sensor assembly. Note that the roof itself, shown in Fig. 11-14, is made from a piece of 18 × 18 in moderately heavy sheet metal bent in a curve. Do not use the no. 28 aluminum sheet metal that is rec-

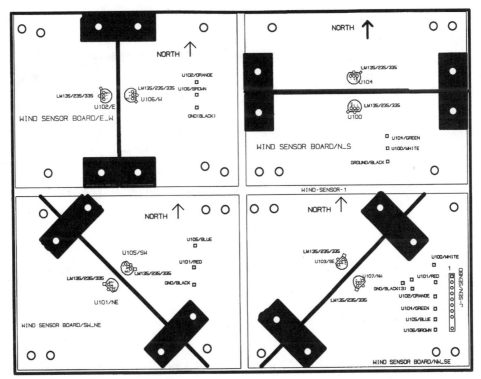

FIGURE 11-12 Component mounting guide for all four sensor boards.

ommended for the windbreaks—it is too light.

The top of the assembly is simply an old PC board with copper completely etched away. This can be considered the secondary roof. As shown in Figs. 11-13 and 11-14, the primary roof for the sensor assembly consists of a curved piece of sheet metal. In the prototype, the sensor assembly is mounted to a surplus 10×12 in PC board. This base board is meant to be attached to a wood board before its final mounting in its permanent location. The $^7/_8$-in wood dowels, which hold up the assembly's roof, are also mounted to this base board.

OPAMP BOARD ASSEMBLY

This board is double-sided. Purchasing a professionally made board is simplest here. See Appendix C. However, if you prefer to make this board yourself, refer to the foil patterns shown in Figs. 11-15 and 11-16. Also refer to Appendix E for tips on replacing a double-sided board with a single-sided one plus jumpers. Figure 11-17 gives the component mounting guide. Figure 11-18 provides a photo of a completed opamp board.

Pots R9, R39, R43, R58, R75, R96, and R99 are listed as 500-Ω pots in the part's list. See Appendix D. This choice assumes that the sensors will be either LM235H or LM335AZ. If you choose the cheaper LM335Z sensor, 1000-Ω pots should be substituted here. (For those who seek perfection and choose the LM135AH, you probably can get away with 200-Ω pots.)

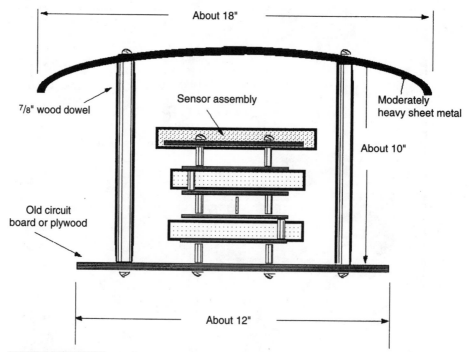

FIGURE 11-13 Drawing of completed sensor assembly with roof shown.

DISPLAY BOARD

This board is a simple, single-sided PC board. Figure 11-19 shows the foil pattern, and Fig. 11-20 provides the component mounting guide. See also Fig. 11-21. This board is also available from Magicland. Again, refer to Appendix C.

Notice that the display board has two connectors for port B. This increases the versatility of the solid-state wind vane because it can be hooked up easily, without disconnecting its own display, to another computer system for automatic recording or even forecasting.

CABLES AND CONNECTORS

You will need the following nine-conductor ribbon cables with suitable 0.1-in single-row connectors with 10 positions. Slot number 9 is left without a contact. A keying plug should be inserted here for polarization purposes. While several manufacturers make suitable connectors, I have used AMP MT connectors. Sources for these connectors include Digi-Key Corporation and Mouser Electronics. Here, you will require, for each connector, a 10-position single-row MODU housing (AMP 1-87499-7, Digi-Key A3019-ND, Mouser 571-1874997), nine contacts (AMP 87667-3, Digi-Key A3000-ND, Mouser 571-876673), a keying plug (AMP 86286-1, Digi-Key A3077-ND, Mouser 538-15-04-9209), and a suitable cable. While special, expensive tools are available to crimp the wire to the contacts, with some care, long-nosed pliers and a soldering iron can be used with great effectiveness.

(A)

(B)

FIGURE 11-14 Photo of completed sensor assembly.

(C)

FIGURE 11-14 (*Continued*) Photo of completed
sensor assembly.

Note:

When connecting to J-PORTE, take special notice of its rather odd pinout arrangements.
Refer to the schematic and component mounting guide.

CONNECTING CABLES

Connect cables as follows:

1. Cable between MagTroll-11's J-MISCL connector and one of the display board's J-
MISCL connector

(D)

FIGURE 11-14 (*Continued*) **Photo of completed sensor assembly.**

2. Cable between the other display board's J-MISCL connector and the opamp board's J-MISCL connector

3. Cable between MagTroll-11's J-PORTE connector and the opamp board's J-PORTE connector

4. Another cable between MagTroll-11's J-PORTB connector and the display board's J-PORTB connector

The length of these cables depends on the case you are using. You also will require a cable between the sensor board and the opamp board (connector J-SEN). This cable must be quite long because you will likely place the sensor assembly on the roof.

Finally, you will need a two-conductor cable to connect MagTroll-11's J1 to the power supply. As the schematic shows, if the power is $+7$ to $+13$ V, connect the positive supply

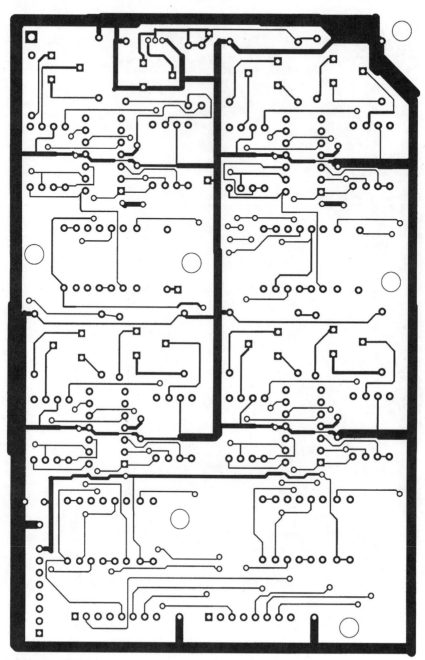

FIGURE 11-15 Solder-side foil pattern for opamp board.

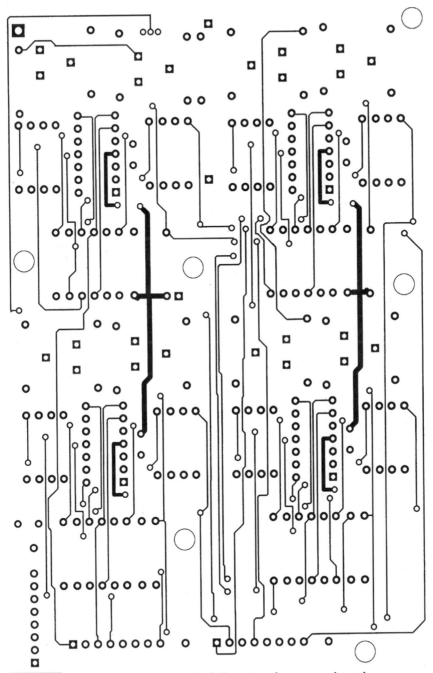

FIGURE 11-16 Component-side foil pattern for opamp board.

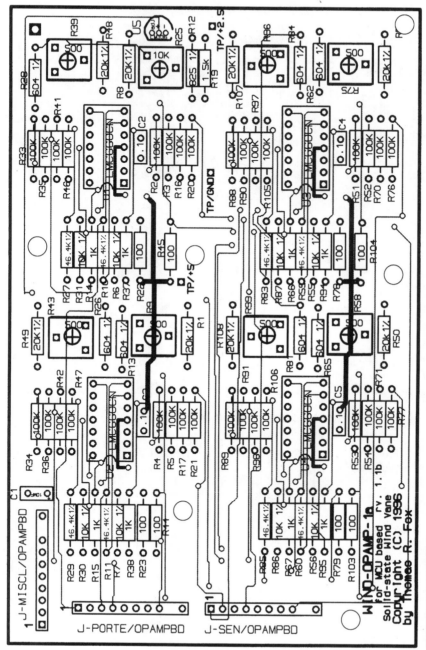

FIGURE 11-17 Component mounting guide for opamp board.

FIGURE 11-18 Photo of opamp board.

lead to pin 1 of J1; if the power supply is a regulated +5 V, connect the positive lead to pin 3. Pin 4, in both cases, is connected to ground.

Note on Power Supplies for MagTroll-11

If an unregulated voltage is used, make sure a jumper is placed at JP1. Because of the large amount of current used by the sensor assembly, make sure a good heat sink is used for MagTroll's voltage regulator IC. It probably would be smarter here to power the system with a separate +5-V regulated supply capable of delivering at least 250 mA. Remember, if you are using a +5-V power source, do not place a jumper at JP1 and connect +5 V to pin 3 of J1. Also, with a +5-V source, the reference voltage IC (U3) cannot be used, so place a jumper between pins 2 and 3 of JP4. Since we are only interested in relative sensor readings, an accurate voltage reference is not absolutely necessary.

SOFTWARE AND MAGTROLL-11 SETUP

Keep in mind that this project, like all projects that make use of an MPU or MCU, is based on software residing somewhere in silicon. This software in silicon is usually referred to as *firmware,* although I sometimes call it *sandware.*

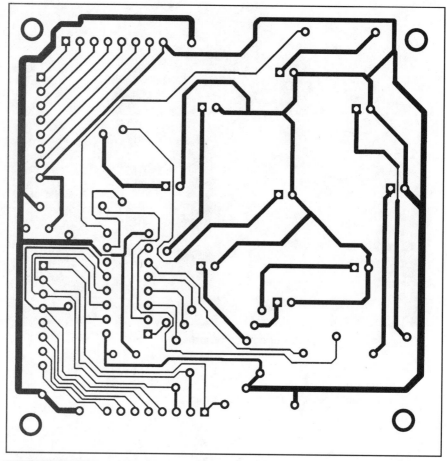

FIGURE 11-19 Foil pattern for display board.

If you are using MagTroll-11 as the controller board, you *must* use an EEPROM-programmed HCx11. You can program the HCx11 yourself with the .S19 code given in the CD-ROM. This disk also contains complete source code as well as other information. As indicated in Appendix C, a preprogrammed MC68HC11A1FN is available at a reasonable price. This is a good deal because you can get the wind vane up and running quickly and easily and still be able to modify the sandware thousands of times.

If you will be using the display board described in this chapter, *you must operate* MagTroll-11 in the single-chip or bootstrap mode. A jumper must be placed at JP6.

If you are using an MC68HC11A1FN, you must set up MagTroll-11 for bootstrap mode. Here, place a jumper between pins 2 and 3 of JP5. You also must place a jumper at JP2 for an automatic jump to the internal EEPROM. Your program must start at $B600 and extend only to $B7FF, at its absolute maximum.

If you are using the MC68HC811E2FN, place a jumper between pins 1 and 2 of JP5 and do not use a jumper at JP2. Here, your program should start at $F800, and the last 2 bytes of memory space, at $FFFE and $FFFF, should contain $F8 and $00, respectively.

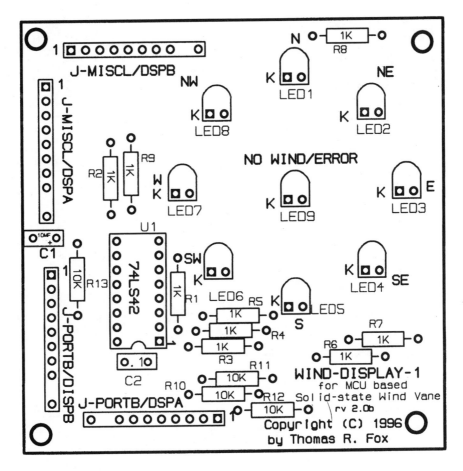

FIGURE 11-20 Component mounting guide for display board.

This will cause an automatic jump to $F800 at reset. Also, store $F800 at the address of the true illegal opcode trap vector at $FFF8 and $FFF9.

Since there is little need for RAM backup for the solid-state wind vane, MagTroll-11's C7 supercapacitor can be changed to a 0.1-µF capacitor.

INITIAL TESTS

Before installing ICs in sockets, connect power to the MagTroll-11, and connect cables to J-MISCL connectors on the opamp and display boards. Check voltages at all IC power pins. For instance, with the voltmeter's negative lead connected to a circuit ground, check the voltage at pin 4 of the opamp's U1. It should show between 4.75 and 5.25 V.

If all voltages appear correct, insert all ICs in their respective sockets. Use care to avoid static damage. With normal humidity, touching a good ground before touching an IC and working slowly are sufficient precautions.

FIGURE 11-21 Photo of display board.

TRYING IT OUT AND CALIBRATING IT

Set jumpers on the MagTroll-11 as described earlier (see Fig. 11-22). Make sure you have installed a cable between the sensor assembly and the opamp board. Power up the system. The center LED should light for about a second. (This may be longer than you would expect because it takes nearly $1/2$ s to register the inputs of 8 sensors 30 times each and to do the necessary data manipulation.) Next, one of the direction LEDs should light at random.

Place a box or bucket over the sensor assembly (to eliminate stray wind currents) and wait at least 5 minutes for the sensors to stabilize. Now using at least a $3^1/_2$-digit digital voltmeter, measure the voltage, with respect to ground, at TP1. If it is not between 2.47 and 2.51 V, adjust R25 until it is. Next, measure the voltage at pin 1 of the opamp board's J-PORTE connector. This will be the reference voltage, so jot it down temporarily. Let's call this V1. Now, measure the voltage at pins 2, 3, 4, 5, 6, 7, and 8 of connector J-PORTE and adjust the respective pots so that these voltages are within 0.01 V of V1 (see Table 11-1). After checking over your adjustments, you may want to place a dab of an easily removable glue on the pot's screw adjustment to keep vibration from changing your adjustment.

FINAL TESTS

Use a fan to determine if the unit is working properly. Place the fan about 3 ft from the sensor assembly. Make sure the assembly is not sitting on the floor or near a piece of furniture—it should be out in the open. Try blowing air at the sensor from several directions to make sure it is working properly. It usually takes about 5 s for the display to respond to

(A)

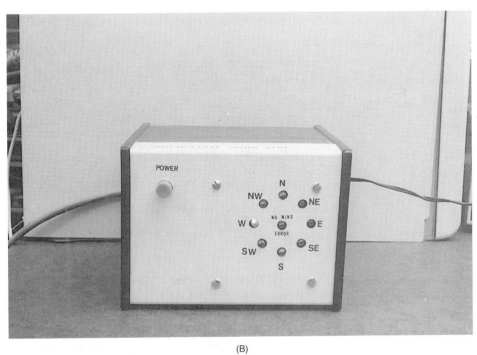

(B)

FIGURE 11-22 Photo of completed wind vane.

(C)

FIGURE 11-22 (*Continued*) Photo of completed wind vane.

TABLE 11-1 OPAMP BOARD POTENTIOMETER ADJUSTMENTS

PIN NUMBER OF J-PORTE	ADJUST POT WITH PART NO.
1	Reference pin—no adjust possible
2	R58
3	R9
4	R75
5	R43
6	R99
7	R39
8	R96

changes of direction. One pleasant, though surprising finding: While the sensors take about 5 minutes to stabilize after being turned on, the solid-state wind vane registers accurately within just a few seconds after being turned on. Thus there is no need to leave the unit on continually, although it draws less than 1 W of current. This adds up to less than $1 per year even if your electricity rates are obscenely high.

RECALIBRATION

Because of aging of components, you may want to readjust the potentiometers after several months of use. Since the sensor assembly is probably in an awkward place to cover with a box, simply do the readjustment before dawn on a day the wind is supposed to be light and variable.

Coming Attractions

Chapter 12 will describe in detail a low-cost "voodoo designed" version of the solid-state wind vane. Instead of eight sensors, this version only uses four, and yet the display shows eight directions. This seeming magic is possible with the help of "intelligent" sandware (firmware, software, or whatever you wish to call it). In addition, hints are given on how to turn this same system into one that includes wind speed, wind chill, and temperature with little modification and, most impressively, little extra cost. Then I will describe the best locations for the sensor assembly and present my view of the future of MCUs.

12

A "VOODOO-DESIGNED" WIND VANE

A CHEAPER MCU-BASED SOLID-STATE WIND VANE

In the last chapter we examined, in detail, the design and construction of a solid-state wind vane that made use of the MagTroll-11. This solid-state wind vane works well. However, it is fairly costly because it uses eight moderately expensive sensors, i.e., National Semiconductor's LM235H.

Since many designers now go along with the saying, "It ain't designed right until its costs have dropped out of sight," we next look at ways to save that green stuff. I already described how using cheaper LM335Z sensors could lower the cost. However, there is a more ingenious way to reduce costs a whole bunch. This method concretely demonstrates the advantage of MCUs. Why not just use *four* sensors instead of *eight,* one each for North, South, East, and West? Before you ridicule this approach and accuse it of bordering on "voodoo design," think about how a color TV or a color monitor can create just about every color there is and yet uses only three different colored phosphors: red, green, and blue.

While there are differences to this analogy with color TV, there are surprising similarities, too. For instance, while no phosphor dots in a picture tube glow yellow when struck by the electron beam, yellow is created by striking both the red and green dots (in a cluster) with the same intensity beam. Now, while no sensor (in the four-sensor wind vane) will pick up a northeast wind directly, what do you suspect would happen theoretically if the wind were exactly northeast and perfectly steady? It is quite obvious that it would appear to the MCU that the wind was blowing from the north and from the east simultaneously. The MCU could then draw the rightful conclusion that the wind was actually from the northeast.

There is another way of looking at this. While the MCU knows for sure that the wind is not from south or west, it cannot tell if it is from the north or east. All one needs is the proper sandware (my flippant name for firmware/software) so that the northeast LED lights when it detects this condition.

An astute observer may notice a slight problem here. If everything is just "perfect," the northeast LED will only light when the wind is exactly midway between north and east. If the wind is slightly to the north of true northeast, the display will show north instead of northeast. Ditto if it is slightly north of exact northwest. Actually, this problem is not as bad as you might first think: Even "steady" real-world winds and voltages vary with time. Regardless, the unit can be improved with ingeniously designed sandware.

A Solid-State Wind Vane with Only Four Sensors

This project truly demonstrates the value of MCUs in product design. With a little creativity, it is possible to produce sandware that will run an inexpensive four-sensor wind vane that will compete with the eight-sensor wind vane described in Chap. 11. Let's not have unrealistic expectations here. The eight-sensor wind vane is inherently more accurate. There isn't much we can do about that. Nonetheless, we can successfully "emulate" an eight-sensor wind vane with a four-sensor one. The fact that, almost invariably, emulations are not as great as the original cannot be denied. The four-sensor wind vane is no exception to this general rule. However, it is good enough for most uses.

BASIC SOFTWARE DESIGN

There are many ways to go about the software design here. I chose to make use of one basic property of statistical analysis: use of the simple mean. As an example, say we divide possible wind directions into 40 different equal numbers. (More commonly, wind directions are divided into 36 equal numbers, each of which is a multiple of 10°—360° total. For example, 90° is an east wind and 40° is a northeast wind.) Since we are only interested in eight directions, we make the following number assignment: North is set equal to 0, Northeast = 5, East = 10, Southeast = 15, South = 20, Southwest = 25, West = 30, and Northwest = 35.

As an illustrative example, say we take two readings from the four sensors. First, the sensors report that the wind is from the east, so our program sets a variable, such as AddEmUp, equal to 10. The following report from the sensor is a south wind, so our program adds 20 to AddEmUp (recall South = 20). Next, the program finds the mean by dividing AddEmUp by 2. Now we have $(10 + 20)/2 = 15$, which we can store at AvgDir. The sandware looks at this 15 in AvgDir and reports to the display that the wind is from the southeast, and the Southeast LED lights. It all seems simple enough.

Too bad things are not as elementary as just explained. What about the northwest direction? What if the wind switches between the North and West sensors? Well, you have $(30 + 0)/2 = 15$, which implies a southeast wind—*not* a northwest one. What happened? Well, we encountered what mathematicians like to call a *discontinuity*. This discontinuity is located at North, where we jumped from 40 to 0.

Have you ever noticed that our measurement of times has a discontinuity? If you doubt this, think about what happens 1 minute after 12:59? Or with 24-hour clocks, what about

1 minute past 23:59? Getting back to our problem with a four-sensor wind vane, you may think that the problem can be solved quickly by setting North = 40 rather than North = 0. We then have (30 + 40)/2 = 35, or northwest. But wait! Hold it! What happens when the wind is from the northeast? We have (10 + 40)/2 = 25 or southwest. Oops! We seem to have a problem. What we did is just move this discontinuity a trifle, but it's still there.

One solution here is to shift the discontinuity back and forth a tiny bit so that we never encounter it. You can look at it as if you assume that the wind is never directly out of the north. It is either a trifle northeast, and thus you set North = 0, or a trifle to the northwest, and thus you set North = 40.

Sneaky, but simple! Not so fast! How do you know which way the wind is coming from before making your calculations? Is not our goal to *find* the direction of the wind? Well, if the first time the sensors measure the wind it is from the east and the next time it is from the north, you can assume that the wind is at least slightly from the northeast. Ditto with the northwest direction. What you do in the program is add a decision loop that sets North = 0 if the last reading was from the east and North = 40 if the last reading was from the west.

The flowchart in Fig. 12-1 describes these things in a more formal way. Also refer to Fig. 12-2, which gives the graphical direction guide for this flowchart. Hopefully, Fig. 12-2 is self-explanatory.

Note that in the preceding example only two readings were taken before we decided which way the wind is blowing. In practice, it is wise to take more than two readings. If AddEmUp is a single byte of storage, we can take up to six readings without any problem. If we took seven readings, we could have an overflow, since $7 \times 40 = 280$, which is a larger number than will fit in 1 byte. For more readings, we would need a word storage space (2 bytes). This is probably unnecessary.

I feel that the sandware for a four-sensor wind vane is a work in process. In other words, while it works, it can stand some improvement. This revelation should stimulate those brilliant minds out there to come up with superior software for a new and improved four-sensor wind vane.

CIRCUIT DESCRIPTION

The display board for the four-sensor solid-state wind vane is identical to that for the eight-sensor wind vane described in Chap. 11. The sensor board for the four-sensor wind vane is similar to that for the eight-sensor wind vane except that it has only four sensors. Sounds logical, doesn't it?

The schematic for the four-sensor opamp board is provided on the CD-ROM. Run the SCHEMAT program to print it out. It, too, is nearly identical to the eight-sensor opamp board described in Chap. 11. The only substantive difference is that it uses only two LMC660CN quad CMOS opamp ICs, while the eight-sensor board uses four. Refer to Chap. 11 for detailed circuit descriptions concerning the display, opamp, and sensor boards.

CONSTRUCTION OF THE SENSOR ASSEMBLY

Unlike the sensor assembly in the preceding chapter, which consisted of four separate boards, this sensor assembly consists of a single PC board. This provides additional substantial cost savings. Refer to the foil pattern in Fig. 12-3 and the component mounting guide in Fig. 12-4. Pay particular notice that the North and South sensors are soldered to

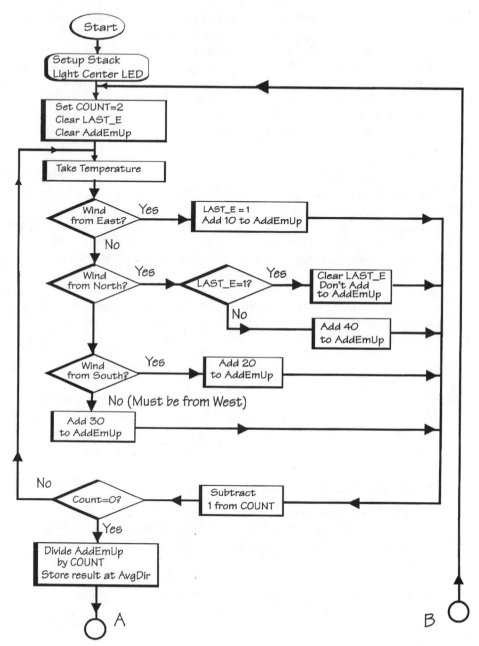

FIGURE 12-1 Flowchart for four-sensor solid-state wind vane.

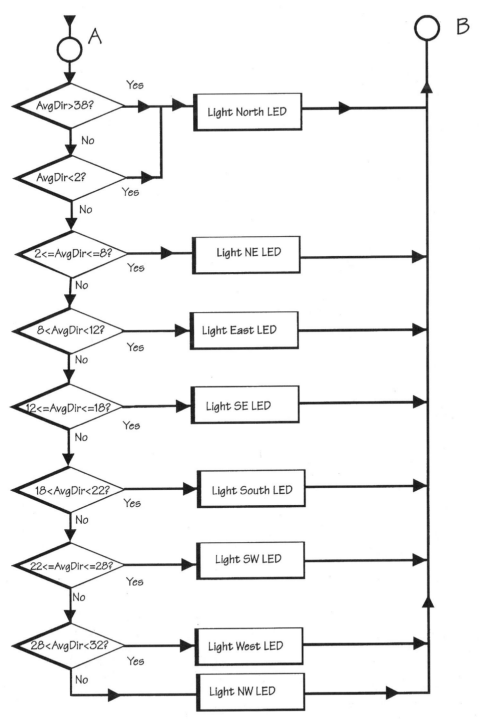

FIGURE 12-1 (*Continued*) Flowchart for four-sensor solid-state wind vane.

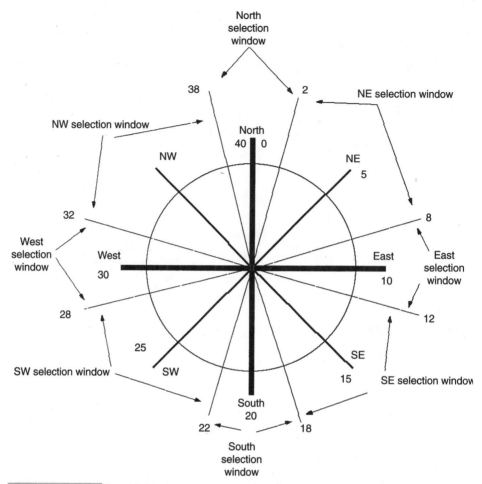

FIGURE 12-2 Graphical direction guide showing software's selection windows based on flowchart in Fig. 12-1. *Note:* The primary direction selection windows are smaller than the secondary selection windows.

the bottom (solder side) of the board. As mentioned in the preceding chapter, use an alligator clip for a heat sink when soldering sensors to the board. Also work as quickly as possible to avoid heat damage to the sensors.

The patterns for the windbreaks are given in Figs. 12-5 and 12-6. The prototype used 28-gauge sheet aluminum, which is available at some building supply stores. Notice the dotted lines on the patterns. Bend the sheet aluminum together along the dotted line. Bend the four small tabs outward. They are used to mount the windbreak to the circuit boards. A $9/_{64}$-in hole is drilled in each tab as shown. Then $1/_4$-in 6-32 hardware is used to mount the windbreak to the circuit boards. The two windbreaks are slightly different, so mount the correct windbreak to each side of the board. Notice that the East-West windbreak is mounted to the top (component side) of the board, and the North-South windbreak is mounted to the bottom (solder side) of the board.

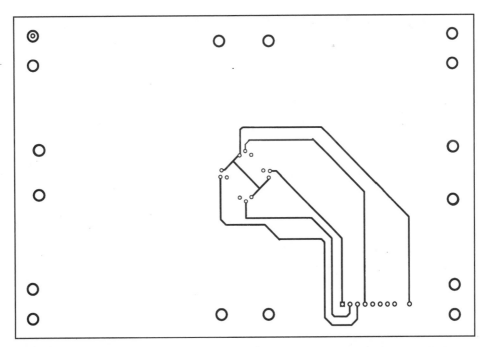

FIGURE 12-3 Foil pattern for sensor board for four-sensor solid-state wind vane.

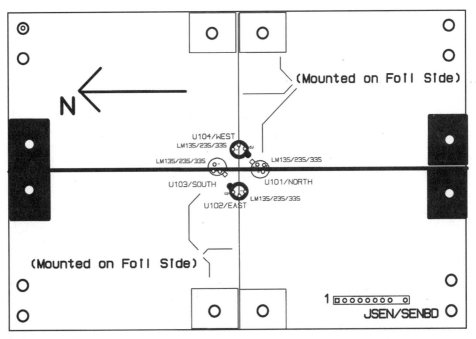

FIGURE 12-4 Component mounting guide for sensor board for four-sensor solid-state wind vane.

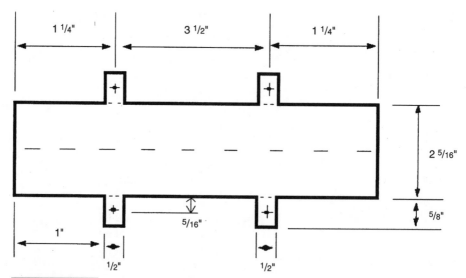

FIGURE 12-5 Construction diagram for North-South windbreak for four-sensor solid-state wind vane.

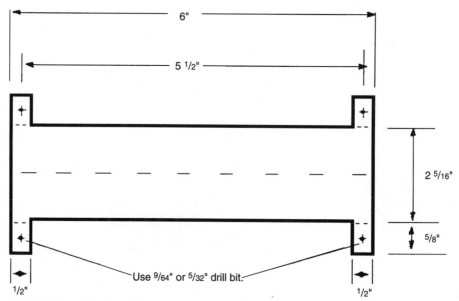

FIGURE 12-6 Construction diagram for East-West windbreak for four-sensor solid-state wind vane.

If there is a chance that the aluminum windbreaks will short out any foil traces on the circuit board, place insulating material over the board before mounting the windbreak to the circuit board. Electrical tape can be used here.

Figure 12-7 does not show the required top horizontal plate. (I show only the bottom plate in the photo.) Here, use spacers to mount an old circuit board or similar material horizontally so that it touches the top of the top windbreak. This plate is necessary so that the wind does not sneak over the top and cool off the sensor on the leeward side of the windbreak. Of course, as mentioned in the preceding chapter, we need a roof to protect the sensor assembly from rain, snow, and especially the sun.

OPAMP BOARD CONSTRUCTION

Here, use the solder-side foil pattern given in Fig. 12-8, the component-side foil pattern given in Fig. 12-9, and the component mounting guide in Fig. 12-10. Notice its relative simplicity compared with the opamp board given in the preceding chapter.

Unlike the board described in the preceding chapter, this PC board is not available from Magicland. However, the PC board given for the eight-sensor wind vane can still be used here. Basically, you only insert half the components listed for the "big" board. Details on using the eight-sensor opamp board in a four-sensor wind vane will be given to those purchasers of the board who request it. See Appendix C.

TESTING IT OUT

Refer to the preceding chapter for details on setting up MagTroll-11 and making connecting cables. Make sure that the MagTroll-11 has installed in it an HCx11 that has its

FIGURE 12-7 Photo of completed sensor assembly for four-sensor solid-state wind vane.

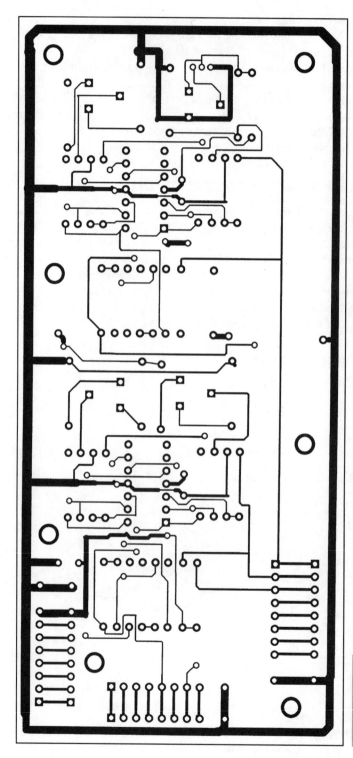

FIGURE 12-8 Solder-side foil pattern for opamp board for four-sensor solid-state wind vane.

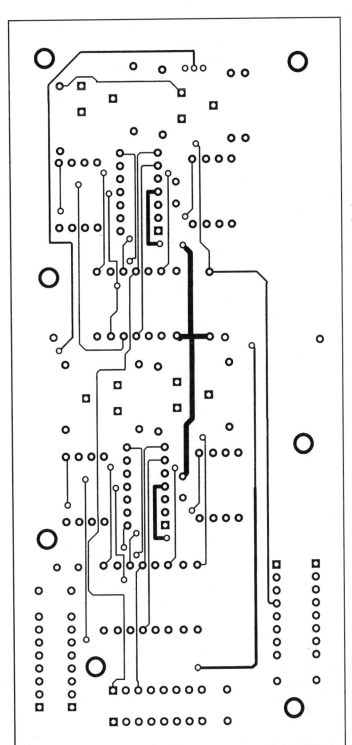

FIGURE 12-9 Component-side foil pattern for opamp board for four-sensor solid-state wind vane.

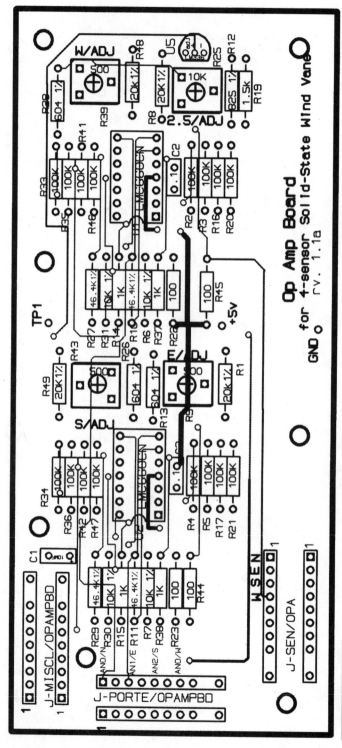

FIGURE 12-10 Component mounting guide for opamp board for four-sensor solid-state wind vane.

EEPROM programmed with appropriate sandware. The source code, assembler listing, and MOT S19 code are included on the CD-ROM. Observe that unlike the .S19 file for the eight-sensor wind vane described in the preceding chapter, there is substantial room in even a 512-byte EEPROM to add additional code.

Since only four sensors are used, the unit does not need as much power as the eight-sensor wind vane. Because of this, you can relax the restrictions suggested in the preceding chapter concerning the power requirements. Refer to Chaps. 10 or 11 concerning power supply connections.

Connect up cables between the MagTroll-11, opamp board, display board, and sensor assembly. Also connect power to the MagTroll-11. Apply power. With WVANE_4A software, the center LED should light for about a second. Next, one of the direction LEDs will light at random. If everything so far appears OK, place a box over the sensor assembly to eliminate stray air currents. Wait 5 minutes before calibration.

Calibration

To calibrate this circuit, you will require a $3\frac{1}{2}$-digit or better digital voltmeter. Connect its ground (black) lead to a circuit board ground. Connect its positive lead to TP1. If the voltage is not between 2.47 and 2.51 V, adjust R25 so that it is. Measure the voltage at pin 1 of the J-PORTE, and jot it down. Next, check pin 3 and adjust R9 so that the voltage is within 0.01 V. Finally, check pin 7 and adjust R39.

TRYING IT OUT

Take the box off the sensor assembly. Use a fan to check out the wind vane's operation. First, check the four main directions—north, east, south, and west. Then check out the other directions: northwest, southeast, southwest, and northeast. With the use of a steady fan, along with WVANE_4A software, you may notice that the display will have a slight tendency to favor the four primary directions over the secondary directions. However, in the real world with shifting winds, the four-sensor wind vane performs better than controlled tests would suggest. Still, there is room for software improvement here. Again, an honest challenge goes out to you to write a superior software routine, a routine that will allow the four-sensor "voodoo designed" wind vane to be practically indistinguishable in operation from its more expensive and conventionally designed cousin—the eight-sensor wind vane.

Location for Sensor Assembly for Both Four- and Eight-Sensor Wind Vanes

The sensor assembly *must* be mounted on a 100-ft tower *at least* 500 ft away from a building or other obstruction. Such a location should only cost a few thousand bucks, but you must add a hundred or so dollars for the 1000-ft cable. Seriously, this description is nearly ideal, but not practical. *The minimum requirement for effective installation of the sensor assembly is a mounting location at least 5 ft off the ground and at least 20 ft from any obstructions.* More ideally, it should be mounted a foot above the highest point of the roof of the building that houses the MCU/display assembly. To eliminate excessively long cables, locate the MCU/display assembly relatively close to a vertical line that intersects the sensor assembly.

A Look at an MCU-Based Device's Versatility—Adding Wind Chill Capability to the Solid-State Wind Vane

If you use an HC811 MCU, with its 2-Kb EEPROM, you can add other features including wind chill measurement, wind speed, and temperature. The exact details are left to you, but let us look at the basic theory and fundamental design. Since the design of such a multipurpose instrument is considerably simpler if a four-sensor wind vane is used, we will restrict our discussion here. But before we "drain our brain" with theory, let's start on the practical side and look at a display board that can be used in such a multipurpose instrument.

THE SCHEMATIC FOR A TEMPERATURE, WIND CHILL, AND WIND SPEED DISPLAY BOARD

One possible schematic for a display that can be used for wind speed, wind chill, and temperature is provided on the CD-ROM. Again, use the SCHEMAT program to print it out. Here, it is assumed that port C is used by MagTroll-11 to output a two-digit BCD (binary-coded decimal) number. Keep in mind that after reset, port C is configured by the HCx11 to function as an input port. To configure port C as an output port, all bits of the DDRC register (data direction register for port C) must be set. Assuming that the equates in the WVANE_4.ASM listing (see CD-ROM) were used in the source code, the following program code segment accomplishes this:

```
. . .
PSHA
LDAA #%1111·1111
STAA DDRC
PULA
. . .
```

U5's purpose here is as a buffer between the MCU and the 7447 TTL ICs U4 and U6. These 7447 ICs are BCD-to-seven-segment decoders/drivers whose outputs are connected to MAN-7 or similar common-anode seven-segment LED displays. The combination of 7447s and MAN-7s can be considered "classic" design. With this combination, the display will show the BCD number that appears at the 7447's input. For instance, with a 0011 binary input (which is 3 in decimal and also hexadecimal), the display will show "3." It does this by lighting display segments a, b, c, d, and g.

Since this is a multipurpose display (it is capable of showing wind speed, wind chill, and temperature), it must have a way of indicating what the display is showing at any instant. Much of this logic is done by the 74HC138 three-line to eight-line decoder U3. (Notice the "logical" designation number—138—for this superuseful IC.) The two LSB inputs of U3 are connected to bits 3 and 4 of port A. U3's other input is connected to ground and thus is permanently at 0. The first three outputs of U3 are connected to discrete LEDs. The remaining outputs are not connected. If you follow through with the logic, you can see that by storing xxx00xxx at PORTA will cause LED3 (the wind chill LED) to light. Note here that "x" stands for "don't care." Thus 11100111 will cause LED3 to light; so will 00000000. Be alert to the fact that only one of these LEDs, connected to the 74HC138, can be made to light at any given instant—but that's all we want here.

The combination of DIS 2 and DIS 3 only displays a positive number from 00 to 99. Now what about a negative number, and how about temperatures above 99°F? The display that does the job here is DIS 1. The logic/decoder circuit that drives DIS 1 is straightforward. It consists of the 74HC00 along with the intelligence of the HCx11.

Notice that the 74HC00 NAND gates here have two purposes. They function as inverters (so a "1" in a bit of port A will cause a segment to light), and they provide more driving current than port A can do all by itself. To display a "−" we light display segment g by storing xx1xxxxx at PORTA. (This set bit causes inverter U1:C's output to be driven low, which allows current to flow through segment g, lighting up the "−" on the display.) The hundred's digit (the leading 1) is lighted in a similar manner. Notice that unlike LED1 through LED3, the "−" and "1" are not mutually exclusive—i.e., they can both be lighted simultaneously. Storing x11xxxxx at PORTA accomplishes this.

Now that you have some idea of how to go about displaying the information, let's take a closer look at how we can obtain the data to be displayed. But let's first look at the wind chill factor.

BASICS ABOUT MEASURING WIND CHILL

The MCU-based solid-state wind vane already has the sensor needed to measure the wind chill factor—actually it has more sensors than we really need here. "What!" You might ask rather vociferously. "Don't you first need the wind speed and the temperature and then look it up somewhere?" True, this is one way to determine the wind chill. But there is a more basic method.

Let's briefly look at the principle behind the wind chill factor. As is apparent to everyone, wind makes one feel cooler. Why? One reason is that it increases the rate of evaporation of sweat. (Remember, the process of evaporation reduces temperature.) But that isn't what "wind chill" is really about, since there is little perspiration once the body is chilled, and with wind chill, we usually talk about chilly conditions. No, there is something else taking place here. The fact is that one's body has an internal 100°F heat source that warms the air immediately around the skin (within a few hundredths of an inch) to a temperature of around 90°F in calm air. This warm air layer is the reason we feel warm, under calm air conditions, when the air temperature is only 75°F—although our bodies are much warmer. When a little breeze comes along and blows some of this warm air away, it is replaced by cooler air, and we feel cooler. The more wind, the less warm air and the more cool air. The more cool air, the cooler we feel. Finally, with a wind speed more than 50 mi/h, the warm air layer has been completely blown away. This is why you do not feel any colder with a 70 mi/h wind than with a 50 mi/h wind—once the layer is gone, it is gone. A sensor that is heated internally mimics the human body. It, too, has heated air immediately surrounding it. It, too, is cooled when wind blows away this warmed air layer. Thus an internally heated sensor can be used to directly measure the wind chill factor. All you need do is calibrate it—this calibration can be done through the software. In an actual system based on the wind vane sensors, the software would first search for the sensor with the lowest temperature—which corresponds to the sensor that receives the most wind. The software would keep tabs of the sensor's data and then compare them with information from a built-in table. The software would then send the appropriate signals to the display board.

AIR TEMPERATURE

Amazingly, it is harder here to measure the air temperature than the wind chill. The reason is that all sensors in the wind vane are heated, and we are looking for an unheated sensor to measure just the air temperature—not the combination air temperature and temperature created by electric power dissipating in the sensor.

We obviously need another sensor—this time an unheated one. (An unheated sensor is actually an impossibility. What we are actually looking for is a minimally heated one. Do you think a usable unheated mercury thermometer is a possibility? If you answered yes, you are wrong, wrong, wrong! Why? You need light to see it, don't you? What does light do when it hits the thermometer? For those who are really interested here, consult a good book about Heisenberg's uncertainty principle.)

As mentioned earlier, the discussion here concerns the four-sensor wind vane. Here, you still have four inputs to the HC811's A/D converter available. All you need do is add another temperature sensor. But be cautious here. Unlike those four wind direction temperature sensors, we wish to heat up this sensor as little as possible. This heating reduction is accomplished by increasing the sensor's current-limiting resistor to 2.2 kΩ. (With the heated sensors, it was 100 Ω.)

The schematic in Fig. 12-11 can be used to input temperature-measuring data to the HC811's internal A/D converter. Referring to this schematic, connect the output (unconnected end of R7) to port E's AN4 input (pin 2 of connector J-PORTE). Adjust R10 so that this output voltage is at 2.6 V when the sensor is at the room temperature of 77°F. When the circuit is adjusted appropriately, the thermometer should be accurate from about −50 to +195°F. Its range is actually greater—from −56 to +199°F. With a little massaging by the software, the display should make sense. Assuming the Fahrenheit system of temperature measurement, the software should be designed so that when the A/D converter puts a 00 into the corresponding result register, the display should show −56; when the result register contains $85 (133 decimal), the display should show 77, and with a $FF (255 decimal) in the result register, the display should show 199.

WIND SPEED

Other than its obvious use as a temperature-measuring sensor, what do you do with the data gathered by this unheated sensor? We use it to find the wind speed. While the details are left to you, let's look at the rough plan of attack.

A complicated formula is available that is used to calculate the wind chill once the wind speed and temperature are known. Well, you do not have to be a math whiz to figure out that if you can calculate the wind chill from the wind speed and temperature, you certainly

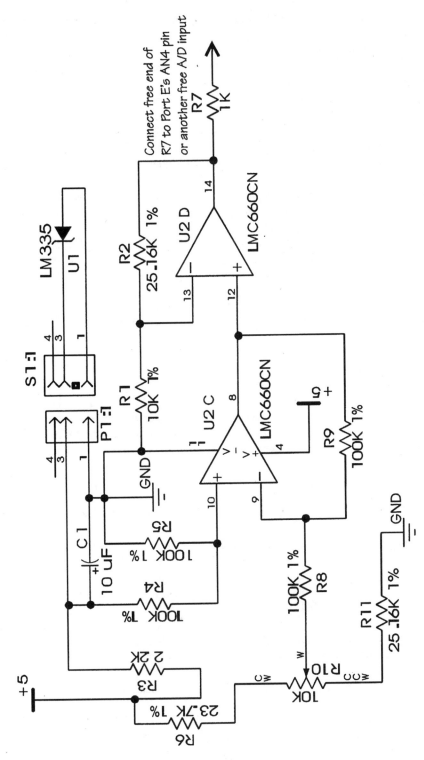

FIGURE 12-11 Schematic showing inputting temperature data for HC811's internal A/D converter.

can figure out the wind speed by knowing the wind chill and temperature. If you can do it, an MCU can do it better and (especially) faster.

If math formulas leave you blank, there is another, less sophisticated way to achieve the same thing. Simply write a software lookup table routine that uses the wind chill chart to look up the wind speed if you know the wind chill and temperature. Try this out on a wind chill chart. It's fun! For instance, referring to the chart in Fig.12-12, say the air temperature is 20°F and the wind chill is −20°F. You can see that the wind speed must be 35 mi/h. Designing software to accomplish this for most wind speeds is not difficult. The software should use interpolation techniques for wind speeds not listed in the chart.

Notice that a heated sensor anemometer has the inherent physical limitation of only recording wind speeds up to about 50 mi/h—once all the heat from a sensor is blown away, there is nothing more to blow away. This fact is obvious by a quick look at any wind chill chart.

While relatively few of you actually will bother writing software to display wind chill, wind speed, or even temperature, the preceding discussion clearly demonstrates the fantastic features of MCUs—their ability to do many amazing things with relatively simple modifications in the software.

Complications with an Eight-Sensor Wind Vane

Everything flows pretty neatly with the four-sensor wind vane. However, the eight-sensor wind vane is already using all eight channels of the HC811's internal A/D converter. If you wish to expand it to measure wind speed and temperature, you will need another unheated sensor connected to an external A/D converter. However, to use an external A/D converter, you will need an 8-bit input port. Here, the HC11's SPI can be used. Refer to Chap. 7 for details.

Using MagTroll-11 in Other Designs

The preceding section has tossed out a few ideas on expanding the solid-state wind vane. But MagTroll-11 and the HCx11 can be used in a vast array of projects far different from the one described here. Of course, as mentioned earlier, the HCx11 was designed originally for use in automotive electronics, but this, too, is just a small part of its capabilities.

As indicated, the MCU-based solid-state wind vane was designed originally as a demonstration project for the MagTroll-11. MagTroll-11 can form the heart (and brains) of a vast number of other projects. While the solid-state wind vane made use of only a small number of HCx11 features—most notably its eight-channel 8-bit A/D converter and 8-bit port B output port (the expanded version also used its C and D ports)—other projects may use many more of its features. A few of these features, such as the SCI and COP watchdog system, were described in Chaps. 5, 6, and 9.

I have completed my discussion of the garden varieties of the HCx11—the MC68HC11A1P, MC68HC11A1FN, and MC68HC811E2FN. The final chapter will look at the MC68HC11A1's sisters, brothers, and a few of its cousins.

Both temperature and wind cause heat loss from body surfaces. A combination of cold and wind makes a body feel colder than the actual temperature. The table below shows, for example, that a temperature of 20°F, plus a wind of 20 miles per hour, causes a body heat loss equal to that of minus 10 degrees with no wind. In other words, the wind makes 20°F feel like −10°F.

Top line of figures shows actual temperatures in degrees Fahrenheit. Column at left shows wind speeds.

MI/H	35	30	25	20	15	10	5	0	-5	-10	-15	-20	-25	-30	-35	-40
5	33	27	21	16	12	7	0	-5	-10	-15	-21	-26	-31	-36	-42	-47
10	22	16	10	3	-3	-9	-15	-22	-27	-34	-40	-46	-52	-58	-64	-71
15	16	9	2	-5	-11	-18	-25	-31	-38	-45	-51	-58	-65	-72	-78	-85
20	12	4	-3	-10	-17	-24	-31	-39	-46	-53	-60	-67	-74	-81	-88	-95
25	8	1	-7	-15	-22	-29	-36	-44	-51	-59	-66	-74	-81	-88	-96	-103
30	6	-2	-10	-18	-25	-33	-41	-49	-56	-64	-71	-79	-86	-93	-101	-109
35	4	-4	-12	-20	-27	-35	-43	-52	-58	-67	-74	-82	-89	-97	-105	-113
40	3	-5	-13	-21	-29	-37	-45	-53	-60	-69	-76	-84	-92	-100	-107	-115
45	2	-6	-14	-22	-30	-38	-46	-54	-62	-70	-78	-85	-93	-102	-109	-117

FIGURE 12-12 Wind chill chart.

13

68HC11A1'S SISTERS, BROTHERS, AND COUSINS

The HC11—A Really Hot MCU

Simply by perusing advertisements in technical and general-interest electronics-related magazines, you can easily draw the rightful conclusion that the HC11 is a popular MCU. Right now, I believe that it is *the 8-bit MCU of choice for many designers.*

The HC11's popularity really is not that new. In fact, back in 1995, this chip was hard to get your hands on because of the unbelievable demand. I discovered, through various sources, that Motorola had suspended order taking for the chip during that period because the demand-supply ratio was obscenely large. One sales agent at a large distributor filled in the rest of the part number of this chip when, in early 1995, he was asked about price and availability. As the buyer started reading off the chip's part number—"MC68HC11…"—the salesperson supplied the rest—"A1FN." He then said: "I bet you've been trying for some time to get this bugger. Well, we got a few. There are 12, 14, 15, no, 17 in stock. I can send them out right away at $15 each." Luckily, the buyer found another supplier at a more reasonable price. This other supplier claimed to have 7000 in stock, but he expected to be cleaned out by the end of the day. Happily, this situation has changed, and most models of the HC11 are no longer hard to get your hands on.

Motorola's Pride of Its Family of Microcontrollers

You may be pleased to discover that Motorola considers the M68HC11 to be the pride of its family of microcontrollers. Apparently, as of mid-1998, Motorola had sold more than 400 million units of the HC11. You would need a huge kid's sandbox to hold all the sand required to make all those chips.

This is all good news! You have no need to fear that your time and effort spent in learning about the HC11 will be useless because Motorola might soon drop the HC11 from its product line. No way will they do that—at least for a long, long time. Also, the M68HC11 family offers architectural compatibility with the M68HC05 family. As you may know, the HC05 can be said to be the "grease monkey" MCU because it is found in so many cars and trucks.

In addition, the MC68HC11 is code-compatible with the M68HC16 family of 16-bit microcontrollers. This means that you can upgrade an HC11 system quickly and easily.

The Sisters and Brothers

This book used three different HC11s: the 48-pin DIP model from the A series (MC68HC11A1P), the 52-pin PLCC model also from the A series (MC68HC11A1FN), and the 52-pin PLCC model from the E-series and noted for its 2-kB EEPROM (MC68HC811E2FN). In other words, we looked at 3 members in 2 series of the HC11 family. As of this writing, the M68HC11 family is made up of more than 60 members in 10 major series. Still, the variety of HC11s is not limited to merely 60+ members! For instance, Motorola's latest price list shows more than 150 different part numbers for the HC11 family.

The Ten Major Series

The following is a list of the 10 major series that make up the HC11 family. I have taken a few descriptions here verbatim, with only minor modifications, from Motorola's own literature.

A series.   The original HC11 belongs to this series. Its features should be well known to you because both the MC68HC11A1P and the MC68HC11A1FN belong to this series. However, the latest Motorola literature mentions a nominal bus speed of 3 MHz. This conflicts with earlier statements, which listed a 2-MHz maximum bus speed. Since most projects in this book used a speed of 1 MHz, this claim of greater speed capability is not really that important to us.

C series.   The 68HC11C0 provides chip selects and memory expansion to 256 kB, 2 pulse width modulation (PWM), and four A/D channels. The multiplexed bus reduces pin count, improves control signals, and reduces chip count.

D series.   This is the "el cheapo" HC11. These chips have fewer peripherals and less memory than the A series. Motorola here seems to want to compete with its own HC05.

E series.   The MC68HC811E2FN belongs to this series. Do not jump to the conclusion that all members of the E series have 2048 bytes of EEPROM. In fact, some E-series members do not have a usable EEPROM at all. The "E" does not stand for "extrabig" EEPROM. However, the "8" shows that the chip does have an EEPROM four times as big as the original HC11. In most other respects, this series is nearly identical to the A series. By the way, when the "8" is replaced with a "7" in part numbers (e.g., MC68HC711), this means that the chip contains a large EPROM (up to 32 kB), which is normally used as a storage space for program instructions.

F series.   The request for high-speed expanded systems required the development of the 68HC11F1. This series has more I/O ports, more RAM, integral chip selects, and a 4-MHz nonmultiplexed bus.

G series.   This is the family to choose if you are into A/D systems bigtime. You get 10-bit A/D resolution plus the most sophisticated timer system in the family.

K series.   The 68HC11K4 offers high speed, large memories, memory management units (MMUs), PWMs, and lots and lots of I/O.

KA.   These variations of the K series offer high integration and large on-chip memories (up to 32 kB of ROM/EPROM) in lower pin-count packages.

L series.   This is the "low power" version of the HC11. It can be operated at frequencies down to dc. Some members of this series have 16 kB of ROM and an additional bidirectional port. Its multiplexed bus can operate up to 3 MHz.

M series.   These enhanced, high-performance MCUs are derived from the 68HC11K4 and include large memory modules, a 16-bit math coprocessor, and four channels of DMA.

P series.   The 68HC11P2 offers a power-saving programmable PLL-based clock circuit along with many I/O pins, large memory, and three SCI ports.

For a list of M68HC11 family devices, see Table 13-1. Also see Fig. 13-1, which shows the M68HC11 part number options. Motorola's Web site provides information on the latest model of the HC11 (see Appendix G). Specifically, the Web address www.mcu.mot-sps.com/lit/sel_guide/sg11mat.htm will provide you with the latest listing for M68HC11 and L68L11 devices.

Cousins of the HC11

Migrating to the HC12 seems natural for an HC11 devotee. This MCU has several features that generally are considered superior to the HC11. For instance, it can operate easily at 4 MHz; it has 1024 bytes of RAM and 4 kB bytes of EEPROM; on-chip memory mapping allows expansion to more than 5 MB of address space; and it has two SCIs and other nifty

TABLE 13-1   M68HC11 FAMILY DEVICES

DEVICE	RAM	ROM	EPROM	EEPROM	COMMENTS
MC68HC11A8	256	8K	0	512	16-bit timer; 8-channel 8-bit A/D, SCI, SPI
MC68HC11A7	256	8K	0	0	
MC68HC11A1	256	0	0	512	
MC68HC11A0	256	0	0	0	
MC68HC11D3	192	4K	0	0	16-bit timer; SCI, SPI
MC68HC711D3	192	0	4K	0	
MC68HC11D0	192	0	0	0	
MC68HC11ED0	512	0	0	0	16-bit timer; SCI, SPI
MC68HC11E9	512	12K	0	512	16-bit timer; SCI, SPI, 8-channel 8-bit A/D
MC68HC711E9	512	0	12K	512	
MC68HC11E8	512	12K	0	0	
MC68HC11E1	512	0	0	512	
MC68HC11E0	512	0	0	0	
MC68HC811E2	256	0	0	2048	16-bit timer; SCI, SPI, 8-channel 8-bit A/D, 2K EEPROM
MC68HC11E20	768	20K	0	512	16-bit timer; SCI, SPI, 8-channel 8-bit A/D, 20-kB ROM EPROM
MC68HC711E20	768	0	20K	512	
MC68HC11F1	1024	0	0	512	Nonmultiplexed bus, 8-channel 8-bit A/D, 4 chip selects, SCI, SPI
MC68HC11G7	512	24K	0	0	Nonmultiplexed bus, 8-channel 10-bit A/D, 4-channel PWM, SCI, SPI, 66 I/O pins
MC68HC11G5	512	16K	0	0	
MC68HC711G5	512	0	16K	0	
MC68HC11G0	512	0	0	0	
MC68HC11K4	768	24K	0	640	Nonmultiplexed bus, memory expansion to 1 MB, 8-channel 8-bit A/D, 4-channel PWM, 4 chip selects
MC68HC711K4	768	0	24K	640	
MC68HC11K3	768	24K	0	0	
MC68HC11K1	768	0	0	640	
MC68HC11K0	768	0	0	0	

DEVICE	RAM	ROM	EPROM	EEPROM	COMMENTS
MC68HC11KA4	768	24K	0	640	Nonmultiplexed bus, 8-channel 8-bit A/D, SCI, SPI, 4-channel PWM
MC68HC711KA4	768	0	24K	640	
MC68HC11KA2	1024	32K	0	640	
MC68HC711KA2	1024	0	32K	640	
MC68HC11L6	512	16K	0	512	Multiplexed bus, 16-bit timer; 8-channel 8-bit A/D, SCI, SPI
MC68HC711L6	512	0	16K	512	
MC68HC11L5	512	16K	0	0	
MC68HC11L1	512	0	0	512	
MC68HC11L0	512	0	0	0	
MC68HC11M2	1280	32K	0	640	Nonmultiplexed bus, 8-channel 8-bit A/D, 4-channel PWM, DMA, on-chip math coprocessor, SCI, 2 SPI
MC68HC711M2	1280	0	32K	640	
MC68HC11N4	768	24K	0	640	Nonmultiplexed bus, 12-channel 8-bit A/D, 2-channel 8-bit D/A, 6-channel PWM, on-chip math coprocessor, SCI, SPI
MC68HC711N4	768	0	24K	640	
MC68HC11P2	1024	32K	0	640	Nonmultiplexed bus, PLL, 8-channel 8-bit A/D, 4-channel PWM, 3 SCI (2 with MI bus), SPI, 62 I/O pins
MC68HC711P2	1024	0	32K	640	

TABLE 13-1   M68HC11 FAMILY DEVICES (CONTINUED)

features. Before you jump from 11 to 12, though, caution. The HC12 is not a snap to use. For instance, the HC12A4 comes in a 112-pin quad flat pack. This physical arrangement itself turns me off because it is a true challenge to solder by hand. Also, since the HC12 is so loaded with features, it is confusing to design with. You thought the HC11 was confusing! However, it is something to look forward to—assuming you need some or all of the goodies that come with it.

The newly famous 68HC16 is a 16-bit version of the HC11. Motorola claims that it is code-compatible with the HC11. In other words, it seems that programs written for the HC11 can be made to work on the HC16 with only minor changes. However, the converse is seldom true. The standard HC16 comes in a 100-pin package, which, like the HC12, is a challenge to use.

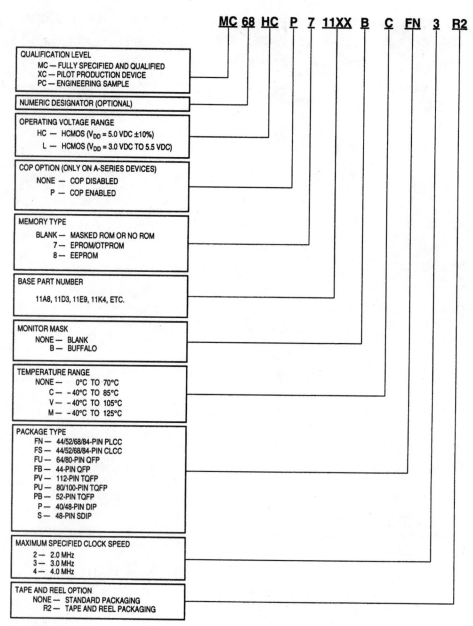

MC 68 HC P 7 11XX B C FN 3 R2

QUALIFICATION LEVEL
   MC — FULLY SPECIFIED AND QUALIFIED
   XC — PILOT PRODUCTION DEVICE
   PC — ENGINEERING SAMPLE

NUMERIC DESIGNATOR (OPTIONAL)

OPERATING VOLTAGE RANGE
   HC — HCMOS (V_{DD} = 5.0 VDC ±10%)
   L — HCMOS (V_{DD} = 3.0 VDC TO 5.5 VDC)

COP OPTION (ONLY ON A-SERIES DEVICES)
   NONE — COP DISABLED
   P — COP ENABLED

MEMORY TYPE
   BLANK — MASKED ROM OR NO ROM
   7 — EPROM/OTPROM
   8 — EEPROM

BASE PART NUMBER
   11A8, 11D3, 11E9, 11K4, ETC.

MONITOR MASK
   NONE — BLANK
   B — BUFFALO

TEMPERATURE RANGE
   NONE —   0°C TO 70°C
   C — –40°C TO 85°C
   V — –40°C TO 105°C
   M — –40°C TO 125°C

PACKAGE TYPE
   FN — 44/52/68/84-PIN PLCC
   FS — 44/52/68/84-PIN CLCC
   FU — 64/80-PIN QFP
   FB — 44-PIN QFP
   PV — 112-PIN TQFP
   PU — 80/100-PIN TQFP
   PB — 52-PIN TQFP
   P — 40/48-PIN DIP
   S — 48-PIN SDIP

MAXIMUM SPECIFIED CLOCK SPEED
   2 — 2.0 MHz
   3 — 3.0 MHz
   4 — 4.0 MHz

TAPE AND REEL OPTION
   NONE — STANDARD PACKAGING
   R2 — TAPE AND REEL PACKAGING

FIGURE 13-1 The meaning of part numbers.

To be honest, I believe that you should put in a few years of snuggling up with the HC11 before you even consider divorcing it for the HC12 and/or HC16.

I cannot end this before mentioning the HC705. Overall, the HC05 series can be looked on as a low-cost, stripped-down version of the HC11. Another way of looking at it is that it is Motorola's answer to Microchip's PIC series of microcontrollers. By the way, the book you are reading is the third in a series on microcontrollers. The first book in the series, *Programming and Customizing the PIC Microcontroller,* was written by Myke Predko. I recommend that you take a serious look at Myke's book, even if you are sure your needs are already met with the HC11. Appreciating other ways of solving similar problems is always wise.

Unlike the HC11, where the first chip of the series is still the most widely used, with the HC05 it seems that the EPROM version, the HC705, is the most popular. One member of the HC05 family, the MC68HC705K1, has the following features:

- 504 bytes of EPROM, including 8 user vector locations
- 32 bytes of user RAM
- 64-bit personality EPROM
- Built-in crystal/resonator/RC clock—no minimum clock speed
- 8-mA sink capability on four I/O pins (the HC11 only has a 1.6-mA sink capability)
- COP system and illegal opcode trap
- A 16-pin DIP model (This is really neat!)
- Multifunction timer and RTI
- External interrupt capability on four I/O pins

While its feature list is impressive, let's face it, there are really three facets of the HC05 series that have made it so popular. They are

1. Low, low cost—some versions are available for a buck and a half in quantities of one.
2. Small size and traditional DIP packaging.
3. Even easier than the HC11 to understand and use.

In other words, it is a kind chip. Kind on the wallet, kind on the hands, kind on the eyes, and kind on the brain.

Disadvantages? The HC05 is only suitable for relatively simple designs. It does not have enough RAM or peripherals to replace the HC11. Nevertheless, for basic designs—well, watch out HC11!

Don't Bother Designing an MCU-Based Blinking Christmas Tree Light!

Remember back in the first six chapters where we spent all that time blinking a LED? Well, we had a purpose there. This purpose was not to impress the wife and kids with a blinking LED. It will not do that. Trust me here! The purpose of the blinking LED was learning how to program and design with the HC11.

This brings up a point. There is such a thing as overkill when designing with MCUs. One example of overkill is to use an MCU in a simple thermostat that turns on a furnace when the temperature drops below the set temperature and shuts it off when the temperature rises 2 degrees above the set temperature. A simple bimetallic thermostat will do this job at far lower cost and more reliably than an MCU-based thermostat. However, if you want to produce an intelligent thermostat that has enough smarts to keep a building at an even temperature and knows which days are holidays so that it can regulate the heat on these days to a minimum, an MCU-based thermostat *may* be justified. However, using an HC16 here is overkill, since an HC11 or even an HC05 will do the job. Remember, in any product, design cost is a major contributor to final cost. This is especially true in products that will likely have relatively small production runs, such as an intelligent thermostat.

A Rosy Future for MCUs Overall

While for a variety of reasons the science fiction version of mobile consumer robotics is perhaps decades away, robots have been spreading seamlessly into everyday life for several years. Look at the electronic bread machine. This MCU-based device is a type of stationary robot. Besides baking a great loaf of bread, several models do many other functions. For instance, my wife uses one that bakes cakes, prepares doughnut dough, and even makes jam! (If you like baked goodies and you do not have a bread machine, you have got to get one yesterday, if not sooner!)

Just about every new electronic device is now MCU-controlled. Look at the ubiquitous VCR. I am looking at another MCU-controlled device as I hit keys—MAG Innovision's MAG-15MX video monitor. Not only does it automatically switch between video modes, but it also has a seven-segment LED display to let you know which mode it is in anytime. (The 17-in version uses an LCD as a secondary display to spell out the mode in plain English—or other language—and other information.) How about the HP-4L printer on which these words were first printed? While its ultrasimple 4-LED display/indicator leaves much to be desired, it, too, uses at least one MCU.

In my view, we have just begun to make use of the MCU. Right now I have thoughts of nearly a dozen projects that would benefit enormously from the addition of MCU intelligence. I am not talking about MCU-controlled toasters here! The only problem is to figure out where to start? Maybe that's another great project, "an MCU-based where-to-start thingumajig"?

BINARY AND HEXADECIMAL NUMBERS: A DIFFERENT WAY OF LOOKING AT THEM

There is absolutely no doubt about it—you *must* completely assimilate the binary number system into your consciousness before it will be possible for you to really and truly understand microcontrollers and how to use them. Happily, the binary number system is as simple as a numbering system can get. Recall that it is a base-2 system. You cannot get any simpler than that. The only reason some people have a problem with it is that it is new and they never spent much time with it.

The following discussion of binary and hexadecimal numbers is different. If you are looking for a textbook description, please check out a number of other good books about computers. Here, however, I have a deep-seated need to take a different look.

Of course, in our quest of MCU knowledge we have come across the hexadecimal numbering system more often than the binary system. However, the *only reason the hexadecimal numbering system is used is because it is a shortcut way of writing binary.* I do not believe this dirty little fact has been emphasized enough in the literature. Because of this, you must be able to eat and sleep the binary system before the hexadecimal system starts to make real sense.

Before going into the binary or base-2 system, it is smart to look at the common garden variety of numbering system known as the *decimal* or *base-10 system.* To "introduce" this system, which is more complicated than the binary system by a factor of five, let's look at several questions from "typical" grade 3 and grade 4 math books.

Problem 1 Write 45,305 in *expanded* form.

Answer 40,000 + 5000 + 300 + 5

Problem 2 In the number 12,056, what place position is the 5? Also, what is the *value* of the 5? Value of the 2?

Answers The 5 is in the tens' position. This means that there are five tens in the number, and the value of the 5 is 50. The 2 is in the thousands' position, and its value is 2000. Remember, there is a definite difference between a digit's place position and its value.

Also, see Fig. A-1, which is a typical illustration from a grade 4 math textbook.

Why Decimal?

Why is our numbering system based on the number 10? Only an intelligent guess can be made here, and this guess has to do with our 10 fingers. If we had only 8 fingers, there is a good chance we would be using the octal (base-8) numbering system. By the way, the octal numbering system is still used by some in the computer field, although "octal users" are a vanishing breed.

Why Binary?

The reason for use of the binary numbering system when dealing with computers is pretty clear-cut. It has to do with electricity, on/off switches, lighted and unlighted light bulbs, relays, high/low voltage levels, and other stuff like that. The very first designers of electronic computers had a sort of tunnel vision and thought the decimal system was the only way to go. They used a set of 10 different *voltage levels* to represent the set of 10 different *symbols* in the base-10 (decimal) numbering system. Nonetheless, it was soon realized that computers could be simpler and more reliable by being designed with the base-2 (binary) numbering system.

To visualize the superiority of the binary system, think of a light bulb. From a *binary*

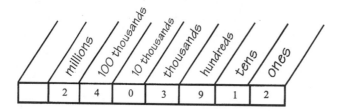

2,403,912

FIGURE A-1 A diagram indicating the meaning of decimal place value. Adapted from a third-grade math book.

perspective, all we care about is whether the bulb is lighted (1) or unlighted (0). Pretty simple, huh? From a decimal perspective, things get a lot more complicated. We then have unlighted (0), lighted so the filament just barely glows (1), lighted so the filament is turning orange (2), ..., lighted so the bulb is very bright but a tinge reddish (8), and finally, when the bulb is completely white hot (9). While this sounds almost ridiculous, it is basically the way decimal-based computers worked. It should be obvious why computers started being, and still are, designed using the binary system in mind. You might be wondering right now why did they ever bother with the decimal system. Well, the superiority of binary is obvious to us now, but at the time, the designers did not have the experience to see this. Experience is important. Trust me.

Practical Look at the Binary Numbering System

Now that you realize that designers chose the binary numbering system for down-to-earth, practical reasons and *not* to confuse those trying to learn about computers, you should feel more comfortable in your quest for knowledge.

Now let's go back to third and fourth grades. Look at Fig. A-2. This shows a problem, similar to problems found early in third grade math books. Complete the table. The first line of the table is done for you to give you an idea what is wanted.

Now we will jump back from base 10 to base 2. Fill in the table in Fig. A-3. Notice the similarity to Fig. A-2. Now, however, the numbers at the left are binary numbers. Here, the first three lines of the table are filled in for you.

The similarity is obvious here between the base-10 and base-2 numbering systems. Both decimal and binary numbering systems use the concept of "place value." Don't jump to the conclusion that this type of numbering system based on the traditional concept of place value is the only one on the block. Think about the Roman system. While in a roundabout way the Roman numbering system can be looked on as a base-10 system, it does not use the same concept of place value that the binary or decimal system uses.

Quick question   Translate the Roman number IX into decimal and binary.

Bytes, Nibbles, and Bits

Everyone has seen the word *byte* in print, although you may be more accustomed to seeing it printed like *BYTE,* which, of course, is the title of a famous monthly computer-oriented publication. Here, however, when I use the term *byte,* I am referring to an 8-bit binary number. Similarly, a *nibble* is simply half a byte, or a 4-bit binary number. By the way, the word *byte* is used approximately 100 times more often than the word *nibble.*

Now that I have defined *byte* and *nibble* in terms of a bit, let's look at a bit. What is it? Well, before I answer this, think about the term *figure* when referring to a six-figure salary. The terms *figure* and *bit* are analogous. The main difference is that *bit* is used

Show each numeral at the left in table form.

	THOUSANDS	HUNDREDS	TENS	ONES
2,403	2	4	0	3
42				
1002				
345				
9,999				
3,040				
7				
200				
1,234				

FIGURE A-2 **A table showing the significance of place value in the decimal system. Adapted from a third-grade math book.**

when talking about binary numbers and *figure* when discussing money in the decimal system. When discussing decimal numbers in general, the term *digit* is usually used in place of *figure*.

Now look at Fig. A-4. This shows the place values of an 8-bit binary number.

Adding Binary Numbers

Before we look at adding binary numbers, let's review adding decimal numbers. Of course, adding numbers such as 23 and 34 is no problem, since all we do is add the numbers in the ones' place and get 7 ones. Next, we add the numbers in the tens' place and get 5 tens. The answer, of course, is 57 (5 tens and 7 ones). This is simple addition, and it is the first type of two-digit addition problems taught to children, usually late in the first grade.

However, if we attempt to add numbers such as 78 and 56 together, things get a bit more complicated. This type of problem is taught to children late in the second grade. As before, we add the numbers in the one's place. Now, however, we get 14. The problem is what do we do with the 1. Well, this 1 stands for 1 ten, so we put this 1 in the tens' place and add it to the 7 in the tens' place to get 8 tens. We then add 8 tens and 5 tens to get 13 tens. Since 13 tens can be written as 1 hundred and 3 tens, we rewrite our answer to show this. Our answer is 1 hundred, 3 tens, and 4 ones, or 134 in standard form.

Now that we have reviewed adding decimal numbers, let's take a serious look at adding bina-

Show each binary number at the left in table form. The first three are filled in for you.

	EIGHTS	FOURS	TWOS	ONES
1111	1	1	1	1
1110	1	1	1	0
1101	1	1	0	1
1100				
1011				
1010				
1001				
1000				
0111				
0110				
0101				
0100				
0011				
0010				
0001				
0000				

FIGURE A-3 A table showing the significance of place value in the binary system.

ry numbers. Let's do an easy one first. What is 0101 + 1010? Of course, it is 1111. What we did is simply add the 1 and 0 together in the one's place and got 1, ditto for the twos', fours', and eights' positions. Now, what happens when we add 1001 and 1001 together? Let's concentrate on the ones' place because that's the toughie. Since we are only restricted to two symbols, 1 and 0, we cannot say $1 + 1 = 2$ because we do not allow 2 to represent the quantity 2. The problem here is analogous to the decimal problem when we added 78 and 56. In decimal, we use 10 symbols, but there is no symbol that represents $8 + 6$. Getting back to binary, since there is no symbol that represents 2, we have to use the concept of place value, and instead of putting down a 2 in the ones place, we put a 1 in the twos' place, so we get $1001 + 1001 = 1010$.

A personal aside When I went to elementary school, this operation of adding that little one to

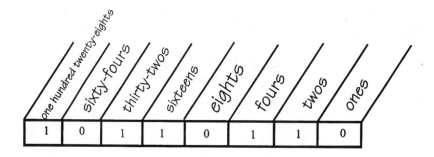

10110110

FIGURE A-4 A diagram indicating the meaning of binary place value.

the number in the next higher place was referred to as *carrying*. Today, this process is referred to as *renaming* in many grade school texts. Both terms, *carrying* and *renaming*, are quite descriptive. *Renaming*, however, is more logical. Nonetheless, I wish to express caution to those who think using logic and basic principles is the best method to use when introducing a new concept to children as well as adults. As an example, consider Ohm's law versus the theory of quantum electrodynamics (QED). Ohm's law is just that, a law. All circuit theory is based on this law. Despite its extreme usefulness, it has no basis in basic physics. Ohm's law makes no judgments and does not offer a hint as to what electrical force really is. Enter QED. This theory, which is said to be the most accurate theory in existence, supposedly explains what electrical force really is (i.e., exchange of virtual photons.) While QED really forms the basis of Ohm's law, electrical engineers do not make use of it (in fact, few have even heard of it) when designing circuits. The reason: It isn't necessary to use QED when designing real-world circuits—Ohm's law and its offspring are good enough. In other words, to an inquisitive electrical engineer (e.g., me), QED can only be looked on as a hobby that makes his or her work more intellectually satisfying. Similarly, the "new math" (e.g., where addition is explained by the joining of sets, etc.) should be treated as a hobby that has the potential of making those specializing in mathematics more intellectually satisfied. This "new math" should only be taught to those who already know the practical, down-to-earth mathematical concepts. The purpose of teaching mathematics is for the student to learn and not merely to be an intellectually satisfying endeavor for the teacher.

By the way, the next section examines binary subtraction. I learned about the concept of "borrowing" when I was in elementary school. Today, the "in word" here is again *renaming*. As before, both the terms *borrowing* and *renaming* are quite descriptive, but *renaming* is more logical. Nonetheless, using logic is not always the best educational aid when a subject is first introduced. The human mind craves concreteness, and only slowly, oh so slowly, does logic seem to take over the gray cells.

Subtracting Binary Numbers

Again, we will first look at subtracting decimal numbers. This time, however, we will look at the "harder" problems first. For instance, let's subtract 49 from 74. First, we subtract the numbers in the ones' place, i.e., 4 take away 9. Since we cannot do this, we have to *rename* ("borrow from") the 7 in the tens' place. We then have only 6 left in the tens' place, but now we have 14 in the ones' place. Since $14 - 9 = 5$, we are left with 5 in the ones' place. Finally, we subtract 4 from 6 to get 2 tens in the tens' position. Since we have 2 tens and 5 ones, the answer is 25.

Once you truly understand what you are doing when subtracting decimal numbers, subtracting binary numbers is as easy as using the Internet.

Example   What is $1010 - 1001$? First, we subtract the numbers in the ones' place, i.e., 0 take away 1. Since this cannot be done without the use of negative numbers, we have to *rename* ("borrow from") that little one in the twos' place. After this, we have nothing in the twos' place, but we have 10 (2 in decimal) ones in the ones' place. Now you have to remember that, in binary, 10 (binary) minus 1 is 1. The remainder of the subtraction process is trivial, so we are left with simply 0001 for the difference.

Remember when you learned arithmetic that you had to remember all those addition and subtraction facts. Binary is much simpler. All you need remember is

In addition: $0 + 1 = 1, 0 + 0 = 0, 1 + 1 = 10$

In subtraction: $0 - 0 = 0, 1 - 0 = 1, 1 - 1 = 0$, and $10 - 1 = 1$

Compare these three addition facts and four subtraction facts you need to remember with the $81+$ addition facts and $81+$ subtraction facts you need to remember to do decimal addition and subtraction. Adding and subtracting really is simple—so simple, in fact, that a tiny piece of ultrapure yet lifeless silicon can do it.

Do the Experiments in This Book

If you really and truly want to get the binary number system ingrained in your gray matter, I strongly recommend that you spend some time on the experiments in this book that deal directly with the binary number system. The following experiments are highly recommended here.

Chapter 1: Do the experiment described in Fig. 1-5.

Chapter 4: Experiment a bit with Mag-11's binary-readout thermometer. This experiment is built into Mag-11 and can be operated as long as the M11DIAG2 firmware is installed.

Chapter 5: Do the binary adder experiment described in Fig. 5-9.

Hexadecimal Numbers

I believe that hexadecimal numbers are one concept that some people in the computer field find repugnant. These people find binary easy to use but seem to be frightened by hexadecimal numbers. There are several reasons why some people find binary numbers simple but are put off by hexadecimal numbers. First, hexadecimal numbers are inherently more complicated. The reason for this should be obvious—they are base-16 numbers. The second, and probably the more significant problem, is that after 9 we run out of "old-time numbers," so we have to substitute some other symbol. Probably because they were already on keyboards, it was decided to substitute the first six letters of the alphabet for these additional six symbols—A, B, C, D, E, and F. Here, 10 is represented by A, 11 by B, 12 by C, 13 by D, 14 by E, and 15 by F.

Quick question   In hexadecimal, how is 16 represented? Simple. By 10!

I may have sent a chill up your spine with this talk of hexadecimal being hard. Well, it's only hard when you start out. It gets easier and easier and easier. Trust me here. Once you get used to it, it is as simple as soldering two shiny copper wires together.

As mentioned, the hexadecimal system is a base-16 system. However, it really and truly is simply a shorthand way of writing and reading binary numbers. Nothing more. Nothing less. If it were not for the fact that computers use the binary numbering system, the hexadecimal system would exist only in footnotes in high school and college textbooks.

Why was hexadecimal chosen for binary shorthand? Well, look at a 4-bit binary number, which is also called a *nibble*, e.g., 1011. This number, 11, can be represented by the hexadecimal number B. *The same is true for every single 4-bit number.* Each one can be represented by a single hexadecimal digit. More than this, two hexadecimal numbers can be used to represent an 8-bit binary number referred to as a *byte*. This is what hexadecimal numbers are all about. Refer to Table A-1 for a listing of 4-bit binary numbers and their corresponding hexadecimal shorthand way of writing them. The decimal equivalents are also provided.

CONVERTING FROM BINARY TO HEXADECIMAL AND BACK AGAIN

This is so simple that it really does not need an explanation. Instead, look at Table A-1. A few minutes of converting from binary to hexadecimal and hexadecimal to binary, and you will quickly discover why hexadecimal numbers are used as a shorthand for binary.

ADDING AND SUBTRACTING HEXADECIMAL NUMBERS

I do not recommend adding and subtracting hexadecimal numbers directly. While some textbooks go through the procedure, it is quickly forgotten because it is done so infrequently. The best way to add and subtract hexadecimal numbers is to convert them first to decimal, then do the operation, and then convert back to hexadecimal. Either that, or use one of the computer programs around that does this for you.

CONVERTING FROM HEXADECIMAL TO DECIMAL AND BACK AGAIN

Many books on computers mention several ways to do this. In my experience, these methods are quickly forgotten. However, you should feel free to try these methods out for yourself. Also refer here to the section "Having More Fun with the RTI" in Chap. 6, which gives my intuitive way of converting from decimal to hexadecimal and back again. Also

TABLE A-1 EQUIVALENT NUMBERS IN BINARY, DECIMAL, AND HEXADECIMAL NUMBERING SYSTEMS		
BINARY	**HEXADECIMAL**	**DECIMAL**
0000	0	0
0001	1	1
0010	2	2
0011	3	3
0100	4	4
0101	5	5
0110	6	6
0111	7	7
1000	8	8
1001	9	9
1010	A	10
1011	B	11
1100	C	12
1101	D	13
1110	E	14
1111	F	15

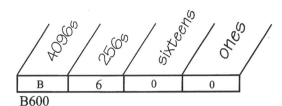

B600

FIGURE A-5 A diagram indicating the meaning of hexadecimal place value.

refer to Fig. A-5, which shows the place values of hexadecimal numbers.

AN EXPERIMENT THAT WILL HELP YOU GET THE FEEL FOR HEXADECIMAL NUMBERS

Again, try out the experiment whose schematic is given in Fig. 7-8 and whose program listing is given in Fig. 7-10.

Summing Up

Today's computer designs use binary numbers. Binary numbers are base-2 numbers, while the common garden variety of numbers (decimal numbers) are base 10. Hexadecimal numbers, which are base 16, are probably used even more frequently when working with MCUs than binary numbers. *Nonetheless, hexadecimal numbers are only a shorthand way of writing binary numbers.*

B

PROGRAMMING EPROMS

A Little Personal History

It was January 1982 when I programmed my first EPROM. That first 1024-byte, 2708 EPROM was programmed with the crudest, "bird's nest" type of breadboarded programmer that probably ever existed. Nonetheless, it worked. And it cost me less than $20. This was at a time when the very, very cheapest EPROM programmers cost nearly as much in dollars as there were bytes in that 2708. The other thing that made up for the shaky operation and messy looks was that it was my own design. Perhaps this was its most salient feature.

Shortly after I programmed my first dozen 2708 EPROMs, the 2708 not only became obsolete, but there came on the scene a slowly growing assortment of relatively inexpensive EPROM programmers. Nearly all these programmers used the intelligence of personal computers and came with delightfully sophisticated software.

In mid-1984 I purchased Ross Custom Electronics Intelliburner for $269 plus $2 shipping and handling. This programmer has worked flawlessly from the time I received it to the present. More than this, the people at Ross Custom Electronics (apparently only two at the time) were really nice and considerate. By the way, Ross Custom Electronics had several other models, referred to as DumBurners, that were even cheaper than the Intelliburner.

The Intelliburner is connected to a PC-compatible computer via its serial interface. While, as its name indicated, the programmer had its own MCU (Intel's 8039), many users bypassed its inherent intelligence and used it in its DumBurner mode. By the way, the simple fact that the Intelliburner still works fine must say something for the reliability of MCUs and EPROMs themselves, since they were used in its design.

While I believe that the Intelliburner and DumBurner are extinct and only preowned models are available (check the Internet as well as Ham Radio Shows), another even more inexpensive programmer, which appeared about 2 years later, is still going strong. It is Needham's Electronics PB-10. The PB-10 is a PC-compatible card that programs 2716 through 8-Mb EPROMs. It is priced at only $139. The PB-10 supports MOT S format object code and comes with Needham's famous EMP programmer control software.

The PB-10 is Needham's oldest and cheapest programmer. The company claims that it has sold over 45,000 units. Its primary disadvantage is that it requires an empty computer slot and the inconvenience that entails. Needham's line now includes the EMP-10, EMP-20, and EMP-30, which are all parallel-port programmers. The EMP-10 costs $220 and the EMP-20 twice this, while the EMP-30 costs $1000.

Introits, Inc., another even smaller company, sells an even cheaper programmer called the Pocket Programmer. The price of this parallel-port programmer is $129.95. Jameco Electronics now sells over 10 different programmers. At least three of them are below $200. BK Precision makes at least one parallel-port programmer for under $200. This, as well as models from Needham's, can be purchased from Digi-Key Corporation.

The preceding is far from being a complete listing of inexpensive sources for EPROM programmers. A quick search of the Internet (keywords *programmer* and *EPROM*) reveals dozens of companies selling low-cost programmers. However, you may feel adventurous and make one yourself. If so, I recommend looking through old issues of *BYTE* and general-interest electronic magazines such as *Radio-Electronics/Electronics Now* for articles on building EPROM programmers. The article by Steve Ciarcia, "Build an Intelligent EPROM Programmer," which appeared in the October 1981 issue of *BYTE,* may be of particular interest. Also look at the book, *Experiments with EPROMS,* by Dave Prochnow. It has several interesting EPROM programmer projects. Like the book you are reading, this book is a McGraw-Hill publication.

One further note: Except for the Intelliburner, I have no hands-on experience with any of the EPROM programmers mentioned. Despite this, I would not hesitate to purchase any of these programmers if I had a need for them (e.g., if the Intelliburner fell off the workbench and one of the kids stepped on it).

Motorola S-Format Object Code

An object code file in MOTOROLA S format is also referred to as a *MOT S file,* an *S-record file,* and an *.S19 file.* Whatever you call it, its the same object code file. By the way, I usually refer to the file itself as an *.S19 file* and refer to the format as the *MOT S format.*

For those who are acquainted with the Intel Hex format, you will quickly see that the MOT S format is just a minor variation. The basic idea is the same. As in the Intel Hex format, *there is no binary data in an .S19 file.* Rather, two ASCII characters are used to represent each byte of binary data. For example, the two ASCII characters "FE" are used to represent the byte "11111110." The first two character positions in the line designate the type field. While there are actually 10 different type fields (S0, S1, S2,..., S9), only two are important to us: S1 and S9. S1 records are the main data records, while an S9 record is used to mark the end of the .S19 file. Each line of an .S19 file must consist of the following:

CHARACTER POSITION	REPRESENTS
1,2	Start of record ("S1")
3,4	Byte count of data for EPROM plus three
5,6,7,8	Base EPROM destination address
9,10	First data byte to be programmed
—	
—	
N-3,N-2	Last data byte to be programmed
N-1,N	Checksum

The checksum in an .S19 file is the ones' complement of the least significant 8 bits of the values of bytes 3 and 4 through N-2. Thus, if the sum were $3FE, the checksum simply would be 01. ($3FE = 0011 1111 1110. The least significant 8 bits is 1111 1110, and its ones' complement is 0000 0001.)

Question   If the sum were $A55, what would be the checksum?

Answer $AA.

The file must end with a null record, usually "S9030000FC." Figure B-1 shows an S1 record of the BLNKLED1 program in Fig. 9-6. The record type, length of data, destination address, object code data, and checksum are indicated.

A PRACTICAL LOOK AT THE .S19 FILE

The .S19 file is created automatically by the AS11 and similar HC11 cross-assemblers. Happily, there is seldom, if ever, a need to manually create an .S19 file. Also, it is not necessary to know anything at all about an .S19, except its name and path, to use it. Of course, I assume here that you never encounter a problem when using it. The .S19 file reminds one of an automobile. It isn't necessary to know anything about the mechanics of a car as long as it starts and runs OK. However, as soon as a problem with a car becomes apparent, then it suddenly becomes necessary for either you or the mechanic to understand what makes the car go. Since there aren't any .S19 file repairpersons listed in the Yellow Pages, this means that it is wise for the .S19 user to understand at least something about its operation.

When a problem occurs either in programming an EPROM or in the operation of an HC11 gizmo that the EPROM controls, take a good look at the .S19 file. Concentrate here on the S1 record. Referring to Fig. B-1, look at the S1 record's destination address. This is often where problems arise. Like all information in this file, the destination address is given in hexadecimal. Notice in Fig. B-1 that the destination address is $B600. Doesn't this look familiar? This is the beginning of the HC11's EEPROM, and many of the programs listed in the book start here.

The S1 record in Fig. B-1 is meant to be programmed into an EEPROM. What if you wanted to program a 32-kB EPROM, such as a 27256, with an .S19 file that had a desti-

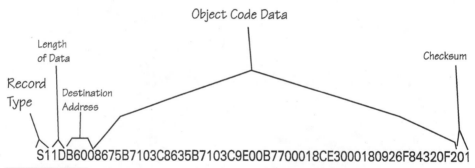

S11DB6008675B7103C8635B7103C9E00B7700018CE3000180926F84320F201

FIGURE B-1   A diagram showing the S1 record file of the BLNKLED1 program.

nation address of $8000? Remember, a 32-kB EPROM's first address is 0000 and its last is $7FFF. Well, there are several ways to do this, but the simplest is to have the EPROM programmer do it for you (Figs. B-2 and B-3). For instance, with the Intelliburner, when the .S19 file starts with a destination address higher than the available address space in the EPROM, the Intelliburner flashes a yellow LED just as soon as the first address space is received. All you need do then is depress the pushbutton switch on the rear of the Intelliburner and the programmer then starts at $8000, which is the same as the destination address. Everything makes sense to the Intelliburner then, and it goes on its merry way programming the EPROM. Other EPROM programmers have similar schemes that do the same thing. Check the documentation.

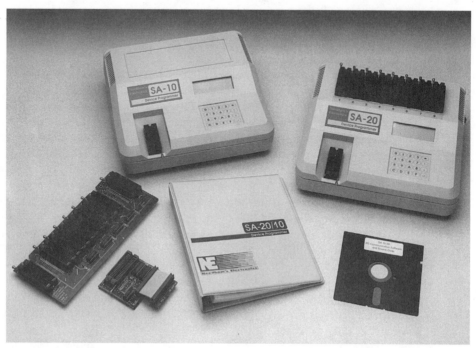

(A)

FIGURE B-2a   Needham's SA-10 and SA-20 EPROM programmers.

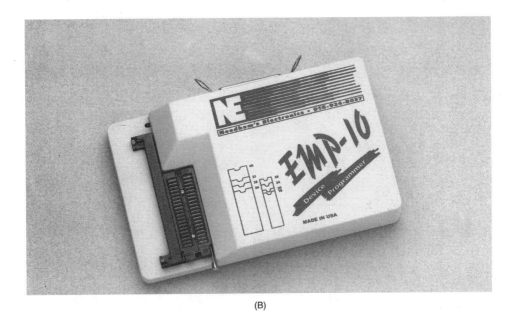

(B)

FIGURE B-2b Needham's EMP-10 EPROM programmer.

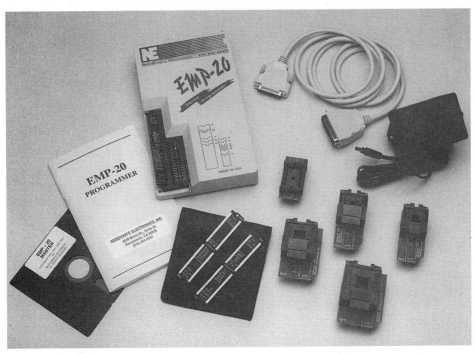

(C)

FIGURE B-2c Needham's EMP-20 EPROM programmer.

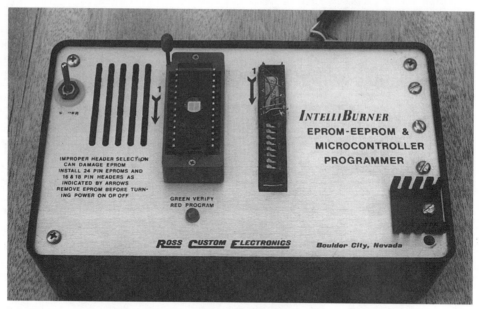

(A)

FIGURE B-3a The Intelliburner (an EPROM programmer).

(B)

FIGURE B-3b The Intelliburner (an EPROM programmer) and Dave Prochinow's book *Experiments with EPROMS*.

SOURCES FOR PRINTED CIRCUIT
BOARDS, KITS, AND PARTS

Magicland is making available professionally made circuit boards, as well as complete kits, for most of the major projects described in this book. All printed circuit boards have plated through holes, and all, except for the Mag-11 boards, have silk screening and solder masking (Fig. C-1).

Mag-11

Mag-11's bareboard and construction guide—part no. MAG-11/BD

Kit for Mag-11 which includes all parts, including the bareboard and construction guide, except for the connectors—part no. MAG-11/KIT

MagPro-11

MagPro-11's bareboard and construction guide—part no. MPRO/BD

Complete kit for MagPro-11, including the bareboard and construction guide, except for the HC11 MCU—part no. MPRO/KIT

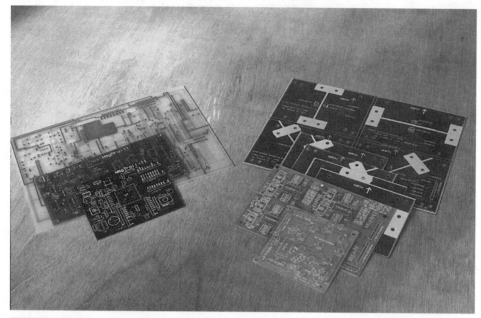

FIGURE C-1 Bare printed circuit boards for Mag-11, MagPro-11, MagTroll-11, and three wind vane boards.

The MC68HC11A1FN MCU with a 512-byte EEPROM—part no. MHC11FN

The MC68HC811E2FN with a 2048-byte EEPROM—part no. MHC811E2FN

MagTroll-11

MagTroll-11's bareboard and construction guide—part no. MTROLL/BD

MagTroll-11 complete kit, including bareboard, construction guide, and one MC68HC11A1FN—part no. MTROLL/KIT

Eight-Sensor Wind Vane

Four bare sensor boards—part no. WV_SEN/BD

Bare opamp board and construction guide—part no. WV_OPAMP/BD

Complete kit for opamp board—part no. WV_OPAMP/KIT

Bare display boards and construction guide—part no. WV_DIS/BD

Complete kit for display board—part no. WV_DIS/KIT

MC68HC11A1FN with WINDVANE firmware in EEPROM—part no. MHC11FN/WV

Four-Sensor Wind Vane

Sensor board

Opamp board/kit—use eight-sensor board/kit and ask for leaflet on converting from eight-sensor operation to four-sensor operation.

Display board—same as eight-sensor wind vane—part no. WV_DIS/BD and part no. WV_DIS/KIT

MC68HC11A1FN with WVANE_4A firmware in EEPROM—part no. MHC11FN/WV_4

For Magicland's latest prices, either write to Magicland at 4380 S. Gordon Avenue, Fremont, MI 49412, or e-mail them a message at *magicland@ncats.net*.
Other vendors that supply parts mentioned in the book include

Digi-Key Corporation
701 Brooks Ave., South
Thief River Falls, MN 56701-0677
1-800-344-4539
Web: *http://www.digikey.com*
Reliable supplier of a wide variety of electronic parts.
Supplier of inexpensive EPROM programmers.
Not a Motorola vendor.

Mouser Electronics
958 N. Main
Mansfield, TX 76063-4827
1-800-346-6873
Web: *http://www.mouser.com*
Another reliable supplier of many types of electronic parts.
Supplier of inexpensive EPROM programmers.
Not a Motorola vendor.

Allied Electronics, Inc.
7410 Pebble Drive
Fort Worth, TX 76118
1-800-433-5700
Web: *http://www.allied.avnet.com*
An oldie but a goodie.
Allied *is* a supplier of Motorola parts.

Jameco
1355 Shoreway Road
Belmont, CA 94002-4100
1-800-831-4242
Web: *http://www.jameco.com*

One of the oldest of the new type of supplier.
A fair selection of common and unique parts.
Supplier of inexpensive EPROM programmers.
Limited supplier of Motorola parts.

Radio Shack Retail Stores
An electronics supermarket that carries a wide supply of parts and tools that are
appreciated by electronic hobbyists. One of the best sources for good-quality solderless breadboards.
Web: *http://www.radioshack.com*

Dick Blick Art Materials
P.O. Box 1267
Galesburg, IL 61402-1267
1-800-447-8192
A source for vellum for making printed circuit boards using the positive photo
method.

FIGURE C-2 Illustrations of two HC11 evaluation boards produced by
Motorola—the M68HC11EVB and the M68HC11EVBU.

FIGURE C-2 (*Continued*) Illustrations of two HC11 evaluation boards produced by Motorola—the M68HC11EVB and the M68HC11EVBU.

Edmund Scientifics
101 East Gloucester Pike
Barrington, NJ 08007-1380
1-800-728-6999
Web: *http://www.edsci.com*
A source for silicon photovoltaic cells and a lot of other neat stuff.

The following vendors supply HC11-related equipment, single-board computers, and software:

Needham's Electronics, Inc.
4630 Beloit Dr. Ste 20
Sacramento, CA 95838
1-916-924-8037
Web: *http://www.needhams.com*
Manufacturer and supplier of low- to midcost EPROM programmers.

Intronics, Inc.
Box 13723
612 Newton St.
Edwardsville, KS 66113
1-913-422-2094
Manufacturer and retailer of inexpensive EPROM programmers.

Zorin
1633 4th Avenue West
Seattle, WA 98119
1-206-282-6061
Web: *http://www.ZORINco.com*
Manufacturer and supplier of HC11 boards and accessories.

EMAC, Inc.
P.O. Box 2042
Carbondale, IL 62902
1-618-529-4525
Manufacturer of HC11-based single-board computers.

Micromint, Inc.
4 Park Street
Vernon, CT 06066
1-860-871-6170
Manufacturer and designer of processor boards including HC11-based ones.

Coactive Aesthetics, Inc.
4000 Bridgeway, Suite 303
Sausalito, CA 94965
1-415-289-1722
e-mail: *coactive@coactive.com*
Sells HC11 boards and software.

Midwest Micro-Tek
2308 East Sixth Street
Brookings, SD 57006
1-605-697-8521
Produces wide range of embedded controllers including HC11-based ones.

TECI
West Shore Road
West Glover, VT 05875
1-802-525-3451
Web: *http://www.tec-I.com*
HC11 development boards and cross-assembler.

P&E Microcomputer Systems, Inc.
P.O. Box 2044
Woburn, MA 01888-0044
1-617-353-9206
Web: *http://www.pemicro.com*
A major, well-known supplier of HC11 software.

Micronix Electronic Systems Inc.
297 Cornett Drive
Red Deer, Alberta T4P 3R9
Canada
PH/FAX: 403-341-5676
Web: *http://www.agt.net/public/micronix*
Supplier of HC11 single-board computers.

Cosmic Software
400 West Cummings Park STE 6000
Woburn, MA 01801-6512
1-781-932-2556
Web: *http://www.cosmic-software.com*
Supplier of HC11-related software.

Avocet Systems
1-800-448-8500 or 1-207-236-9055
Web: *http://avocetsystems.com*
Supplier of development software for most MCUs including the HC11.

PARTS LISTS

Mag-11, MagPro-11, MagTroll-11, Eight-Sensor Solid-State Wind Vane, Four-Sensor Solid-State Wind Vane, and Temperature and Wind Chill/Speed Display Board

Complete Parts List for Mag-11 Single-Board Computer, Version 1e

C1—1000-μF, 25-V electrolytic capacitor (axial lead)

C2—47-μF, 16-V electrolytic capacitor (radial lead)

C3—10-μF, 16-V tantalum capacitor (radial lead)

C4, C11—1-μF, 100-V, metallized polyester capacitor

C5—0.01-μF, 50-V monolithic ceramic capacitor

C6, C9, C10, C18, C20-C32—0.1-μF, 50-V monolithic ceramic capacitor

C7, C8—15-pF, 500-, dipped mica capacitor

C12–C16—22-μF, 25-V electrolytic capacitor (radial lead)

C17—100,000-μF, 5.5-V Panasonic gold capacitor (DK #P6952)

C19—1000-pF, 50-V monolithic ceramic capacitor

D1, D4—1N5817 Schottky barrier rectifier

D2—1N4001 silicon rectifier

D3—1N4733A 5.1-V, 1-W zener diode

J1–J6—20-connector, dual-row female .1"X.1" matrix (DK #929975-01-36-ND, cut to length; see instruction guide)

J7—DB-9 right-angle male PC board mount (DK #409M-ND)

JP1, JP2, JP4–JP7, JP12—3-post male with 0.1-in centers (DK #929834-01-36-ND, cut to length)

JP8, JP10, JP11, JP14—2-post male with 0.1-in centers (DK #929834-01-36-ND, cut to length)

LED1, LED2, LED4, LED5, LED9, LED10—green, low-current, high-efficiency LEDs (T-1 case)

LED3, LED6–LED8, LED11—red, low-current, high-efficiency LEDs (T-1 case)

Q1—2N2222A NPN transistor

R1—330-Ω, 1/4-W, 5% resistor

R2, R6—1-kΩ, PC pot (DK #D4AA13) *Note:* R3 is not used. It is replaced by the resistance in RN4

R4—1-MΩ, 1/4-W, 5% resistor

R5, R11—1-kΩ, 1/4-W, 5% resistor

R7—150-Ω, 1/4-W, 5% resistor

R8—10-MΩ, 1/4-W, 5% resistor

R9—4.7-kΩ, 1/4-W, 5% resistor

R10—500-Ω PC pot (DK #D4AA52)

R12—3.3-kΩ, 1/4-W, 5% resistor (optional)

R13—10-kΩ, PC pot (optional) (DK #D4AA14)

R14—4.7-Ω, 1/4-W, 5% resistor

R15—1.5-kΩ, 1/4-W, 5% resistor

R16, R18—10-kΩ, 1/4-W, 5% resistor

R17—1.2-kΩ, 1/4-W, 5% resistor

R19—7.32-kΩ (or 6.04-kΩ; see construction guide), 1% resistor (substitute for the 7.27-kΩ resistor)

RTH—1-kΩ thermistor (Fenwall's JB31J1 or equal)

S1—10-circuit DIP switch

S2—4-circuit DIP switch

S3—PC mount SPST pushbutton switch (DK #P8035). Cut off diagonal pins; see construction guide.

SO1—compatible 2-circuit socket and cable for power connections

RN1—9 × 4.7-kΩ resistor network

RN2, RN3—5 × 10-kΩ resistor network

RN4—9 × 1.2-kΩ resistor network

RN5—9 × 10-kΩ resistor network

U1—MC68HC11A1P MCU IC

U2—MAX690CPA or MAX690ACPA IC

U3—MAX232 IC

U4—LM2931T low I/O DIF +5-V voltage regulator

U5—74HC245 CMOS IC

U6, U11—74HC373 CMOS IC

U7—27C256 EPROM (programmed with either M11DIAG2 or BUFFALO)

U8—74HC138 CMOS IC

U9—6264LP 8-kB RAM (optional)

U10—27C256 32-kB EPROM or 32-kB RAM (optional)

U12—74HC259 CMOS IC

U13—74HC4078 CMOS IC

U14, U15—74HCT04 CMOS IC (The 74HCT04 has been substituted for the 74HC04 and 74HC14.)

U16—74HC32 CMOS IC

U17—74HC11 CMOS IC

U18—LM336-5.0

XTAL1—4-MHz or 8-MHz crystal (4-MHz is preferred.)

1—48-pin IC socket (Two 24-pin sockets are used.)

3—28-pin IC socket

3—20-pin IC socket

3—16-pin IC socket

5—14-pin IC socket

1—8-pin IC socket

10 shorting jumpers (DK# S-9000), optional heat sink for U4, circuit board (Magicland part no. MAG-11/BD)

Parts List for Magpro-11

C1, C2—27-μF monolithic ceramic capacitors

C3, C4—10-μF, 16-V radial lead tantalum capacitors

C5–C8—22-μF, 25-V radial electrolytic capacitors

CU1–CU8—0.1-μF monolithic ceramic capacitors

D1—1N4729A, 1-W, 3.6-V zener diode

D2—1N5817 Schottky barrier rectifier

J1—4 ckt, 0.1-in male header (Digi-Key part no. A1912-ND)

J2—male printed circuit DB9 connector

JP1–JP3—2 ckt, 0.1-in male header (Digi-Key part no. A1911-ND)

JP4—6 ckt, 0.1-in male header (Digi-Key part no. A1913-ND). Cut 6 ckt header in half. Use only one half.

LED1, LED4—low-power yellow LED

LED2—standard green LED

LED3—low-power red LED

R1—10-MΩ, 1/4-W, 5% resistor

R2—2.2-kΩ, 1/4-W, 5% resistor

R3—4.7-Ω, 1/4-W, 5% resistor

R4, R10, R12, R15, R17—4.7-kΩ, 1/4-W, 5% resistor

R5—330-Ω, 1/4-W, 5% resistor

R6—6.04-kΩ, 1% resistor

R7—7.32-kΩ, 1% resistor

R8—100-Ω, 1/4-W, 5% resistor

R9, R13—1-kΩ, 1/4-W, 5% resistor

R11, R14—100-kΩ, 1/4-W, 5% resistor

R16—5-kΩ minipot (Digi-Key part no. D4AA53)

RN1—4.7-kΩ, 9-element SIP common terminal resistor network

RN2—10-kΩ, 5-element SIP common terminal resistor network

S1—SPST PC pushbutton switch (Digi-Key part no. P8035S)

U1—52-pin solder-tail PLCC socket (Digi-Key part no. A418-ND)

Note: Socket at U1 accepts either a MC68HC11A1FN or a MC68HC811E2FN MCU. See text.

U2—27C256 EPROM programmed with BUFFALO 3.4 firmware

U3—74HC373 Octal-D tri-state latch

U4—74HC00 quad NAND

U5—6264 8-kB $\times$ 8 static RAM

U6—MAX709LCPA MPU supervisor

U7—MAX232CPE +5-V pwrd dual RS-232 transmitters and receivers

U8—LM324N quad opamp

XTAL1—4-MHz crystal

PC board (Magicland part no. MPRO/BD), power cable with socket that mates with J1 [socket: Digi-Key part no. A3013-ND (housing), Digi-Key part no. A3000-ND (contacts), A3077-ND (keying plug)], shorting jumpers, solder, etc.

Note:

DK stands for Digi-Key Corporation. These part numbers are listed as a convenience to the builder only and are not meant to be recommendations of Digi-Key or Magicland over similar companies.

Parts List for Magtroll-11

C1—1000-μF, 16-V axial lead electrolytic capacitor (see text)

C2, C3—27-pF monolithic ceramic capacitors

C4—100-μF, 16-V axial lead electrolytic capacitor

C5–C10—0.1-μF monolithic ceramic capacitors

D1—1N914 silicon diode

J1—4-pin male 0.1-in center headers (1 pin removed)

J-MISCL, J-PORTA–J-PORTE (6 required)—10-pin male 0.1-in center headers (1 pin removed—save 2 pins for TP1,TP2)

JP1, JP2, JP3—2-pin male 0.1-in center headers

JP4, JP5—3-in male 0.1-in center headers

LED1—yellow, low-power LED

R1—10-MΩ, 1/4-W, 5% resistor

R2—3.3-kΩ, 1/4-W, 5% resistor

R3—10-kΩ mini-pc pot (Digi-Key part no. D4AA14 or Mouser Electronics 594-63M103)

R4—53.6-kΩ, 1% resistor

R5—10-kΩ, 1% resistor

R6—470-kΩ, 1/4-W, 5% resistor

R7—2.2-kΩ, 1/4-W, 5% resistor

R8—1-kΩ, 1/4-W, 5% resistor

R9, R10—4.7-kΩ, 1/4-W, 5% resistor

RN1—4.7-kΩ, 10-pin/9-res, bused SIP resistor network

RN2—10-kΩ, 6-pin/5-res, bused SIP resistor network

S1—SPST PC switch (Digi-Key part no. P8035S-ND or equal)

TP1, TP2—pins that were removed from J-MISCL–J-PORTE or even cut off leads from a resistor.

U1—MC68HC11A1FN MCU or MC68HC811E2FN MCU. See text.

U2—MAX690ACPA MPU supervisor

U3—LM336-5 5-V IC voltage reference

U4—LM7805CTB or LM2931T 5-V voltage regulator

XTAL1—4-MHz or 8-MHz crystal

Printed circuit board (Magicland part no. MTROLL/BD), shorting plugs (jumpers), IC sockets, etc.

Note:

DK stands for Digi-Key Corporation. These part numbers are listed as a convenience to the builder only and are not meant as a recommendation of Digi-Key, Mouser Electronics, or Magicland over similar companies.

Parts List for the Sensor Assembly for the Eight-Sensor Solid-State Wind Vane

J-SEN/ASSEMBLY—10-pin male 0.1-in center headers (pin 9 removed). Right-angle type recommended. (Mouser's part no. 538-22-05-2101)

U100–U107—National Semiconductor's LM135/LM235/LM335 series of precision temperature sensors. LM235H especially recommended for its excellent performance/price ratio. The LM135AH is tops for performance here, but price is too high. The inexpensive LM335Z can be used but requires 1-kΩ pots be used in place of 500-Ω pots on the opamp board.

Wire—7 different colors of hookup wire (white, black, brown, orange, green, red, and blue)

North-South PC board, East-West PC board, NE-SW PC board, and SE-NW PC board (Magicland part no. WV_SEN/BD)

Hardware   1-in 6-32 spacers, 6-32 screws and nuts
Additional parts for roof   18 $\times$ 18-in 20-gauge sheet metal, 4 10-in lengths of 7/8-in wood dowels, screws, etc.

Parts List for the Opamp Board for the Solid-State Wind Vane

C1—10-μF, 20-V dc electrolytic capacitor

C2–C5 (4 reqd.)—0.1-μF, 50-V monolithic ceramic capacitor

J-MISCL, J-PORTE, J-SEN (3 reqd.) headers (1 pin removed)—10-pin male 0.1-in center (save 2 pins for TP1, TP2)

R1, R8, R48–R50, R57, R107, R108 (8 reqd.)—20-kΩ, 1% resistors

R2–R5, R16, R17, R20, R21, R33–R36, R41, R42, R46, R47, R51–R54, R71, R77, R88–R91, R97, R98, R105, R106 (32 reqd.)—100-kΩ, 1%, resistors

R6, R7, R30, R31, R55, R56, R86, R87 (8 reqd.)—10-kΩ, 1% resistors

R9, R39, R43, R58, R75, R96, R99 (7 reqd.)—500-Ω minipot (DK part no. D4AA52 or Mouser's part no. 594-63M500)

R10, R11, R27, R29, R59, R60, R83, R85 (8 reqd.)—46.4-kΩ, 1% resistors

R12—825-Ω, 1% resistor

R13, R26, R28, R62, R65 ,R81, R84 (7 reqd.)—604-Ω, 1% resistors

R14, R15, R37, R38, R66, R67, R94, R95 (8 reqd.)—1-kΩ, 1% resistors

R19—1.5-kΩ, 5% resistor

R22, R23, R44, R45, R78, R79, R103, R104 (8 reqd.)—100-Ω, 1% resistor

R25—10-kΩ mini-pc pot (DK part no. D4AA14 or Mouser's part no. 594-63M103)

U1–U4 (4 reqd.)—LMC660CN CMOS quad opamp

U5—LM336-2.5 IC voltage reference

PC board (Magicland part no. WV_OPAMP/KIT), hardware, etc.

Note:

DK stands for Digi-Key Corporation. These part numbers are listed as a convenience to the builder only and are not meant as a recommendation of Digi-Key, Mouser Electronics, or Magicland over similar companies.

Parts List of the Display Board for the Solid-State Wind Vane (Eight-Sensor and Four-Sensor Versions)

C1—10-μF, 20-V electrolytic capacitor

C2—0.1-μF monolithic ceramic capacitor

J-MISCL/DISPA, J-MISCL/DISPB, J-PORTB/DISPA, J-PORTB/DISPB (4reqd.)—10-pin male 0.1-in center headers (1 pin removed). Right-angle type recommended. (Mouser's part no. 538-22-05-2101)

LED1, LED9—red superbright LED

LED2, LED4, LED6, LED8 (4 reqd.)—yellow superbright LED

LED3, LED5, LED7 (3 reqd.)—green superbright LED

R1–R9 (8 reqd.)—1-kΩ, 5% resistors

R10–R13—10-kΩ, 5% resistors

U1—74LS42 CMOS BCD to DEC decoder

PC board (Magicland part no. WV_DIS/BD), hardware, etc.

Note:

These part numbers are listed as a convenience to the builder only and are not meant as a recommendation of Digi-Key, Mouser Electronics, or Magicland over similar companies.

Parts List for Four-Sensor Sensor Assembly

J-SEN/ASSEMBLY—10-pin male 0.1-in center headers (pin 9 removed). Right-angle type recommended. (Mouser's part no. 538-22-05-2101)

U100–U103—National Semiconductor's LM135/LM235/LM335 series of precision temperature sensors. LM235H especially recommended for best performance/price ratio. The LM135AH is tops for performance here, but price is too high. The inexpensive LM335Z can be used but requires 1-kΩ pots be used in place of 500-Ω pots on the opamp board.

PC board (check with Magicland for availability)

North-South windbreak, East-West windbreak

Hardware 1-in 6-32 spacers, 6-32 screws and nuts.
Additional parts for roof 18 $\times$ 18-in 20-gauge sheet metal, 4 10-in lengths of 7/8-in wood dowels, screws, etc.

Parts List for Four-Sensor Opamp Board

C1—10-μF, 20-V tantalum capacitor

C2, C3—0.1-μF, 50-V monolithic ceramic capacitor

J-MISCL/OPA, J-PORTE/OPA, J-SEN/OPA—10-pin male 0.1-in center headers (1 pin removed). Right-angle type recommended. Mouser Electronics part no. 538-22-05-2101

R1, R8, R48, R49—20-kΩ, 1% resistors

R2–R5, R16, R17, R20, R21, R33–R36, R41, R42, R47, (16 reqd.)—100-kΩ, 1% resistors

R6, R7, R30, R31—10-kΩ, 1% resistors

R9, R39, R43—500-Ω minipot (DK part no. D4AA52 or Mouser part no. 594-63M500)

R10, R11, R27, R29—46.4-kΩ, 1% resistors

R12—825-Ω, 1% resistor

R13, R26, R28—604-Ω, 1% resistors

R14, R15, R37, R38—1-kΩ, 5% resistors

R19—1.5-kΩ, 5% resistor

R22, R23, R44, R45—100-Ω, 1% resistor

R25—10-kΩ minipot (DK part no. D4AA14 or Mouser part no. 594-63M103)

U1, U2—LMC660CN CMOS quad opamp

U3—LM336-2.5 IC voltage reference

PC board (Magicland's part no. WV_OPAMP/BD can be adapted for use), solder, etc.

Note:

DK stands for Digi-Key Corporation. These part numbers are listed as a convenience to the builder only and are not meant as recommendations of Digi-Key, Mouser Electronics, or Magicland over similar companies.

Note:

To simplify the revision process, the following reference designators have been eliminated: R18, R24, R32, R40

Parts List for Temperature and Wind Chill/Speed Display Board

DIS1, DIS2, DIS3—7-segment, common-anode LED displays

J-PORTA, J-PORTC (or J-PORTB; see text)—10-pin male 0.1-in center headers (1 pin removed). Right-angle type recommended. Mouser part no. 538-22-05-2101

LED1, LED2, LED3—low-power LEDs (any color)

R1–R3, R7–R20 (17 needed)—330-Ω, 1/4-W, 5% resistors

R4, R5, R6—1-kΩ, 1/4-W, 5% resistors

U1—74HC00 high-speed CMOS quad NAND IC

U2—74HC138 high-speed CMOS 3 line to 8 line decoder

U3, U4—7447 TTL BCD-to-7-seg decoder/driver IC

U5—74LS541 low-power Schottky octal buffer with tristate outputs

PC board, solder, etc.

Note:

DK stands for Digi-Key Corporation. These part numbers are listed as a convenience to the builder only and are not meant as recommendations of Digi-Key, Mouser Electronics, or Magicland over similar companies.

E

Dynamite Hints on Making
PC Boards

The positive photographic method is the simplest, and one of the best, methods of making printed circuit boards using laser printer–produced art. Keep in mind here that the bottom (solder side) of the board is made by placing the art printed side down (in contact with the board). The top (component side) is made by placing the artwork with printed side up.

I made single-sided prototypes of Mag-11, MagPro-11, MagTroll-11, and the opamp board for the solid-state wind vane using this method. The Okidata 400 laser-type LED printer and the HP LaserJet 4L were both used here with excellent results.

Laser printers will print on suitable clear transparencies. However, some nearly transparent vellums seem to do a better job and are easier to touch up with a pen. The main difference is that vellums require the developing light be left on for a 10 to 30 percent longer time period than with a transparency. The exact time period depends on the light-transmission characteristics of the vellum used. Make sure when printing the artwork that you choose the maximum darkness setting of your printer. Also, manually feed the vellum and set the printer up so that the finished copy feeds out the back and not the top.

I have used several different types of vellums. Most work OK. One was found superior. It was Micro 100% rag vellum tracing paper, available through the Dick Blick Art Catalog (cat. no. 931218). With this vellum and GC Electronics presensitized circuit boards with positive etch resist, it was found that an exposure time of around 15 minutes, using a 275-W sunlamp, with reflector, at a distance of 12 in, produced good boards. However, to allow for variations, it is a good idea to try a small sample test yourself before you attempt to make a large, expensive board. Figure E-1 shows the setup for making PC boards using the positive photographic method.

One problem was sometimes detected. The corners of thin traces did not always come out well. They sometimes left a tiny break in the foil that did not pass a "continuity" check. The solution to this problem was simple. The vellum with artwork printed on it was placed

FIGURE E-1 Setup for making PC boards using the positive photographic method.

printed side down on a light table, and a fine pen was used to touch up the corners, even if not needed. Even though I worked slowly, it took less than 30 minutes to go over all corners. (Another possible solution to this potential problem is to simply use a slightly shorter exposure time.)

The real tests come after etching. The boards usually turn out nearly perfectly. No shorts or breaks are usually detected either visually or with an ohmmeter. No "Mickey Mouse type" of repair has to be done. I am not sure where the problem of imperfect corners originated in the software or the printers, but a fine pen with permanent black ink quickly and easily solved the problem, *and* one should check over the artwork anyway before using it to make a PC board.

Printing the Foil Patterns

To simplify the process for you, I have created a menu-driven program called FOIL that will print out most foil patterns described in this book. However, you must have access to a laser printer to print out the foil patterns.

To print out a foil pattern, run the FOIL program from the DOS prompt by first selecting the FOIL folder directory and then type "FOIL." Windows 95/98 users can run the program simply by double clicking the FOIL icon (which is in the FOIL folder) in My Computer. The simple program is menu-driven, so you should not have a problem.

Tips in Making Boards That Have Both a Solder-Side and a Component-Side Foil Pattern on a Single-Sided Board

Most projects in this book use two-sided boards. The creation of quality two-sided boards is beyond the capability of most hobbyists. However, it is possible to substitute a board with foil only on the solder side and jumper wires on the component side. Also, additional jumpers are usually added to the solder side. I have done this a number of times—even with boards more complicated than Mag-11. The following is some "stuff" I have learned and am anxious to share with other avid project builders.

1. Create a good-quality single-sided board from the solder side of the foil pattern.
2. Locate all vias (holes that do not have any component leads inserted in them). Tack solder jumpers, using no. 30 wire-wrap wire on the component side connecting the appropriate vias. *Do not connect a jumper wire on the component side to any hole a component lead is inserted into or you will have problems inserting the component into the board.*
3. Mount all sockets and components, and carefully solder them to the board. *Do not install ICs yet.*

4. On the solder side, tack solder the remaining jumper wires.
5. Use a tiny dab of glue from a glue gun to tack down any long jumper wires. Short wires (under an inch or so) should not be glued down to the board.

Notice that the preceding procedure not only takes up valuable time but leaves plenty of room for slipups. To be honest, if a professionally made double-sided board with plated through holes is available at a reasonable price, it is often wise to purchase the board rather than going through the agony of the preceding procedure. However, such a time-consuming procedure is well suited to one- or two-of-a-kind prototypes.

HANDLING STATIC-SENSITIVE

PARTS

All ICs can be damaged by static electricity. However, normally, we are more concerned here with MOS and CMOS and relatives (e.g., EPROMs, RAMs, MPUs) than bipolar ICs (e.g., TTL or LSTTL). Most ICs used in the projects and experiments in this book are in the first group—extremely sensitive to static.

Keep in mind that one does not have to "feel" a static shock to damage an IC. The minimum precautionary requirements are listed here. With average humidity, they are generally sufficient. If the humidity is low, extra precautions should be taken. One precaution that is often recommended is a grounded wrist strap. While such a device will protect ICs quite well, there is the danger of being "fried" yourself if you inadvertently touch a hot ac line. Because of this, I only recommend grounded wrist straps for the professional.

1. Be calmly seated for at least 5 minutes before touching any sensitive electronic part.
2. Touch a good ground with one hand and the work surface and circuit board with the other. With one hand still touching a ground, pick up the protective package containing the IC with the other. This should lower the potential voltage to an acceptable value. Let go of the ground, and touch the circuit board and work surface again.
3. Slowly take the IC from its protective package, and without letting go of it until it is installed in a socket, straighten the pins, if necessary. Now install the IC in the socket or board.

Keep in mind that the board itself also should be treated with care, since even installed ICs can be damaged if the board experiences a moderate jolt of static electricity.

Hint:

From personal experience, I have found that EPROMs, EEPROMs, and RAMs are unusually sensitive to static electricity. However, with even these, I have only rarely damaged an IC by giving it a shock. Much more often, damaged ICs are found on the floor with pins all bent and a suspicious looking sneaker mark on their backs.

Quickie Troubleshoot Guide

You will mostly likely have problems with some experiments or projects in this book. This is to be expected and should be viewed positively. As mentioned before, I doubt if real learning can take place without mistakes being made. However, be cautious because you do not want those mistakes to turn into a catastrophe. Catastrophes have happened in the past when hobbyists worked with vacuum tubes that required voltages over 300 V to work right. You are fortunate. All you need deal with to work on experiments and projects in this book is 13 V maximum, and usually only 5 V. While under ideal circumstances 13 V will give you a tickle, it will not give you a jolt like 300 V will. However, minor catastrophes still do happen. However, these catastrophes are now relegated to the destruction of an expensive IC or a minor burn from a drop of hot solder. Still, be cautious!

As mentioned earlier, it is always wise to power a board, at least the first time, with a power supply that has a current monitor. If you do not have such a power supply with a current monitor, consider using the supply described in Fig. 1-2 the first time you apply power to the board. If either the power supply's meter or lightbulb indicates excessive current, shut off power at once and attempt to determine the problem. Quick action can save money.

Assuming that you have purchased a professionally made printed circuit board, the most likely cause of a malfunctioning board is improper insertion of an IC in a socket. Watch out for bent IC pins. Sometimes the pin is bent so badly that detection is really difficult. The next most likely cause of improper operation is a poor or missing soldered connection. Another likely possibility is a solder bridge. Other possibilities include wrong placement of parts and improper orientation of parts such as diodes, ICs, and electrolytic capacitors. If you have any doubts as to the location of parts, it is best to check the location against the schematic using an ohmmeter as a guide tool. Far down on the list of possibilities are defective ICs and possible breaks in the foil. Be assured that the inherent design of the book's major projects (Mag-11, MagPro-11, MagTroll-11, wind vane) is not at fault. I have built and tested each of these projects successfully a number of times and did not discover a problem in the design or circuit board.

G

HC11-RELATED INTERNET
RESOURCES

Once you really get serious with the HC11, you will need, in addition to this book, the *M68HC11 Reference Manual,* which I have referred to as the "HC11 bible," and the *MC68HC11A8 Technical Data Book.* These two publications are included on the CD-ROM in PDF format. See the CD-ROM's DATA_SHT folder/directory.

You also will want the data sheet for the MC68HC811E2. Happily, this publication, as well as the previous two, are now available at no cost from Motorola's Web site: *http://motsps.com/home/lit_ord.html.* They are also available by calling the Motorola Literature Distribution Center at 800-441-2447 or 303-675-2140. However, at times there is a charge for this phone service. If you wish, you can call them and see.

To order the literature described, use the order numbers below:

M68HC11 Reference Manual—order no. M68HC11RM/AD

MC68HC11A8 Technical Data Book—order no. MC68HC11A8/D

MC68HC811E2—order no. BR586/D

You also will likely want to obtain a copy of the *Technical Literature Guide*—order no. BR101/D.

Motorola has a Web site where it answers frequently asked questions (FAQs) about the HC11. I have visited this site a number of times and found it only fair. The main problem, I believe, is that its answers often lack depth and sometimes commit the sin of omission. However, it still is an interesting Web site. Its URL is *http://www.mcu.motsps.com/hc11/faq.*

Of course, the more general Web address *http://www.mcu.motsps.com/hc11* will direct you to other Motorola Web pages that deal with the HC11.

A Little History

Those who started using computers after the Internet became so popular (roughly 1996–1997) may not be acquainted with bulletin board services (BBSs). Some BBSs are still in existence, but most have given up and migrated to the Web. One of the very best BBSs of all time was Motorola's FREEWARE. This BBS had just about everything an electronics nut like me could ask for. Sadly, it is now gone. However, a remnant of FREE-WARE still exists on Motorola's Web site. This remnant, apparently now referred to as the *freeweb,* can be reached at *http://www.mcu.motsps.com/freeweb/index.html.* The freeweb *is good, but it is no FREEWARE BBS!* Trust me! Nonetheless, it still provides an abundance of information including various versions of the BUFFALO monitor's source code. Check it out. You may not get a true taste of the "good old days," but you may get at least a whiff.

Other Web Sites of Interest to HC11 Addicts

While Motorola Web sites will get you the most info on the HC11, there are many others you will likely want to visit. First, visit Seattle Robotics at *http://www.seattlerobotics.org.* You will find a wealth of MCU information here, and much of it is HC11/HC12/HC16-related. If you go to their links page at *http://www.seattlerobotics.org/websites.html,* you will be able to go directly to a multitude of HC11-related home pages, including at least two, Karl Lunt's and Kevin Ross's, that offer much of interest to the HC11 user.

Other interesting HC11-related sites include

University of Alberta EE FTP service.   A bigtime FTP site that provides information on many Motorola products, including the HC11. This Web site seems closer to the old Motorola FREEWARE site than Motorola's new freeweb. A University of Alberta server: *ftp://nyquist.ee.ualberta.ca/pub/motorola.*

List of HC11 resources including freeware, shareware, and cashware. *http://www.geocities.com/ResearchTriangle/1495/ee_hc11.html.*

Phillip Musumeci home page.   Support for HC11 systems: *http://mirriwinni.csc.rmit.cdu.au/~phillip/f1/index.html.*

Sylvain Bissonnette's home page.   More support for the HC11. See "Girls Love the HC11 Too!" *http://pages.infinit.net/hc1x/.*

Roger's home page.   More HC1 support, including additional links: *http://www.czl.com/~rsch/index.html.*

Viper's HC11 page.   A great source for even more HC11 Web sites: *http://soli.inav.net/~sviper/hc11/hc11.html.*

HC11-Related Magazines with Web Sites

Micro Control Journal is an electronic magazine (i.e., only published on the Web) devoted to microcontrollers. For a free trial subscription, contact them at *http://www.mc journal.com/.*

 Print magazines, with Web sites, that often contain articles/columns on the HC11 include

Circuit Cellar Ink
P.O. Box 698
Holmes, PA 19043-9613
1-800-269-6301
Web: *http://www.circellar.com/*

Nuts & Volts
430 Princeland Court
Corona, CA 91719
1-800-783-4624
Web: *http://www.nutsvolts.com*

CONTACTING THE AUTHOR

You can reach the author through e-mail at: *tfox@ncats.net* or *tomfox@ncats.net*

DATA SHEETS AND ALL

This appendix is sort of a catchall for quite important stuff like HC11 pin assignments, a brief synopsis of the HC11 instruction set/cycle-by-cycle operation, HC11 electrical characteristics, data sheets for National Semiconductor's LM135/235/335, LM2931, and LMC660 series of ICs, information about the data sheets on the CD-ROM, and tips on obtaining data sheets off the World Wide Web.

I wish to thank Motorola and National Semiconductor for granting permission to reprint their copyrighted literature. Also, I wish to take this time to thank Maxim for granting permission to include several of its data sheets on the CD-ROM.

The CD-ROM's Data Sheets

All data sheets on the CD-ROM are in the DATA_SHT folder/directory in PDF format. You will need Adobe's Acrobat Reader to display/print these data sheets. If you do not already have Acrobat Reader on your hard disk, see the next section for details on obtaining it. By the way, the DATA_SHT folder/directory is continually expanding. For instance, it now contains the entire "HC11 bible."

Caution:

Some semiconductor companies (notably Maxim) have strongly advised me to only use the latest data sheets available. In turn, I strongly advise you to only use the latest data sheets available, especially when designing with Maxim parts. Today, with the availability of the World Wide Web, it seems that it is wise to download the latest data sheets directly from the manufacturer's Web site.

Tips on Obtaining Data Sheets Off the World Wide Web

Of course, you will need Adobe's Acrobat Reader to display/print any data sheets you download off the Web. However, some downloads now require that the latest version of the Acrobat Reader be on your hard disk *before* you finish the download procedure. The Acrobat4 folder/directory on the CD-ROM contains the AR40ENG installation program that is used to install Acrobat Reader 4.0 on your hard disk. Acrobat Reader 4.0 requires Windows 95/98. All you need do to start the installation process is double click the AR40ENG file (in the Acrobat4 folder/directory) on Windows 95/98 MyComputer. If you use Windows 3.x, you can obtain Acrobat Reader 3.02 from Adobe's Web site at *http://www.adobe.com.*

Nearly all semiconductor companies now have Web sites where you can download data sheets, price lists, application notes, and other information. The URLs for four company Web sites are given below:

Motorola: *http://www.mcu.motsps.com/.* The URL that will take you right to Motorola's *HC11 User Manual* download page is *http://www.mcu.motsps.com/cgi-bin /Lit?Table*=HC11&Type=USER.

Maxim Semiconductor: Data sheets: *http://www.maxim-ic.com/DataSheets.html.*

National Semiconductor: Data sheets: *http://www.national.com/catalog/.*

Dallas Semiconductor: Data sheets: *http://www.dalsemi.com/DocControl/index.html.*

Personal aside from the author   I hesitated including the 19-page synopsis of the HC11's instruction set and and cycle-by-cycle operation in this appendix because you can obtain a free copy of the HC11's *Technical Data Book* directly from Motorola. However, somehow this book would not seem complete without it.

Listing of Data Sheets (In Order of Appearance)

Figure H-1—Pin assignments for the 48-pin dual-in-line (DIP) packaged HC11 (e.g., MC68HC11A1P)

Figure H-2—Pin assignments for the 52-pin plastic leaded chip carrier (PLCC) packaged HC11 (e.g., MC68HC11A1FN and MC68HC811E2FN)

Figure H-3—HC11's electrical characteristics

Figure H-4—Synopsis of the HC11 instruction set and cycle-by-cycle operation (18 pages)

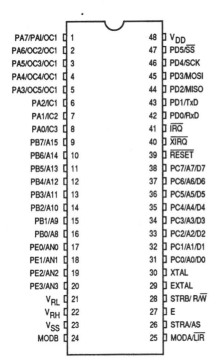

FIGURE H-1 Pin assignments for the 48-pin DIP-packaged HC11. (*Copyright of Motorola. Used by permission.*)

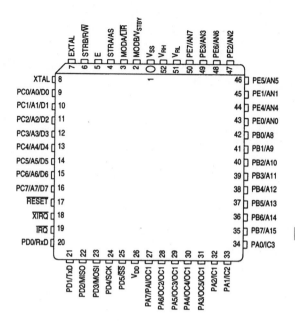

FIGURE H-2 Pin assignments for the 52-pin plastic leaded chip carrier (PLCC) packaged HC11. (*Copyright of Motorola. Used by permission.*)

Maximum Rating

Rating	Symbol	Value	Unit
Supply Voltage	V_{DD}	− 0.3 to + 7.0	V
Input Voltage	V_{in}	− 0.3 to + 7.0	V
Operating Temperature Range MC68HC11A8 MC68HC11A8C MC68HC11A8V MC68HC11A8M MC68L11A8	T_A	T_L to T_H 0 to 70 − 40 to 85 − 40 to 105 − 40 to 125 − 20 to 70	°C
Storage Temperature Range	T_{stg}	− 55 to 150	°C
Current Drain per Pin* Excluding V_{DD}, V_{SS}, V_{RH}, and V_{RL}	I_D	25	mA

*One pin at a time, observing maximum power dissipation limits.

Internal circuitry protects the inputs against damage caused by high static voltages or electric fields; however, normal precautions are necessary to avoid application of any voltage higher than maximum-rated voltages to this high-impedance circuit. Extended operation at the maximum ratings can adversely affect device reliability. Tying unused inputs to an appropriate logic voltage level (either GND or V_{DD}) enhances reliability of operation.

(A)

FIGURE H-3 HC11's electrical characteristics. (*Copyright of Motorola. Used by permission.*)

Thermal Characteristics

Characteristic	Symbol	Value	Unit
Average Junction Temperature	T_J	$T_A + (P_D \times \Theta_{JA})$	°C
Ambient Temperature	T_A	User-determined	°C
Package Thermal Resistance (Junction-to-Ambient) 52-Pin Plastic Quad Pack (PLCC) 48-Pin Plastic Dual In-Line Package (DIP)	Θ_{JA}	50 40	°C/W
Total Power Dissipation	P_D	$P_{INT} + P_{I/O}$ $K + (T_J + 273°C)$ (Note 1)	W
Device Internal Power Dissipation	P_{INT}	$I_{DD} \times V_{DD}$	W
I/O Pin Power Dissipation	$P_{I/O}$ (Note 2)	User-determined	W
A Constant	K	$P_D \times (T_A + 273°C) +$ $\Theta_{JA} \times P_D2$ (Note 3)	W · °C

NOTES:
1. This is an approximate value, neglecting $P_{I/O}$.
2. For most applications $P_{I/O} \ll P_{INT}$ and can be neglected.
3. K is a constant pertaining to the device. Solve for K with a known T_A and a measured P_D (at equilibrium). Use this value of K to solve for P_D and T_J iteratively for any value of T_A.

(B)

FIGURE H-3 (*Continued*) HC11's electrical characteristics.

DC Electrical Characteristics

$V_{DD} = 5.0$ Vdc $\pm$ 10%, $V_{SS} = 0$ Vdc, $T_A = T_L$ to T_H, unless otherwise noted

Characteristics		Symbol	Min	Max	Unit
Output Voltage (Note 1)	All Outputs except XTAL	V_{OL}	—	0.1	V
$I_{Load} = \pm 10.0$ µA	All Outputs Except XTAL, $\overline{RESET}$, and MODA	V_{OH}	$V_{DD} - 01$	—	
Output High Voltage (Note 1) $I_{Load} = -0.8$ mA, $V_{DD} = 4.5$ V	All Outputs Except XTAL, $\overline{RESET}$, and MODA	V_{OH}	$V_{DD} - 0.8$	—	V
Output Low Voltage $I_{Load} = 1.6$ mA	All Outputs Except XTAL	V_{OL}	—	0.4	V
Input High Voltage	All Inputs Except $\overline{RESET}$	V_{IH}	$0.7 \times V_{DD}$	$V_{DD} + 0.3$	V
	$\overline{RESET}$		$0.8 \times V_{DD}$	$V_{DD} + 0.3$	
Input Low Voltage	All Inputs	V_{IL}	$V_{SS} - 0.3$	$0.2 \times V_{DD}$	V
I/O Ports, Three-State Leakage $V_{in} = V_{IH}$ or V_{IL}	PA7, PC0-PC7, PD0-PD5, AS/STRA, MODA/$\overline{LIR}$, $\overline{RESET}$	I_{OZ}	—	±10	µA
Input Leakage Current (Note 2)		I_{in}			µA
$V_{in} = V_{DD}$ or V_{SS}	PA0-PA2, $\overline{IRQ}$, $\overline{XIRQ}$		—	±1	
$V_{in} = V_{DD}$ or V_{SS}	MODB/V_{STBY}		—	±10	
RAM Standby Voltage	Power down	V_{SB}	4.0	V_{DD}	V
RAM Standby Current	Power down	I_{SB}	—	10	µA
Total Supply Current (Note 3)		I_{DD}			mA
RUN:					
Single-Chip Mode		dc – 2 MHz	—	15	
		3 MHz	—	27	
Expanded Multiplexed Mode		dc – 2 MHz	—	27	
		3 MHz	—	35	
WAIT:		W_{IDD}			mA
All Peripheral Functions Shut Down					
Single-Chip Mode		dc – 2 MHz	—	6	
		3 MHz	—	15	
Expanded Multiplexed Mode		dc – 2 MHz	—	10	
		3 MHz	—	20	
STOP:		S_{IDD}			µA
No Clocks, Single-Chip Mode		dc – 2 MHz	—	50	
		3 MHz	—	150	
Input Capacitance PA0-PA2, PE0-PE7, $\overline{IRQ}$, $\overline{XIRQ}$, EXTAL		C_{in}	—	8	pF
PA7, PC0-PC7, PD0-PD5, AS/STRA, MODA/$\overline{LIR}$, $\overline{RESET}$			—	12	
Power Dissipation		P_D			mW
	Single-Chip Mode	2 MHz	—	85	
	Expanded Multiplexed Mode		—	150	
	Single-Chip Mode	3 MHz	—	150	
	Expanded Multiplexed Mode		—	195	

NOTES:

1. V_{OH} specification for $\overline{RESET}$ and MODA is not applicable because they are open-drain pins. V_{OH} specification not applicable to ports C and D in wired-OR mode.
2. Refer to A/D specification for leakage current for port E.
3. EXTAL is driven with a square wave, and
   $t_{cyc} = 500$ ns for 2 MHz rating;
   $t_{cyc} = 333$ ns for 3 MHz rating.
   $V_{IL} \leq 0.2$ V
   $V_{IH} \geq V_{DD} - 0.2$ V
   No dc loads.

(C)

FIGURE H-3 (*Continued*) HC11's electrical characteristics.

DC Electrical Characteristics (MC68L11A8)

$V_{DD} = 3.0$ Vdc to 5.5 Vdc, $V_{SS} = 0$ Vdc, $T_A = T_L$ to T_H, unless otherwise noted

Characteristics		Symbol	Min	Max	Unit
Output Voltage (Note 1)	All Outputs except XTAL	V_{OL}	—	0.1	V
	All Outputs Except XTAL, RESET, and MODA	V_{OH}	$V_{DD} - 01$	—	
$I_{Load} = \pm 10.0\ \mu A$					
Output High Voltage (Note 1)	All Outputs Except XTAL, RESET, and MODA	V_{OH}	$V_{DD} - 0.8$	3	V
$I_{Load} = -0.8$ mA, $V_{DD} = 4.5$ V					
Output Low Voltage	All Outputs Except XTAL	V_{OL}	—	0.4	V
$I_{Load} = 1.6$ mA					
Input High Voltage	All Inputs Except RESET	V_{IH}	$0.7 \times V_{DD}$	$V_{DD} + 0.3$	V
	RESET		$0.8 \times V_{DD}$	$V_{DD} + 0.3$	
Input Low Voltage	All Inputs	V_{IL}	$V_{SS} - 0.3$	$0.2 \times V_{DD}$	V
I/O Ports, Three-State Leakage	PA7, PC0-PC7, PD0-PD5,	I_{OZ}	—	±10	µA
$V_{in} = V_{IH}$ or V_{IL}	AS/STRA, MODA/LIR, RESET				
Input Leakage Current (Note 2)		I_{in}			µA
$V_{in} = V_{DD}$ or V_{SS}	PA0-PA2, IRQ, XIRQ		—	±1	
$V_{in} = V_{DD}$ or V_{SS}	MODB/V_{STBY}		—	±10	
RAM Standby Voltage	Power down	V_{SB}	2.0	V_{DD}	V
RAM Standby Current	Power down	I_{SB}	—	10	µA
Total Supply Current (Note 3)					mA
RUN:		I_{DD}			
Single-Chip Mode		dc – 1 MHz	—	4	
		2MHz	—	8	
Expanded Multiplexed Mode		dc – 1 MHz	—	7	
		2 MHz	—	14	
WAIT:		W_{IDD}			mA
All Peripheral Functions Shut Down					
Single-Chip Mode		dc – 1 MHz	—	3	
		2 MHz	—	6	
Expanded Multiplexed Mode		dc – 1 MHz	—	2.5	
		2 MHz	—	5	
STOP:		S_{IDD}			µA
No Clocks, Single-Chip Mode		dc – 1 MHz	—	25	
		2 MHz	—	25	
Input Capacitance PA0-PA2, PE0-PE7, IRQ, XIRQ, EXTAL		C_{in}	—	8	pF
PA7, PC0-PC7, PD0-PD5, AS/STRA, MODA/LIR, RESET			—	12	
Power Dissipation		P_D			mW
	Single-Chip Mode 1 MHz		—	12	
	Expanded Multiplexed Mode		—	21	
	Single-Chip Mode 2 MHz		—	24	
	Expanded Multiplexed Mode		—	42	

NOTES:
1. V_{OH} specification for RESET and MODA is not applicable because they are open-drain pins. V_{OH} specification not applicable to ports C and D in wired-OR mode.
2. Refer to A/D specification for leakage current for port E.
3. EXTAL is driven with a square wave, and
   $t_{cyc} = 1000$ ns for 1MHz rating;
   $t_{cyc} = 500$ ns for 2 MHz rating.
   $V_{IL} \leq 0.2$ V
   $V_{IH} \geq V_{DD} - 0.2$ V
   No dc loads.

(D)

FIGURE H-3   (*Continued*) HC11's electrical characteristics.

INSTRUCTION SET

The central processing unit (CPU) in the MC68HC11A8 is basically a proper extension of the MC6801 CPU. In addition to its ability to execute all M6800 and M6801 instructions, the MC68HC11A8 CPU has a paged operation code (opcode) map with a total of 91 new opcodes. Major functional additions include a second 16-bit index register (Y register), two types of 16-by-16 divide instructions, STOP and WAIT instructions, and bit manipulation instructions.

Table 10-1 shows all MC68HC11A8 instructions in all possible addressing modes. For each instruction, the operand construction is shown as well as the total number of machine code bytes and execution time in CPU E-clock cycles. Notes are provided at the end of Table 10-1 which explain the letters in the Operand and Execution Time columns for some instructions. Definitions of "Special Ops" found in the Boolean Expression column are found in Figure 10-2.

Tables 10-2 through 10-8 provide a detailed description of the information present on the address bus, data bus, and the read/write (R/W̄) line during each cycle of each instruction. The information is useful in comparing actual with expected results during debug of both software and hardware as the program is executed. The information is categorized in groups according to addressing mode and number of cycles per instruction. In general, instructions with the same address mode and number of cycles execute in the same manner. Exceptions are indicated in the table.

(A)

FIGURE H-4 **Synopsis of the HC11 instruction set and cycle-by-cycle operation.** (*Copyright of Motorola. Used by permission.*)

Table 10-1. MC68HC11A8 Instructions, Addressing Modes, and Execution Times (Sheet 1 of 7)

Source Form(s)	Operation	Boolean Expression		Machine Coding (Hexadecimal) Opcode	Operand(s)	Bytes	Cycle	Cycle by Cycle*	Condition Codes S X H I N Z V C
ABA	Add Accumulators	A + B → A		1B		1	2	2-1	- - ↕ - ↕ ↕ ↕ ↕
ABX	Add B to X	IX + 00:B → IX	INH	3A		1	3	2-2	- - - - - - - -
ABY	Add B to Y	IY + 00:B → IY	INH	18 3A		2	4	2-4	- - - - - - - -
ADCA(opr)	Add with Carry to A	A + M + C → A	A IMM	89	ii	2	2	3-1	- - ↕ - ↕ ↕ ↕ ↕
			A DIR	99	dd	2	3	4-1	
			A EXT	B9	hh ll	3	4	5-2	
			A IND,X	A9	ff	2	4	6-2	
			A IND,Y	18 A9	ff	3	5	7-2	
ADCB (opr)	Add with Carry to B	B + M + C → B	B IMM	C9	ii	2	2	3-1	- - ↕ - ↕ ↕ ↕ ↕
			B DIR	D9	dd	2	3	4-1	
			B EXT	F9	hh ll	3	4	5-2	
			B IND,X	E9	ff	2	4	6-2	
			B IND,Y	18 E9	ff	3	5	7-2	
ADDA (opr)	Add Memory to A	A + M → A	A IMM	8B	ii	2	2	3-1	- - ↕ - ↕ ↕ ↕ ↕
			A DIR	9B	dd	2	3	4-1	
			A EXT	BB	hh ll	3	4	5-2	
			A IND,X	AB	ff	2	4	6-2	
			A IND,Y	18 AB	ff	3	5	7-2	
ADDB(opr)	Add Memory to B	B + M → B	B IMM	CB	ii	2	2	3-1	- - ↕ - ↕ ↕ ↕ ↕
			B DIR	DB	dd	2	3	4-1	
			B EXT	FB	hh ll	3	4	5-2	
			B IND,X	EB	ff	2	4	6-2	
			B IND,Y	18 EB	ff	3	5	7-2	
ADDD (opr)	Add 16-Bit to D	D + M:M + 1 → D	IMM	C3	jj kk	3	4	3-3	- - - - ↕ ↕ ↕ ↕
			DIR	D3	dd	2	5	4-7	
			EXT	F3	hh ll	3	6	5-10	
			IND,X	E3	ff	2	6	6-10	
			IND,Y	18 E3	ff	3	7	7-8	
ANDA (opr)	AND A with Memory	A•M → A	A IMM	84	ii	2	2	3-1	- - - - ↕ ↕ 0 -
			A DIR	94	dd	2	3	4-1	
			A EXT	B4	hh ll	3	4	5-2	
			A IND,X	A4	ff	2	4	6-2	
			A IND,Y	18 A4	ff	3	5	7-2	
ANDB (opr)	AND B with Memory	B•M → B	B IMM	C4	ii	2	2	3-1	- - - - ↕ ↕ 0 -
			B DIR	D4	dd	2	3	4-1	
			B EXT	F4	hh ll	3	4	5-2	
			B IND,X	E4	ff	2	4	6-2	
			B IND,Y	18 E4	ff	3	5	7-2	
ASL (opr)	Arithmetic Shift Left	□←□□□□□□□□←0 C b7 b0	EXT	78	hh ll	3	6	5-8	- - - - ↕ ↕ ↕ ↕
			IND,X	68	ff	2	6	6-3	
			IND,Y	18 68	ff	3	7	7-3	
ASLA			A INH	48		1	2	2-1	
ASLB			B INH	58		1	2	2-1	
ASLD	Arithmetic Shift Left Double	□←□□ - - - □□←0 C b15 b0	INH	05		1	3	2-2	- - - - ↕ ↕ ↕ ↕
ASR (opr)	Arithmetic Shift Right	□→□□□□□□□□→□ b7 b0 C	EXT	77	hh ll	3	6	5-8	- - - - ↕ ↕ ↕ ↕
			IND,X	67	ff	2	6	6-3	
			IND,Y	18 67	ff	3	7	7-3	
ASRA			A INH	47		1	2	2-1	
ASRB			B INH	57		1	2	2-1	
BCC (rel)	Branch if Carry Clear	? C = 0	REL	24	rr	2	3	8-1	- - - - - - - -
BCLR (opr) (msk)	Clear Bit(s)	M•(mm̄) → M	DIR	15	dd mm	3	6	4-10	- - - - ↕ ↕ 0 -
			IND,X	1D	ff mm	3	7	6-13	
			IND,Y	18 1D	ff mm	4	8	7-10	
BCS (rel)	Branch if Carry Set	? C = 1	REL	25	rr	2	3	8-1	- - - - - - - -
BEQ (rel)	Branch if = Zero	? Z = 1	REL	27	rr	2	3	8-1	- - - - - - - -

*Cycle-by-cycle number provides a reference to Tables 10-2 through 10-8 which detail cycle-by-cycle operation.
  Example: Table 10-1 Cycle-by-Cycle column reference number 2-4 equals Table 10-2 line item 2-4.

(B)

FIGURE H-4 *(Continued)* Synopsis of the HC11 instruction set and cycle-by-cycle operation.

Table 10-1. MC68HC11A8 Instructions, Addressing Modes, and Execution Times (Sheet 2 of 7)

Source Form(s)	Operation	Boolean Expression	Addressing Mode for Operand	Opcode	Operand(s)	Bytes	Cycle	Cycle by Cycle*	S	X	H	I	N	Z	V	C
BGE (rel)	Branch if ≥ Zero	? N ⊕ V = 0	REL	2C	rr	2	3	8-1	-	-	-	-	-	-	-	-
BGT (rel)	Branch if > Zero	? Z + (N ⊕ V) = 0	REL	2E	rr	2	3	8-1	-	-	-	-	-	-	-	-
BHI (rel)	Branch if Higher	? C + Z = 0	REL	22	rr	2	3	8-1	-	-	-	-	-	-	-	-
BHS (rel)	Branch if Higher or Same	? C = 0	REL	24	rr	2	3	8-1	-	-	-	-	-	-	-	-
BITA (opr)	Bit(s) Test A with Memory	A•M	A IMM	85	ii	2	2	3-1	-	-	-	-	↕	↕	0	-
			A DIR	95	dd	2	3	4-1								
			A EXT	B5	hh ll	3	4	5-2								
			A IND,X	A5	ff	2	4	6-2								
			A IND,Y	18 A5	ff	3	5	7-2								
BITB (opr)	Bit(s) Test B with Memory	B•M	B IMM	C5	ii	2	2	3-1	-	-	-	-	↕	↕	0	-
			B DIR	D5	dd	2	3	4-1								
			B EXT	F5	hh ll	3	4	5-2								
			B IND,X	E5	ff	2	4	6-2								
			B IND,Y	18 E5	ff	3	5	7-2								
BLE (rel)	Branch if ≤ Zero	? Z + (N ⊕ V) = 1	REL	2F	rr	2	3	8-1	-	-	-	-	-	-	-	-
BLO (rel)	Branch if Lower	? C = 1	REL	25	rr	2	3	8-1	-	-	-	-	-	-	-	-
BLS (rel)	Branch if Lower or Same	? C + Z = 1	REL	23	rr	2	3	8-1	-	-	-	-	-	-	-	-
BLT (rel)	Branch If < Zero	? N ⊕ V = 1	REL	2D	rr	2	3	8-1	-	-	-	-	-	-	-	-
BMI (rel)	Branch if Minus	? N = 1	REL	2B	rr	2	3	8-1	-	-	-	-	-	-	-	-
BNE (rel)	Branch if Not = Zero	? Z = 0	REL	26	rr	2	3	8-1	-	-	-	-	-	-	-	-
BPL (rel)	Branch if Plus	? N = 0	REL	2A	rr	2	3	8-1	-	-	-	-	-	-	-	-
BRA (rel)	Branch Always	? 1 = 1	REL	20	rr	2	3	8-1	-	-	-	-	-	-	-	-
BRCLR(opr) (msk) (rel)	Branch if Bit(s) Clear	? M • mm = 0	DIR	13	dd mm rr	4	6	4-11	-	-	-	-	-	-	-	-
			IND,X	1F	ff mm rr	4	7	6-14								
			IND,Y	18 1F	ff mm rr	5	8	7-11								
BRN (rel)	Branch Never	? 1 = 0	REL	21	rr	2	3	8-1	-	-	-	-	-	-	-	-
BRSET(opr) (msk) (rel)	Branch if Bit(s) Set	? (M̄) • mm = 0	DIR	12	dd mm rr	4	6	4-11	-	-	-	-	-	-	-	-
			IND,X	1E	ff mm rr	4	7	6-14								
			IND,Y	18 1E	ff mm rr	5	8	7-11								
BSET(opr) (msk)	Set Bit(s)	M + mm → M	DIR	14	dd mm	3	6	4-10	-	-	-	-	↕	↕	0	-
			IND,X	1C	ff mm	3	7	6-13								
			IND,Y	18 1C	ff mm	4	8	7-10								
BSR (rel)	Branch to Subroutine	See Special Ops	REL	8D	rr	2	6	8-2	-	-	-	-	-	-	-	-
BVC (rel)	Branch if Overflow Clear	? V = 0	REL	28	rr	2	3	8-1	-	-	-	-	-	-	-	-
BVS (rel)	Branch if Overflow Set	? V = 1	REL	29	rr	2	3	8-1	-	-	-	-	-	-	-	-
CBA	Compare A to B	A – B	INH	11		1	2	2-1	-	-	-	-	↕	↕	↕	↕
CLC	Clear Carry Bit	0 → C	INH	0C		1	2	2-1	-	-	-	-	-	-	-	0
CLI	Clear Interrupt Mask	0 → I	INH	0E		1	2	2-1	-	-	-	0	-	-	-	-
CLR (opr)	Clear Memory Byte	0 → M	EXT	7F	hh ll	3	6	5-8	-	-	-	-	0	1	0	0
			IND,X	6F	ff	2	6	6-3								
			IND,Y	18 6F	ff	3	7	7-3								
CLRA	Clear Accumulator A	0 → A	A INH	4F		1	2	2-1	-	-	-	-	0	1	0	0
CLRB	Clear Accumulator B	0 → B	B INH	5F		1	2	2-1	-	-	-	-	0	1	0	0
CLV	Clear Overflow Flag	0 → V	INH	0A		1	2	2-1	-	-	-	-	-	-	0	-
CMPA (opr)	Compare A to Memory	A – M	A IMM	81	ii	2	2	3-1	-	-	-	-	↕	↕	↕	↕
			A DIR	91	dd	2	3	4-1								
			A EXT	B1	hh ll	3	4	5-2								
			A IND,X	A1	ff	2	4	6-2								
			A IND,Y	18 A1	ff	3	5	7-2								

*Cycle-by-cycle number provides a reference to Tables 10-2 through 10-8 which detail cycle-by-cycle operation.
  Example: Table 10-1 Cycle-by-Cycle column reference number 2-4 equals Table 10-2 line item 2-4.

(C)

FIGURE H-4   *(Continued)* Synopsis of the HC11 instruction set and cycle-by-cycle operation.

Table 10-1. MC68HC11A8 Instructions, Addressing Modes, and Execution Times (Sheet 3 of 7)

Source Form(s)	Operation	Boolean Expression	Addressing Mode for Operand	Machine Coding (Hexadecimal) Opcode	Operand(s)	Bytes	Cycle	Cycle by Cycle*	Condition Codes S X H I N Z V C
CMPB (opr)	Compare B to Memory	B – M	B IMM	C1	ii	2	2	3-1	- - - - ↕ ↕ ↕ ↕
			B DIR	D1	dd	2	3	4-1	
			B EXT	F1	hh ll	3	4	5-2	
			B IND,X	E1	ff	2	4	6-2	
			B IND,Y	18 E1	ff	3	5	7-2	
COM (opr)	1's Complement Memory Byte	$FF – M → M	EXT	73	hh ll	3	6	5-8	- - - - ↕ ↕ 0 1
			IND,X	63	ff	2	6	6-3	
			IND,Y	18 63	ff	3	7	7-3	
COMA	1's Complement A	$FF – A → A	A INH	43		1	2	2-1	- - - - ↕ ↕ 0 1
COMB	1's Complement B	$FF – B → B	B INH	53		1	2	2-1	- - - - ↕ ↕ 0 1
CPD (opr)	Compare D to Memory 16-Bit	D – M:M + 1	IMM	1A 83	jj kk	4	5	3-5	- - - - ↕ ↕ ↕ ↕
			DIR	1A 93	dd	3	6	4-9	
			EXT	1A B3	hh ll	4	7	5-11	
			IND,X	1A A3	ff	3	7	6-11	
			IND,Y	CD A3	ff	3	7	7-8	
CPX (opr)	Compare X to Memory 16-Bit	IX – M:M + 1	IMM	8C	jj kk	3	4	3-3	- - - - ↕ ↕ ↕ ↕
			DIR	9C	dd	2	5	4-7	
			EXT	BC	hh ll	3	6	5-10	
			IND,X	AC	ff	2	6	6-10	
			IND,Y	CD AC	ff	3	7	7-8	
CPY (opr)	Compare Y to Memory 16-Bit	IY – M:M + 1	IMM	18 8C	jj kk	4	5	3-5	- - - - ↕ ↕ ↕ ↕
			DIR	18 9C	dd	3	6	4-9	
			EXT	18 BC	hh ll	4	7	5-11	
			IND,X	1A AC	ff	3	7	6-11	
			IND,Y	18 AC	ff	3	7	7-8	
DAA	Decimal Adjust A	Adjust Sum to BCD	INH	19		1	2	2-1	- - - - ↕ ↕ ↕ ↕
DEC (opr)	Decrement Memory Byte	M – 1 → M	EXT	7A	hh ll	3	6	5-8	- - - - ↕ ↕ ↕ -
			IND,X	6A	ff	2	6	6-3	
			IND,Y	18 6A	ff	3	7	7-3	
DECA	Decrement Accumulator A	A – 1 → A	A INH	4A		1	2	2-1	- - - - ↕ ↕ ↕ -
DECB	Decrement Accumulator B	B – 1 → B	B INH	5A		1	2	2-1	- - - - ↕ ↕ ↕ -
DES	Decrement Stack Pointer	SP – 1 → SP	INH	34		1	3	2-3	- - - - - - - -
DEX	Decrement Index Register X	IX – 1 → IX	INH	09		1	3	2-2	- - - - - ↕ - -
DEY	Decrement Index Register Y	IY – 1 → IY	INH	18 09		2	4	2-4	- - - - - ↕ - -
EORA (opr)	Exclusive OR A with Memory	A ⊕ M → A	A IMM	88	ii	2	2	3-1	- - - - ↕ ↕ 0 -
			A DIR	98	dd	2	3	4-1	
			A EXT	88	hh ll	3	4	5-2	
			A IND,X	A8	ff	2	4	6-2	
			A IND,Y	18 A8	ff	3	5	7-2	
EORB (opr)	Exclusive OR B with Memory	B ⊕ M → B	B IMM	C8	ii	2	2	3-1	- - - - ↕ ↕ 0 -
			B DIR	D8	dd	2	3	4-1	
			B EXT	F8	hh ll	3	4	5-2	
			B IND,X	E8	ff	2	4	6-2	
			B IND,Y	18 E8	ff	3	5	7-2	
FDIV	Fractional Divide 16 by 16	D/IX → IX; r → D	INH	03		1	41	2-17	- - - - - ↕ ↕ ↕
IDIV	Integer Divide 16 by 16	D/IX → IX; r → D	INH	02		1	41	2-17	- - - - - ↕ 0 ↕
INC (opr)	Increment Memory Byte	M + 1 → M	EXT	7C	hh ll	3	6	5-8	- - - - ↕ ↕ ↕ -
			IND,X	6C	ff	2	6	6-3	
			IND,Y	18 6C	ff	3	7	7-3	
INCA	Increment Accumulator A	A + 1 → A	A INH	4C		1	2	2-1	- - - - ↕ ↕ ↕ -
INCB	Increment Accumulator B	B + 1 → B	B INH	5C		1	2	2-1	- - - - ↕ ↕ ↕ -
INS	Increment Stack Pointer	SP + 1 → SP	INH	31		1	3	2-3	- - - - - - - -

*Cycle-by-cycle number provides a reference to Tables 10-2 through 10-8 which detail cycle-by-cycle operation.
   Example: Table 10-1 Cycle-by-Cycle column reference number 2-4 equals Table 10-2 line item 2-4.

(D)

FIGURE H-4   (*Continued*) Synopsis of the HC11 instruction set and cycle-by-cycle operation.

Table 10-1. MC68HC11A8 Instructions, Addressing Modes, and Execution Times (Sheet 4 of 7)

Source Form(s)	Operation	Boolean Expression	Addressing Mode for Operand	Opcode	Operand(s)	Bytes	Cycle	Cycle by Cycle*	S	X	H	I	N	Z	V	C
INX	Increment Index Register X	IX + 1 → IX	INH	08		1	3	2-2	-	-	-	-	-	↕	-	-
INY	Increment Index Register Y	IY + 1 → IY	INH	18 08		2	4	2-4	-	-	-	-	-	↕	-	-
JMP (opr)	Jump	See Special Ops	EXT	7E	hh ll	3	3	5-1	-	-	-	-	-	-	-	-
			IND,X	6E	ff	2	3	6-1								
			IND,Y	18 6E	ff	3	4	7-1								
JSR (opr)	Jump to Subroutine	See Special Ops	DIR	9D	dd	2	5	4-8	-	-	-	-	-	-	-	-
			EXT	BD	hh ll	3	6	5-12								
			IND,X	AD	ff	2	6	6-12								
			IND,Y	18 AD	ff	3	7	7-9								
LDAA (opr)	Load Accumulator A	M → A	A IMM	86	ii	2	2	3-1	-	-	-	-	↕	↕	0	-
			A DIR	96	dd	2	3	4-1								
			A EXT	B6	hh ll	3	4	5-2								
			A IND,X	A6	ff	2	4	6-2								
			A IND,Y	18 A6	ff	3	5	7-2								
LDAB (opr)	Load Accumulator B	M → B	B IMM	C6	ii	2	2	3-1	-	-	-	-	↕	↕	0	-
			B DIR	D6	dd	2	3	4-1								
			B EXT	F6	hh ll	3	4	5-2								
			B IND,X	E6	ff	2	4	6-2								
			B IND,Y	18 E6	ff	3	5	7-2								
LDD (opr)	Load Double Accumulator D	M → A, M + 1 → B	IMM	CC	jj kk	3	3	3-2	-	-	-	-	↕	↕	0	-
			DIR	DC	dd	2	4	4-3								
			EXT	FC	hh ll	3	5	5-4								
			IND,X	EC	ff	2	5	6-6								
			IND,Y	18 EC	ff	3	6	7-6								
LDS (opr)	Load Stack Pointer	M:M + 1 → SP	IMM	8E	jj kk	3	3	3-2	-	-	-	-	↕	↕	0	-
			DIR	9E	dd	2	4	4-3								
			EXT	BE	hh ll	3	5	5-4								
			IND,X	AE	ff	2	5	6-6								
			IND,Y	18 AE	ff	3	6	7-6								
LDX (opr)	Load Index Register X	M:M + 1 → IX	IMM	CE	jj kk	3	3	3-2	-	-	-	-	↕	↕	0	-
			DIR	DE	dd	2	4	4-3								
			EXT	FE	hh ll	3	5	5-4								
			IND,X	EE	ff	2	5	6-6								
			IND,Y	CD EE	ff	3	6	7-6								
LDY (opr)	Load Index Register Y	M:M + 1 → IY	IMM	18 CE	jj kk	4	4	3-4	-	-	-	-	↕	↕	0	-
			DIR	18 DE	dd	3	5	4-5								
			EXT	18 FE	hh ll	4	6	5-6								
			IND,X	1A EE	ff	3	6	6-7								
			IND,Y	18 EE	ff	3	6	7-6								
LSL (opr)	Logical Shift Left	C ◄─[b7 ─── b0]◄─ 0	EXT	78	hh ll	3	6	5-8	-	-	-	-	↕	↕	↕	↕
			IND,X	68	ff	2	6	6-3								
			IND,Y	18 68	ff	3	7	3-7								
LSLA			A INH	48		1	2	2-1								
LSLB			B INH	58		1	2	2-1								
LSLD	Logical Shift Left Double	C ◄─[b15]···[b0]◄─ 0	INH	05		1	3	2-2	-	-	-	-	↕	↕	↕	↕
LSR (opr)	Logical Shift Right	0 ─►[b7 ─── b0]─► C	EXT	74	hh ll	3	6	5-8	-	-	-	-	0	↕	↕	↕
			IND,X	64	ff	2	6	6-3								
			IND,Y	18 64	ff	3	7	7-3								
LSRA			A INH	44		1	2	2-1								
LSRB			B INH	54		1	2	2-1								
LSRD	Logical Shift Right Double	0 ─►[b15]···[b0]─► C	INH	04		1	3	2-2	-	-	-	-	0	↕	↕	↕
MUL	Multiply 8 by 8	AxB → D	INH	3D		1	10	2-13	-	-	-	-	-	-	-	↕

*Cycle-by-cycle number provides a reference to Tables 10-2 through 10-8 which detail cycle-by-cycle operation.
   Example: Table 10-1 Cycle-by-Cycle column reference number 2-4 equals Table 10-2 line item 2-4.

(E)

FIGURE H-4 (*Continued*) Synopsis of the HC11 instruction set and cycle-by-cycle operation.

Table 10-1. MC68HC11A8 Instructions, Addressing Modes, and Execution Times (Sheet 5 of 7)

Source Form(s)	Operation	Boolean Expression	Addressing Mode for Operand	Opcode	Operand(s)	Bytes	Cycle	Cycle by Cycle*	S	X	H	I	N	Z	V	C
NEG (opr)	2's Complement Memory Byte	0 − M → M	EXT	70	hh ll	3	6	5-8	-	-	-	-	↕	↕	↕	↕
			IND,X	60	ff	2	6	6-3								
			IND,Y	18 60	ff	3	7	7-3								
NEGA	2's Complement A	0 − A → A	A INH	40		1	2	2-1	-	-	-	-	↕	↕	↕	↕
NEGB	2's Complement B	0 − B → B	B INH	50		1	2	2-1	-	-	-	-	↕	↕	↕	↕
NOP	No Operation	No Operation	INH	01		1	2	2-1	-	-	-	-	-	-	-	-
ORAA (opr)	OR Accumulator A (Inclusive)	A + M → A	A IMM	8A	ii	2	2	3-1	-	-	-	-	↕	↕	0	-
			A DIR	9A	dd	2	3	4-1								
			A EXT	BA	hh ll	3	4	5-2								
			A IND,X	AA	ff	2	4	6-2								
			A IND,Y	18 AA	ff	3	5	7-2								
ORAB (opr)	OR Accumulator B (Inclusive)	B + M → B	B IMM	CA	ii	2	2	3-1	-	-	-	-	↕	↕	0	-
			B DIR	DA	dd	2	3	4-1								
			B EXT	FA	hh ll	3	4	5-2								
			B IND,X	EA	ff	2	4	6-2								
			B IND,Y	18 EA	ff	3	5	7-2								
PSHA	Push A onto Stack	A → Stk, SP = SP−1	A INH	36		1	3	2-6	-	-	-	-	-	-	-	-
PSHB	Push B onto Stack	B → Stk, SP = SP−1	B INH	37		1	3	2-6	-	-	-	-	-	-	-	-
PSHX	Push X onto Stack (Lo First)	IX → Stk, SP = SP−2	INH	3C		1	4	2-7	-	-	-	-	-	-	-	-
PSHY	Push Y onto Stack (Lo First)	IY → Stk, SP = SP−2	INH	18 3C		2	5	2-8	-	-	-	-	-	-	-	-
PULA	Pull A from Stack	SP = SP + 1, A←Stk	A INH	32		1	4	2-9	-	-	-	-	-	-	-	-
PULB	Pull B from Stack	SP = SP + 1, B←Stk	B INH	33		1	4	2-9	-	-	-	-	-	-	-	-
PULX	Pull X from Stack (Hi First)	SP = SP + 2, IX←Stk	INH	38		1	5	2-10	-	-	-	-	-	-	-	-
PULY	Pull Y from Stack (Hi First)	SP = SP + 2, IY←Stk	INH	18 38		2	6	2-11	-	-	-	-	-	-	-	-
ROL (opr)	Rotate Left		EXT	79	hh ll	3	6	5-8	-	-	-	-	↕	↕	↕	↕
			IND,X	69	ff	2	6	6-3								
			IND,Y	18 69	ff	3	7	7-3								
ROLA			A INH	49		1	2	2-1								
ROLB			B INH	59		1	2	2-1								
ROR (opr)	Rotate Right		EXT	76	hh ll	3	6	5-8	-	-	-	-	↕	↕	↕	↕
			IND,X	66	ff	2	6	6-3								
			IND,Y	18 66	ff	3	7	7-3								
RORA			A INH	46		1	2	2-1								
RORB			B INH	56		1	2	2-1								
RTI	Return from Interrupt	See Special Ops	INH	3B		1	12	2-14	↕	↕	↕	↕	↕	↕	↕	↕
RTS	Return from Subroutine	See Special Ops	INH	39		1	5	2-12	-	-	-	-	-	-	-	-
SBA	Subtract B from A	A − B → A	INH	10		1	2	2-1	-	-	-	-	↕	↕	↕	↕
SBCA (opr)	Subtract with Carry from A	A − M − C → A	A IMM	82	ii	2	2	3-1	-	-	-	-	↕	↕	↕	↕
			A DIR	92	dd	2	3	4-1								
			A EXT	B2	hh ll	3	4	5-2								
			A IND,X	A2	ff	2	4	6-2								
			A IND,Y	18 A2	ff	3	5	7-2								
SBCB (opr)	Subtract with Carry from B	B − M − C → B	B IMM	C2	ii	2	2	3-1	-	-	-	-	↕	↕	↕	↕
			B DIR	D2	dd	2	3	4-1								
			B EXT	F2	hh ll	3	4	5-2								
			B IND,X	E2	ff	2	4	6-2								
			B IND,Y	18 E2	ff	3	5	7-2								
SEC	Set Carry	1 → C	INH	0D		1	2	2-1	-	-	-	-	-	-	-	1
SEI	Set Interrupt Mask	1 → I	INH	0F		1	2	2-1	-	-	-	1	-	-	-	-
SEV	Set Overflow Flag	1 → V	INH	0B		1	2	2-1	-	-	-	-	-	-	1	-

*Cycle-by-cycle number provides a reference to Tables 10-2 through 10-8 which detail cycle-by-cycle operation.
  Example: Table 10-1 Cycle-by-Cycle column reference number 2-4 equals Table 10-2 line item 2-4.

(F)

FIGURE H-4 *(Continued)* Synopsis of the HC11 instruction set and cycle-by-cycle operation.

Table 10-1. MC68HC11A8 Instructions, Addressing Modes, and Execution Times (Sheet 6 of 7)

Source Form(s)	Operation	Boolean Expression	Addressing Mode for Operand	Machine Coding (Hexadecimal) Opcode	Operand(s)	Bytes	Cycle	Cycle by Cycle*	Condition Codes S X H I N Z V C
STAA (opr)	Store Accumulator A	A → M	A DIR	97	dd	2	3	4-2	- - - - ↕ ↕ 0 -
			A EXT	B7	hh ll	3	4	5-3	
			A IND,X	A7	ff	2	4	6-5	
			A IND,Y	18 A7	ff	3	5	7-5	
STAB (opr)	Store Accumulator B	B → M	B DIR	D7	dd	2	3	4-2	- - - - ↕ ↕ 0 -
			B EXT	F7	hh ll	3	4	5-3	
			B IND,X	E7	ff	2	4	6-5	
			B IND,Y	18 E7	ff	3	5	7-5	
STD (opr)	Store Accumulator D	A → M, B → M + 1	DIR	DD	dd	2	4	4-4	- - - - ↕ ↕ 0 -
			EXT	FD	hh ll	3	5	5-5	
			IND,X	ED	ff	2	5	6-8	
			IND,Y	18 ED	ff	3	6	7-7	
STOP	Stop Internal Clocks		INH	CF		1	2	2-1	- - - - - - - -
STS (opr)	Store Stack Pointer	SP → M:M + 1	DIR	9F	dd	2	4	4-4	- - - - ↕ ↕ 0 -
			EXT	BF	hh ll	3	5	5-5	
			IND,X	AF	ff	2	5	6-8	
			IND,Y	18 AF	ff	3	6	7-7	
STX (opr)	Store Index Register X	IX → M:M + 1	DIR	DF	dd	2	4	4-4	- - - - ↕ ↕ 0 -
			EXT	FF	hh ll	3	5	5-5	
			IND,X	EF	ff	2	5	6-8	
			IND,Y	CD EF	ff	3	6	7-7	
STY (opr)	Store Index Register Y	IY → M:M + 1	DIR	18 DF	dd	3	5	4-6	- - - - ↕ ↕ 0 -
			EXT	18 FF	hh ll	4	6	5-7	
			IND,X	1A EF	ff	3	6	6-9	
			IND,Y	18 EF	ff	3	6	7-7	
SUBA (opr)	Subtract Memory from A	A – M → A	A IMM	80	ii	2	2	3-1	- - - - ↕ ↕ ↕ ↕
			A DIR	90	dd	2	3	4-1	
			A EXT	B0	hh ll	3	4	5-2	
			A IND,X	A0	ff	2	4	6-2	
			A IND,Y	18 A0	ff	3	5	7-2	
SUBB (opr)	Subtract Memory from B	B – M → B	B IMM	C0	ii	2	2	3-1	- - - - ↕ ↕ ↕ ↕
			B DIR	D0	dd	2	3	4-1	
			B EXT	F0	hh ll	3	4	5-2	
			B IND,X	E0	ff	2	4	6-2	
			B IND,Y	18 E0	ff	3	5	7-2	
SUBD (opr)	Subtract Memory from D	D – M:M + 1 → D	IMM	83	jj kk	3	4	3-3	- - - - ↕ ↕ ↕ ↕
			DIR	93	dd	2	5	4-7	
			EXT	B3	hh ll	3	6	5-10	
			IND,X	A3	ff	2	6	6-10	
			IND,Y	18 A3	ff	3	7	7-8	
SWI	Software Interrupt	See Special Ops	INH	3F		1	14	2-15	- - - 1 - - - -
TAB	Transfer A to B	A → B	INH	16		1	2	2-1	- - - - ↕ ↕ 0 -
TAP	Transfer A to CC Register	A → CCR	INH	06		1	2	2-1	↕ ↕ ↕ ↕ ↕ ↕ ↕ ↕
TBA	Transfer B to A	B → A	INH	17		1	2	2-1	- - - - ↕ ↕ 0 -
TEST	TEST (Only in Test Modes)	Address Bus Counts	INH	00		1	**	2-20	- - - - - - - -
TPA	Transfer CC Register to A	CCR → A	INH	07		1	2	2-1	- - - - - - - -
TST (opr)	Test for Zero or Minus	M – 0	EXT	7D	hh ll	3	6	5-9	- - - - ↕ ↕ 0 0
			IND,X	6D	ff	2	6	6-6	
			IND,Y	18 6D	ff	3	7	7-4	
TSTA		A – 0	A INH	4D		1	2	2-1	- - - - ↕ ↕ 0 0
TSTB		B – 0	B INH	5D		1	2	2-1	- - - - ↕ ↕ 0 0
TSX	Transfer Stack Pointer to X	SP + 1 → IX	INH	30		1	3	2-3	- - - - - - - -
TSY	Transfer Stack Pointer to Y	SP + 1 → IY	INH	18 30		2	4	2-5	- - - - - - - -

*Cycle-by-cycle number provides a reference to Tables 10-2 through 10-8 which detail cycle-by-cycle operation.
Example: Table 10-1 Cycle-by-Cycle column reference number 2-4 equals Table 10-2 line item 2-4.

(G)

FIGURE H-4 (*Continued*) Synopsis of the HC11 instruction set and cycle-by-cycle operation.

Table 10-1. MC68HC11A8 Instructions, Addressing Modes, and Execution Times (Sheet 7 of 7)

Source Form(s)	Operation	Boolean Expression	Addressing Mode for Operand	Machine Coding (Hexadecimal) Opcode	Machine Coding (Hexadecimal) Operand(s)	Bytes	Cycle	Cycle by Cycle*	Condition Codes S X H I N Z V C
TXS	Transfer X to Stack Pointer	IX − 1 → SP	INH	35		1	3	2-2	- - - - - - - -
TYS	Transfer Y to Stack Pointer	IY − 1 → SP	INH	18 35		2	4	2-4	- - - - - - - -
WAI	Wait for Interrupt	Stack Regs & WAIT	INH	3E		1	***	2-16	- - - - - - - -
XGDX	Exchange D with X	IX → D, D → IX	INH	8F		1	3	2-2	- - - - - - - -
XGDY	Exchange D with Y	IY → D, D → IY	INH	18 8F		2	4	2-4	- - - - - - - -

*Cycle-by-cycle number provides a reference to Tables 10-2 through 10-8 which detail cycle-by-cycle operation.
   Example: Table 10-1 Cycle-by-Cycle column reference number 2-4 equals Table 10-2 line item 2-4.

**Infinity or Until Reset Occurs

***12 Cycles are used beginning with the opcode fetch. A wait state is entered which remains in effect for an integer number of MPU E-clock cycles (n) until an interrupt is recognized. Finally, two additional cycles are used to fetch the appropriate interrupt vector (14 + n total).

dd	=	8-Bit Direct Address ($0000 –$00FF) (High Byte Assumed to be $00)
ff	=	8-Bit Positive Offset $00 (0) to $FF (255) (Is Added to Index)
hh	=	High Order Byte of 16-Bit Extended Address
ii	=	One Byte of Immediate Data
jj	=	High Order Byte of 16-Bit Immediate Data
kk	=	Low Order Byte of 16-Bit Immediate Data
ll	=	Low Order Byte of 16-Bit Extended Address
mm	=	8-Bit Bit Mask (Set Bits to be Affected)
rr	=	Signed Relative Offset $80 (− 128) to $7F (+ 127)
		(Offset Relative to the Address Following the Machine Code Offset Byte)

(H)

FIGURE H-4   (*Continued*) Synopsis of the HC11 instruction set and cycle-by-cycle operation.

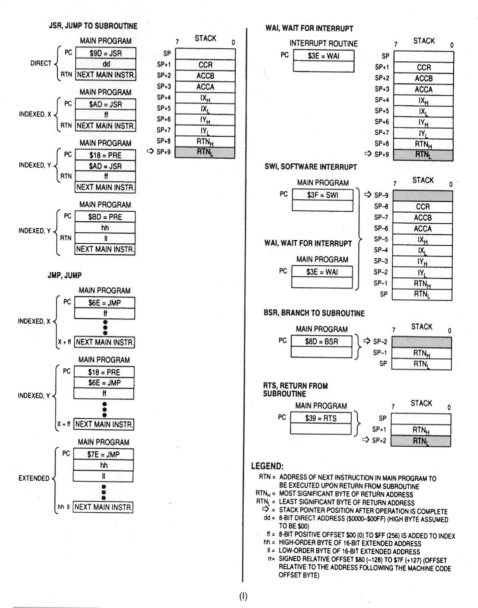

(I)

FIGURE H-4 (*Continued*) Synopsis of the HC11 instruction set and cycle-by-cycle operation.

Table 10-2. Cycle-by-Cycle Operation — Inherent Mode (Sheet 1 of 3)

Reference Number*	Address Mode and Instructions	Cycles	Cycle #	Address Bus	R/W Line	Data Bus
2-1	ABA, ASLA, ASLB, ASRA, ASRB, CBA, CLC, CLI, CLRA, CLRB, CLV, COMA, COMB, DAA, DECA, DECB, INCA, INCB, LSLA, LSLB, LSRA, LSRB, NEGA, NEGB, NOP, ROLA, ROLB, RORA, RORB, SBA, SEC, SEI, SEV, STOP, TAB, TAP, TBA, TPA, TSTA, TSTB	2	1	Opcode Address	1	Opcode
			2	Opcode Address + 1	1	Irrelevant Data
2-2	ABX, ASLD, DEX, INX, LSLD, LSRD, TXS, XGDX	3	1	Opcode Address	1	Opcode
			2	Opcode Address + 1	1	Irrelevant Data
			3	$FFFF	1	Irrelevant Data
2-3	DES, INS, TSX	3	1	Opcode Address	1	Opcode
			2	Opcode Address + 1	1	Irrelevant Data
			3	Previous SP Value	1	Irrelevant Data
2-4	ABY, DEY, INY, TYS, XGDY	4	1	Opcode Address	1	Opcode (Page Select Byte) ($18)
			2	Opcode Address + 1	1	Opcode (Second Byte)
			3	Opcode Address + 2	1	Irrelevant Data
			4	$FFFF	1	Irrelevant Data
2-5	TSY	4	1	Opcode Address	1	Opcode (Page Select Byte) ($18)
			2	Opcode Address + 1	1	Opcode (Second Byte) ($30)
			3	Opcode Address + 2	1	Irrelevant Data
			4	Stack Pointer	1	Irrelevant Data
2-6	PSHA, PSHB	3	1	Opcode Address	1	Opcode
			2	Opcode Address + 1	1	Irrelevant Data
			3	Stack Pointer	0	Accumulator Data
2-7	PSHX	4	1	Opcode Address	1	Opcode ($3C)
			2	Opcode Address + 1	1	Irrelevant Data
			3	Stack Pointer	0	IXL (Low Byte) to Stack
			4	Stack Pointer − 1	0	IXH (High Byte) to Stack
2-8	PSHY	5	1	Opcode Address	1	Opcode (Page Select Byte) ($18)
			2	Opcode Address + 1	1	Opcode (Second Byte) ($3C)
			3	Opcode Address + 2	1	Irrelevant Data
			4	Stack Pointer	0	IXL (Low Byte) to Stack
			5	Stack Pointer − 1	0	IXH (High Byte) to Stack
2-9	PULA, PULB	4	1	Opcode Address	1	Opcode
			2	Opcode Address + 1	1	Irrelevant Data
			3	Stack Pointer	1	Irrelevant Data
			4	Stack Pointer + 1	1	Operand Data from Stack
2-10	PULX	5	1	Opcode Address	1	Opcode ($38)
			2	Opcode Address + 1	1	Irrelevant Data
			3	Stack Pointer	1	Irrelevant Data
			4	Stack Pointer + 1	1	IXH (High Byte) from Stack
			5	Stack Pointer + 2	1	IXL (Low Byte) from Stack
2-11	PULY	6	1	Opcode Address	1	Opcode (Page Select Byte) ($18)
			2	Opcode Address + 1	1	Opcode (Second Byte) ($38)
			3	Opcode Address + 2	1	Irrelevant Data
			4	Stack Pointer	1	Irrelevant Data
			5	Stack Pointer + 1	1	IYH (High Byte) from Stack
			6	Stack Pointer + 2	1	IYH (Low Byte) from Stack

*The reference number is given to provide a cross-reference to Table 10-1

(J)

FIGURE H-4 (*Continued*) Synopsis of the HC11 instruction set and cycle-by-cycle operation.

Table 10-2. Cycle-by-Cycle Operation — Inherent Mode (Sheet 2 of 3)

Reference Number*	Address Mode and Instructions	Cycles	Cycle #	Address Bus	R/W Line	Data Bus
2-12	RTS	5	1	Opcode Address	1	Opcode ($39)
			2	Opcode Address + 1	1	Irrelevant Data
			3	Stack Pointer	1	Irrelevant Data
			4	Stack Pointer +1	1	Address of Next Instruction (High Byte)
			5	Stack Pointer + 2	1	Address of Next Instruction (Low Byte)
2-13	MUL	10	1	Opcode Address	1	Opcode ($3D)
			2	Opcode Address + 1	1	Irrelevant Data
			3	$FFFF	1	Irrelevant Data
			4	$FFFF	1	Irrelevant Data
			5	$FFFF	1	Irrelevant Data
			6	$FFFF	1	Irrelevant Data
			7	$FFFF	1	Irrelevant Data
			8	$FFFF	1	Irrelevant Data
			9	$FFFF	1	Irrelevant Data
			10	$FFFF	1	Irrelevant Data
2-14	RTI	12	1	Opcode Address	1	Opcode ($3B)
			2	Opcode Address + 1	1	Irrelevant Data
			3	Stack Pointer	1	Irrelevant Data
			4	Stack Pointer + 1	1	Condition Code Register from Stack
			5	Stack Pointer + 2	1	B Accumulator from Stack
			6	Stack Pointer + 3	1	A Accumulator from Stack
			7	Stack Pointer + 4	1	IXH (High Byte) from Stack
			8	Stack Pointer + 5	1	IXL (Low Byte) from Stack
			9	Stack Pointer + 6	1	IYH (High Byte) from Stack
			10	Stack Pointer + 7	1	IYL (Low Byte) from Stack
			11	Stack Pointer + 8	1	Address of Next Instruction (High Byte)
			12	Stack Pointer + 9	1	Address of Next Instruction (Low Byte)
2-15	SWI	14	1	Opcode Address	1	Opcode ($3F)
			2	Opcode Address + 1	1	Irrelevant Data
			3	Stack Pointer	0	Return Address (Low Byte)
			4	Stack Pointer − 1	0	Return Address (High Byte)
			5	Stack Pointer − 2	0	IYL (Low Byte) to Stack
			6	Stack Pointer − 3	0	IYH (High Byte) to Stack
			7	Stack Pointer − 4	0	IXL (Low Byte) to Stack
			8	Stack Pointer − 5	0	IXH (High Byte) to Stack
			9	Stack Pointer − 6	0	A Accumulator to Stack
			10	Stack Pointer − 7	0	B Accumulator to Stack
			11	Stack Pointer − 8	0	Condition Code Register to Stack
			12	Stack Pointer − 8	1	Irrelevant Data
			13	Address of SWI Vector (First Location)	1	SWI Service Routine Address (High Byte)
			14	Address of Vector + 1 (Second Location)	1	SWI Service Routine Address (Low Byte)
2-16	WAI	14 + n	1	Opcode Address	1	Opcode ($3E)
			2	Opcode Address + 1	1	Irrelevant Data
			3	Stack Pointer	0	Return Address (Low Byte)
			4	Stack Pointer − 1	0	Return Address (High Byte)
			5	Stack Pointer − 2	0	IYL (Low Byte) to Stack
			6	Stack Pointer − 3	0	IYH (High Byte) to Stack
			7	Stack Pointer − 4	0	IXL (Low Byte) to Stack
			8	Stack Pointer − 5	0	IXH (High Byte) to Stack
			9	Stack Pointer − 6	0	A Accumulator to Stack

*The reference number is given to provide a cross-reference to Table 10-1

(K)

FIGURE H-4 *(Continued)* Synopsis of the HC11 instruction set and cycle-by-cycle operation.

Table 10-2. Cycle-by-Cycle Operation — Inherent Mode (Sheet 3 of 3)

Reference Number*	Address Mode and Instructions	Cycles	Cycle #	Address Bus	R/W Line	Data Bus
2-16 (Continued)	WAI	14 + n	10	Stack Pointer – 7	0	B Accumulator to Stack
			11	Stack Pointer – 8	0	Condition Code Register to Stack
			12 to			
			n + 12	Stack Pointer – 8	1	Irrelevant Data
			n + 13	Address of Vector (First Location)	1	Service Routine Address (High Byte)
			n + 14	Address of Vector + 1 (Second Location)	1	Service Routine Address (Low Byte)
2-17	FDIV, IDIV	41	1	Opcode Address	1	Opcode
			2	Opcode Address + 1	1	Irrelevant Data
			3 – 41	$FFFF	1	Irrelevant Data
2-18	Page 1 Illegal Opcodes	15	1	Opcode Address	1	Opcode (Illegal)
			2	Opcode Address + 1	1	Irrelevant Data
			3	$FFFF	1	Irrelevant Data
			4	Stack Pointer	0	Return Address (Low Byte)
			5	Stack Pointer – 1	0	Return Address (High Byte)
			6	Stack Pointer – 2	0	IYL (Low Byte) to Stack
			7	Stack Pointer – 3	0	IYH (High Byte) to Stack
			8	Stack Pointer – 4	0	IXL (Low Byte) to Stack
			9	Stack Pointer – 5	0	IXH (High Byte) to Stack
			10	Stack Pointer – 6	0	A Accumulator
			11	Stack Pointer – 7	0	B Accumulator
			12	Stack Pointer – 8	0	Condition Code Register to Stack
			13	Stack Pointer – 8	1	Irrelevant Data
			14	Address of Vector (First Location)	1	Service Routine Address (High Byte)
			15	Address of Vector + 1 (Second Location)	1	Service Routine Address (Low Byte)
2-19	Pages 2, 3, or 4 Illegal Opcodes	16	1	Opcode Address	1	Opcode (Legal Page Select)
			2	Opcode Address + 1	1	Opcode (Illegal Second Byte)
			3	Opcode Address + 2	1	Irrelevant Data
			4	$FFFF	1	Irrelevant Data
			5	Stack Pointer	0	Return Address (Low Byte)
			6	Stack Pointer – 1	0	Return Address (High Byte)
			7	Stack Pointer – 2	0	IYL (Low Byte) to Stack
			8	Stack Pointer – 3	0	IYH (High Byte) to Stack
			9	Stack Pointer – 4	0	IXL (Low Byte) to Stack
			10	Stack Pointer – 5	0	IXH (High Byte) to Stack
			11	Stack Pointer – 6	0	A Accumulator
			12	Stack Pointer – 7	0	B Accumulator
			13	Stack Pointer – 8	0	Condition Code Register to Stack
			14	Stack Pointer – 8	1	Irrelevant Data
			15	Address of Vector (First Location)	1	Service Routine Address (High Byte)
			16	Address of Vector + 1 (Second Location)	1	Service Routine Address (Low Byte)
2-20	TEST	Infinite	1	Opcode Address	1	Opcode ($00)
			2	Opcode Address + 1	1	Irrelevant Data
			3	Opcode Address + 1	1	Irrelevant Data
			4	Opcode Address + 2	1	Irrelevant Data
			5 – n	Previous Address + 1	1	Irrelevant Data

*The reference number is given to provide a cross-reference to Table 10-1

(L)

FIGURE H-4 (*Continued*) Synopsis of the HC11 instruction set and cycle-by-cycle operation.

Table 10-3. Cycle-by-Cycle Operation — Immediate Mode

Reference Number*	Address Mode and Instructions	Cycles	Cycle #	Address Bus	R/W Line	Data Bus
3-1	ADCA, ADCB, ADDA, ADDB, ANDA, ANDB, BITA, BITB, CMPA, CMPB, EORA, EORB, LDAA, LDAB, ORAA, ORAB, SBCA, SBCB, SUBA, SUBB,	2	1	Opcode Address	1	Opcode
			2	Opcode Address + 1	1	Operand Data
3-2	LDD, LDS, LDX	3	1	Opcode Address	1	Opcode
			2	Opcode Address + 1	1	Operand Data (High Byte)
			3	Opcode Address + 2	1	Operand Data (Low Byte)
3-3	ADDD, CPX, SUBD	4	1	Opcode Address	1	Opcode
			2	Opcode Address + 1	1	Operand Data (High Byte)
			3	Opcode Address + 2	1	Operand Data (Low Byte)
			4	$FFFF	1	Irrelevant Data
3-4	LDY	4	1	Opcode Address	1	Opcode (Page Select Byte) ($18)
			2	Opcode Address + 1	1	Opcode (Second Byte) ($EC)
			3	Opcode Address + 2	1	Operand Data (High Byte)
			4	Opcode Address + 3	1	Operand Data (Low Byte)
3-5	CPD, CPY	5	1	Opcode Address	1	Opcode (Page Select Byte)
			2	Opcode Address + 1	1	Opcode (Second Byte)
			3	Opcode Address + 2	1	Operand Data (High Byte)
			4	Opcode Address + 3	1	Operand Data (Low Byte)
			5	$FFFF	1	Irrelevant Data

*The reference number is given to provide a cross-reference to Table 10-1

(M)

Table 10-4. Cycle-by-Cycle Operation — Direct Mode (Sheet 1 of 2)

Reference Number*	Address Mode and Instructions	Cycles	Cycle #	Address Bus	R/W Line	Data Bus
4-1	ADCA, ADCB, ADDA, ADDB, ANDA, ANDB, BITA, BITB, CMPA, CMPB, EORA, EORB, LDAA, LDAB, ORAA, ORAB, SBCA, SBCB, SUBA, SUBB	3	1	Opcode Address	1	Opcode
			2	Opcode Address + 1	1	Operand Address (Low Byte) (High Byte Assumed to be $00)
			3	Operand Address	1	Operand Data
4-2	STAA, STAB	3	1	Opcode Address	1	Opcode
			2	Opcode Address + 1	1	Operand Address (Low Byte) (High Byte Assumed to be $00)
			3	Operand Address	0	Data from Accumulator
4-3	LDD, LDS, LDX	4	1	Opcode Address	1	Opcode
			2	Opcode Address + 1	1	Operand Address (Low Byte) (High Byte Assumed to be $00)
			3	Operand Address	1	Operand Data (High Byte)
			4	Operand Address + 1	1	Operand Data (Low Byte)
4-4	STD, STS, STX	4	1	Opcode Address	1	Opcode
			2	Opcode Address + 1	1	Operand Address (Low Byte) (High Byte Assumed to be $00)
			3	Operand Address	0	Register Data (High Byte)
			4	Operand Address + 1	0	Register Data (Low Byte)
4-5	LDY	5	1	Opcode Address	1	Opcode (Page Select Byte) ($18)
			2	Opcode Address + 1	1	Opcode (Second Byte) ($DE)
			3	Opcode Address + 2	1	Operand Address (Low Byte) (High Byte Assumed to be $00)
			4	Operand Address	1	Operand Data (High Byte)
			5	Operand Address + 1	1	Operand Data (Low Byte)

*The reference number is given to provide a cross-reference to Table 10-1

(N)

FIGURE H-4 (*Continued*) Synopsis of the HC11 instruction set and cycle-by-cycle operation.

Table 10-4. Cycle-by-Cycle Operation — Direct Mode (Sheet 2 of 2)

Reference Number*	Address Mode and Instructions	Cycles	Cycle #	Address Bus	R/W Line	Data Bus
4-6	STY	5	1	Opcode Address	1	Opcode (Page Select Byte) ($18)
			2	Opcode Address + 1	1	Opcode (Second Byte) ($DF)
			3	Opcode Address + 2	1	Operand Address (Low Byte)
						(High Byte Assumed to be $00)
			4	Operand Address	0	Register Data (High Byte)
			5	Operand Address + 1	0	Register Data (Low Byte)
4-7	ADDD, CPX, SUBD	5	1	Opcode Address	1	Opcode
			2	Opcode Address + 1	1	Operand Address (Low Byte)
						(High Byte Assumed to be $00)
			3	Operand Address	1	Operand Data (High Byte)
			4	Operand Address + 1	1	Operand Data (Low Byte)
			5	$FFFF	1	Irrelevant Data
4-8	JSR	5	1	Opcode Address	1	Opcode ($9D)
			2	Opcode Address + 1	1	Subroutine Address (Low Byte)
						(High Byte Assumed to be $00)
			3	Subroutine Address	1	First Subroutine Opcode
			4	Stack Pointer	0	Return Address (Low Byte)
			5	Stack Pointer – 1	0	Return Address (High Byte)
4-9	CPD, CPY	6	1	Opcode Address	1	Opcode (Page Select Byte)
			2	Opcode Address + 1	1	Opcode (Second Byte)
			3	Opcode Address + 2	1	Operand Address (Low Byte)
						(High Byte Assumed to be $00)
			4	Operand Address	1	Operand Data (High Byte)
			5	Operand Address + 1	1	Operand Data (Low Byte)
			6	$FFFF	1	Irrelevant Data
4-10	BCLR, BSET	6	1	Opcode Address	1	Opcode
			2	Opcode Address + 1	1	Operand Address (Low Byte)
						(High Byte Assumed to be $00)
			3	Operand Address	1	Original Operand Data
			4	Opcode Address + 2	1	Mask Byte
			5	$FFFF	1	Irrelevant Data
			6	Operand Address	0	Result Operand Data
4-11	BRCLR, BRSET	6	1	Opcode Address	1	Opcode
			2	Opcode Address + 1	1	Operand Address (Low Byte)
						(High Byte Assumed to be $00)
			3	Operand Address	1	Original Operand Data
			4	Opcode Address + 2	1	Mask Byte
			5	Opcode Address + 3	1	Branch Offset
			6	$FFFF	1	Irrelevant Data

*The reference number is given to provide a cross-reference to Table 10-1

(O)

Table 10-5. Cycle-by-Cycle Operation — Extended Mode (Sheet 1 of 2)

Reference Number*	Address Mode and Instructions	Cycles	Cycle #	Address Bus	R/W Line	Data Bus
5-1	JMP	3	1	Opcode Address	1	Opcode ($7E)
			2	Opcode Address + 1	1	Jump Address (High Byte)
			3	Opcode Address + 2	1	Jump Address (Low Byte)
5-2	ADCA, ADCB, ADDA, ADDB, ANDA, ANDB, BITA, BITB, CMPA, CMPB, EORA, EORB, LDAA, LDAB, ORAA ORAB, SBCA, SBCB, SUBA, SUBB	4	1	Opcode Address	1	Opcode
			2	Opcode Address + 1	1	Operand Address (High Byte)
			3	Opcode Address + 2	1	Operand Address (Low Byte)
			4	Operand Address	1	Operand Data

*The reference number is given to provide a cross-reference to Table 10-1

(P)

FIGURE H-4 *(Continued)* Synopsis of the HC11 instruction set and cycle-by-cycle operation.

Table 10-5. Cycle-by-Cycle Operation — Extended Mode (Sheet 2 of 2)

Reference Number*	Address Mode and Instructions	Cycles	Cycle #	Address Bus	R/W Line	Data Bus
5-3	STAA, STAB	4	1	Opcode Address	1	Opcode
			2	Opcode Address + 1	1	Operand Address (High Byte)
			3	Opcode Address + 2	1	Operand Address (Low Byte)
			4	Operand Address	0	Accumulator Data
5-4	LDD, LDS, LDX	5	1	Opcode Address	1	Opcode
			2	Opcode Address + 1	1	Operand Address (High Byte)
			3	Opcode Address + 2	1	Operand Address (Low Byte)
			4	Operand Address	1	Operand Data (High Byte)
			5	Operand Address + 1	1	Operand Data (Low Byte)
5-5	STD, STS, STX	5	1	Opcode Address	1	Opcode
			2	Opcode Address + 1	1	Operand Address (High Byte)
			3	Opcode Address + 2	1	Operand Address (Low Byte)
			4	Operand Address	0	Register Data (High Byte)
			5	Operand Address + 1	0	Register Data (Low Byte)
5-6	LDY	6	1	Opcode Address	1	Opcode (Page Select Byte) ($18)
			2	Opcode Address + 1	1	Opcode (Second Byte) ($FE)
			3	Opcode Address + 2	1	Operand Address (High Byte)
			4	Opcode Address + 3	1	Operand Address (Low Byte)
			5	Operand Address	1	Operand Data (High Byte)
			6	Operand Address + 1	1	Operand Data (Low Byte)
5-7	STY	6	1	Opcode Address	1	Opcode (Page Select Byte) ($18)
			2	Opcode Address + 1	1	Opcode (Second Byte) ($FF)
			3	Opcode Address + 2	1	Operand Address (High Byte)
			4	Opcode Address + 3	1	Operand Address (Low Byte)
			5	Operand Address	0	Register Data (High Byte)
			6	Operand Address + 1	0	Register Data (Low Byte)
5-8	ASL, ASR, CLR, COM, DEC, INC, LSL, LSR, NEG, ROL, ROR	6	1	Opcode Address	1	Opcode
			2	Opcode Address + 1	1	Operand Address (High Byte)
			3	Opcode Address + 2	1	Operand Address (Low Byte)
			4	Operand Address	1	Original Operand Data
			5	$FFFF	1	Irrelevant Data
			6	Operand Address	0	Result Operand Data
5-9	TST	6	1	Opcode Address	1	Opcode ($7D)
			2	Opcode Address + 1	1	Operand Address (High Byte)
			3	Opcode Address + 2	1	Operand Address (Low Byte)
			4	Operand Address	1	Original Operand Data
			5	$FFFF	1	Irrelevant Data
			6	$FFFF	1	Irrelevant Data
5-10	ADDD, CPX, SUBD	6	1	Opcode Address	1	Opcode
			2	Opcode Address + 1	1	Operand Address (High Byte)
			3	Opcode Address + 2	1	Operand Address (Low Byte)
			4	Operand Address	1	Operand Data (High Byte)
			5	Operand Address + 1	1	Operand Data (Low Byte)
			6	$FFFF	1	Irrelevant Data
5-11	CPD, CPY	7	1	Opcode Address	1	Opcode (Page Select Byte)
			2	Opcode Address + 1	1	Opcode (Second Byte)
			3	Opcode Address + 2	1	Operand Address (High Byte)
			4	Opcode Address + 3	1	Operand Address (Low Byte)
			5	Operand Address	1	Operand Data (High Byte)
			6	Operand Address + 1	1	Operand Data (Low Byte)
			7	$FFFF	1	Irrelevant Data
5-12	JSR	6	1	Opcode Address	1	Opcode ($BD)
			2	Opcode Address + 1	1	Subroutine Address (High Byte)
			3	Opcode Address + 2	1	Subroutine Address (Low Byte)
			4	Subroutine Address	1	First Opcode in Subroutine
			5	Stack Pointer	0	Return Address (Low Byte)
			6	Stack Pointer – 1	0	Return Address (High Byte)

*The reference number is given to provide a cross-reference to Table 10-1

(Q)

FIGURE H-4　(*Continued*) Synopsis of the HC11 instruction set and cycle-by-cycle operation.

Table 10-6. Cycle-by-Cycle Operation — Indexed X Mode (Sheet 1 of 2)

Reference Number*	Address Mode and Instructions	Cycles	Cycle #	Address Bus	R/W̄ Line	Data Bus
6-1	JMP	3	1	Opcode Address	1	Opcode ($6E)
			2	Opcode Address + 1	1	Index Offset
			3	$FFFF	1	Irrelevant Data
6-2	ADCA, ADCB, ADDA, ADDB, ANDA, ANDB, BITA, BITB, CMPA, CMPB, EORA, EORB, LDAA, LDAB, ORAA, ORAB, SBCA, SBCB, SUBA, SUBB	4	1	Opcode Address	1	Opcode
			2	Opcode Address + 1	1	Index Offset
			3	$FFFF	1	Irrelevant Data
			4	(IX) + Offset	1	Operand Data
6-3	ASL, ASR, CLR, COM, DEC, INC, LSL, LSR, NEG, ROL, ROR	6	1	Opcode Address	1	Opcode
			2	Opcode Address + 1	1	Index Offset
			3	$FFFF	1	Irrelevant Data
			4	(IX) + Offset	1	Original Operand Data
			5	$FFFF	1	Irrelevant Data
			6	(IX) + Offset	0	Result Operand Data
6-4	TST	6	1	Opcode Address	1	Opcode ($6D)
			2	Opcode Address + 1	1	Index Offset
			3	$FFFF	1	Irrelevant Data
			4	(IX) + Offset	1	Original Operand Data
			5	$FFFF	1	Irrelevant Data
			6	$FFFF	1	Irrelevant Data
6-5	STAA, STAB	4	1	Opcode Address	1	Opcode
			2	Opcode Address + 1	1	Index Offset
			3	$FFFF	1	Irrelevant Data
			4	(IX) + Offset	0	Accumulator Data
6-6	LDD, LDS, LDX	5	1	Opcode Address	1	Opcode
			2	Opcode Address + 1	1	Index Offset
			3	$FFFF	1	Irrelevant Data
			4	(IX) + Offset	1	Operand Data (High Byte)
			5	(IX) + Offset + 1	1	Operand Data (Low Byte)
6-7	LDY	6	1	Opcode Address	1	Opcode (Page Select Byte) ($1A)
			2	Opcode Address + 1	1	Opcode (Second Byte) ($EE)
			3	Opcode Address + 2	1	Index Offset
			4	$FFFF	1	Irrelevant Data
			5	(IX) + Offset	1	Operand Data (High Byte)
			6	(IX) + Offset + 1	1	Operand Data (Low Byte)
6-8	STD, STS, STX	5	1	Opcode Address	1	Opcode
			2	Opcode Address + 1	1	Index Offset
			3	$FFFF	1	Irrelevant Data
			4	(IX) + Offset	0	Register Data (High Byte)
			5	(IX) + Offset + 1	0	Register Data (Low Byte)
6-9	STY	6	1	Opcode Address	1	Opcode (Page Select Byte) ($1A)
			2	Opcode Address + 1	1	Opcode (Second Byte) ($EF)
			3	Opcode Address + 2	1	Index Offset
			4	$FFFF	1	Irrelevant Data
			5	(IX) + Offset	0	Register Data (High Byte)
			6	(IX) + Offset + 1	0	Register Data (Low Byte)
6-10	ADDD, CPX, SUBD	6	1	Opcode Address	1	Opcode
			2	Opcode Address + 1	1	Index Offset
			3	$FFFF	1	Irrelevant Data
			4	(IX) + Offset	1	Operand Data (High Byte)
			5	(IX) + Offset + 1	1	Operand Data (Low Byte)
			6	$FFFF	1	Irrelevant Data

*The reference number is given to provide a cross-reference to Table 10-1

(R)

FIGURE H-4 *(Continued)* Synopsis of the HC11 instruction set and cycle-by-cycle operation.

Table 10-6. Cycle-by-Cycle Operation — Indexed X Mode (Sheet 2 of 2)

Reference Number*	Address Mode and Instructions	Cycles	Cycle #	Address Bus	R/W̄ Line	Data Bus
6-11	CPD, CPY	7	1	Opcode Address	1	Opcode (Page Select Byte)
			2	Opcode Address + 1	1	Opcode (Second Byte)
			3	Opcode Address + 2	1	Index Offset
			4	$FFFF	1	Irrelevant Data
			5	(IX) + Offset	1	Operand Data (High Byte)
			6	(IX) + Offset + 1	1	Operand Data (Low Byte)
			7	$FFFF	1	Irrelevant Data
6-12	JSR	6	1	Opcode Address	1	Opcode ($AD)
			2	Opcode Address + 1	1	Index Offset
			3	$FFFF	1	Irrelevant Data
			4	(IX) + Offset	1	First Opcode in Subroutine
			5	Stack Pointer	0	Return Address (Low Byte)
			6	Stack Pointer − 1	0	Return Address (High Byte)
6-13	BCLR, BSET	7	1	Opcode Address	1	Opcode
			2	Opcode Address + 1	1	Index Offset
			3	$FFFF	1	Irrelevant Data
			4	(IX) + Offset	1	Original Operand Data
			5	Opcode Address + 2	1	Mask Byte
			6	$FFFF	1	Irrelevant Data
			7	(IX) + Offset	0	Result Operand Data
6-14	BRCLR, BRSET	7	1	Opcode Address	1	Opcode
			2	Opcode Address + 1	1	Index Offset
			3	$FFFF	1	Irrelevant Data
			4	(IX) + Offset	1	Original Operand Data
			5	Opcode Address + 2	1	Mask Byte
			6	Opcode Address + 3	1	Branch Offset
			7	$FFFF	1	Irrelevant Data

*The reference number is given to provide a cross-reference to Table 10-1

(S)

Table 10-7. Cycle-by-Cycle Operation — Indexed Y Mode (Sheet 1 of 2)

Reference Number*	Address Mode and Instructions	Cycles	Cycle #	Address Bus	R/W̄ Line	Data Bus
7-1	JMP	4	1	Opcode Address	1	Opcode (Page Select Byte) ($18)
			2	Opcode Address + 1	1	Opcode (Second Byte) ($6E)
			3	Opcode Address + 2	1	Index Offset
			4	$FFFF	1	Irrelevant Data
7-2	ADCA, ADCB, ADDA, ADDB, ANDA ANDB, BITA, BITB, CMPA, CMPB, EORA, EORB, LDAA, LDAB, ORAA, ORAB, SBCA, SBCB, SUBA, SUBB,	5	1	Opcode Address	1	Opcode (Page Select Byte) ($18)
			2	Opcode Address + 1	1	Opcode (Second Byte)
			3	Opcode Address + 2	1	Index Offset
			4	$FFFF	1	Irrelevant Data
			5	(IY) + Offset	1	Operand Data
7-3	ASL, ASR, CLR, COM, DEC, INC, LSL, LSR, NEG, ROL, ROR	7	1	Opcode Address	1	Opcode (Page Select Byte)
			2	Opcode Address + 1	1	Opcode (Second Byte)
			3	Opcode Address + 2	1	Index Offset
			4	$FFFF	1	Irrelevant Data
			5	(IY) + Offset	1	Original Operand Data
			6	$FFFF	1	Irrelevant Data
			7	(IY) + Offset	0	Result Operand Data

*The reference number is given to provide a cross-reference to Table 10-1

(T)

FIGURE H-4 (*Continued*) Synopsis of the HC11 instruction set and cycle-by-cycle operation.

Table 10-7. Cycle-by-Cycle Operation — Indexed Y Mode (Sheet 2 of 2)

Reference Number*	Address Mode and Instructions	Cycles	Cycle #	Address Bus	R/W̄ Line	Data Bus
7-4	TST	7	1	Opcode Address	1	Opcode (Page Select Byte) ($18)
			2	Opcode Address + 1	1	Opcode (Second Byte) ($6D)
			3	Index Offset	1	Index Offset
			4	$FFFF	1	Irrelevant Data
			5	(IY) + Offset	1	Original Operand Data
			6	$FFFF	1	Irrelevant Data
			7	$FFFF	1	Irrelevant Data
7-5	STAA, STAB	5	1	Opcode Address	1	Opcode (Page Select Byte) ($18)
			2	Opcode Address + 1	1	Opcode (Second Byte)
			3	Index Offset	1	Index Offset
			4	$FFFF	1	Irrelevant Data
			5	(IY) + Offset	0	Accumulator Data
7-6	LDD, LDS, LDX, LDY	6	1	Opcode Address	1	Opcode (Page Select Byte)
			2	Opcode Address + 1	1	Opcode (Second Byte)
			3	Index Offset	1	Index Offset
			4	$FFFF	1	Irrelevant Data
			5	(IY) + Offset	1	Operand Data (High Byte)
			6	(IY) + Offset + 1	1	Operand Data (Low Byte)
7-7	STD, STS, STX, STY	6	1	Opcode Address	1	Opcode (Page Select Byte)
			2	Opcode Address + 1	1	Opcode (Second Byte)
			3	Index Offset	1	Index Offset
			4	$FFFF	1	Irrelevant Data
			5	(IY) + Offset	0	Register Data (High Byte)
			6	(IY) + Offset + 1	0	Register Data (Low Byte)
7-8	ADDD, CPD, CPX, CPY, SUBD	7	1	Opcode Address	1	Opcode (Page Select Byte)
			2	Opcode Address + 1	1	Opcode (Second Byte)
			3	Index Offset	1	Index Offset
			4	$FFFF	1	Irrelevant Data
			5	(IY) + Offset	1	Operand Data (High Byte)
			6	(IY) + Offset + 1	1	Operand Data (Low Byte)
			7	$FFFF	1	Irrelevant Data
7-9	JSR	7	1	Opcode Address	1	Opcode (Page Select Byte) ($18)
			2	Opcode Address + 1	1	Opcode (Second Byte) ($AD)
			3	Index Offset	1	Index Offset
			4	$FFFF	1	Irrelevant Data
			5	(IY) + Offset	1	First Opcode in Subroutine
			6	Stack Pointer	0	Return Address (Low Byte)
			7	Stack Pointer − 1	0	Return Address (High Byte)
7-10	BCLR, BSET	8	1	Opcode Address	1	Opcode (Page Select Byte) ($18)
			2	Opcode Address + 1	1	Opcode (Second Byte)
			3	Index Offset	1	Index Offset
			4	$FFFF	1	Irrelevant Data
			5	(IY) + Offset	1	Original Operand Data
			6	Opcode Address + 3	1	Mask Byte
			7	$FFFF	1	Irrelevant Data
			8	(IY) + Offset	0	Result Operand Data
7-11	BRCLR, BRSET	8	1	Opcode Address	1	Opcode (Page Select Byte) ($18)
			2	Opcode Address + 1	1	Opcode (Second Byte)
			3	Index Offset	1	Index Offset
			4	$FFFF	1	Irrelevant Data
			5	(IY) + Offset	1	Original Operand Data
			6	Opcode Address + 3	1	Mask Byte
			7	Opcode Address + 4	1	Branch Offset
			8	$FFFF	1	Irrelevant Data

*The reference number is given to provide a cross-reference to Table 10-1

(U)

FIGURE H-4   (*Continued*) Synopsis of the HC11 instruction set and cycle-by-cycle operation.

Table 10-8. Cycle-by-Cycle Operation — Relative Mode

Reference Number*	Address Mode and Instructions	Cycles	Cycle #	Address Bus	R/W Line	Data Bus
8-1	BCC, BCS, BEQ, BGE, BGT, BHI, BHS, BLE, BLO, BLS, BLT, BMI, BNE, BPL, BRA, BRN, BVC, BVS,	3	1	Opcode Address	1	Opcode
			2	Opcode Address + 1	1	Branch Offset
			3	$FFFF	1	Irrelevant Data
8-2	BSR	6	1	Opcode Address	1	Opcode ($8D)
			2	Opcode Address + 1	1	Branch Offset
			3	$FFFF	1	Irrelevant Data
			4	Subroutine Address	1	Opcode of Next Instruction
			5	Stack Pointer	0	Return Address (Low Byte)
			6	Stack Pointer – 1	0	Return Address (High Byte)

*The reference number is given to provide a cross-reference to Table 10-1

(V)

FIGURE H-4 (*Continued*) Synopsis of the HC11 instruction set and cycle-by-cycle operation.

**National
Semiconductor
Corporation**

LM135/LM235/LM335, LM135A/LM235A/LM335A
Precision Temperature Sensors

General Description

The LM135 series are precision, easily-calibrated, integrated circuit temperature sensors. Operating as a 2-terminal zener, the LM135 has a breakdown voltage directly proportional to absolute temperature at +10 mV/°K. With less than 1Ω dynamic impedance the device operates over a current range of 400 μA to 5 mA with virtually no change in performance. When calibrated at 25°C the LM135 has typically less than 1°C error over a 100°C temperature range. Unlike other sensors the LM135 has a linear output.

Applications for the LM135 include almost any type of temperature sensing over a −55°C to +150°C temperature range. The low impedance and linear output make interfacing to readout or control circuitry especially easy.

The LM135 operates over a −55°C to +150°C temperature range while the LM235 operates over a −40°C to +125°C temperature range. The LM335 operates from −40°C to +100°C. The LM135/LM235/LM335 are available packaged in hermetic TO-46 transistor packages while the LM335 is also available in plastic TO-92 packages.

Features

- Directly calibrated in °Kelvin
- 1°C initial accuracy available
- Operates from 400 μA to 5 mA
- Less than 1Ω dynamic impedance
- Easily calibrated
- Wide operating temperature range
- 200°C overrange
- Low cost

Schematic Diagram

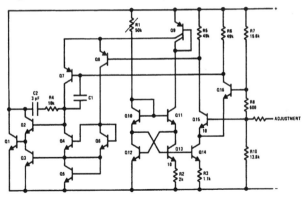

TL/H/5698−1

Connection Diagrams

**TO-92
Plastic Package**

BOTTOM VIEW

**Order Number LM335Z or LM335AZ
See NS Package Number Z03A**

**TO-46
Metal Can Package***

BOTTOM VIEW

TL/H/5698−8

*Case is connected to negative pin

**Order Number LM135H, LM235H,
LM335H, LM135AH, LM235AH or LM335AH
See NS Package Number H03H**

(A)

FIGURE H-5 National Semiconductor data sheets for the LM135/235/335 series of precision temperature sensors. (*Copyright© 1999 National Semiconductor Corporation; Santa Clara, CA; All rights reserved.*)

Absolute Maximum Ratings

If Military/Aerospace specified devices are required, contact the National Semiconductor Sales Office/Distributors for availability and specifications. (Note 4)

Reverse Current ..... 15 mA

Forward Current ..... 10 mA

Storage Temperature

TO-46 Package ..... $-60°C$ to $+180°C$

TO-92 Package ..... $-60°C$ to $+150°C$

Specified Operating Temp. Range

	Continuous	Intermittent (Note 2)
LM135, LM135A	$-55°C$ to $+150°C$	150°C to 200°C
LM235, LM235A	$-40°C$ to $+125°C$	125°C to 150°C
LM335, LM335A	$-40°C$ to $+100°C$	100°C to 125°C

Lead Temp. (Soldering, 10 seconds)

TO-92 Package: ..... 260°C

TO-46 Package: ..... 300°C

Temperature Accuracy LM135/LM235, LM135A/LM235A (Note 1)

Parameter	Conditions	LM135A/LM235A			LM135/LM235			Units
		Min	Typ	Max	Min	Typ	Max	
Operating Output Voltage	$T_C = 25°C$, $I_R = 1$ mA	2.97	2.98	2.99	2.95	2.98	3.01	V
Uncalibrated Temperature Error	$T_C = 25°C$, $I_R = 1$ mA		0.5	1		1	3	°C
Uncalibrated Temperature Error	$T_{MIN} \leq T_C \leq T_{MAX}$, $I_R = 1$ mA		1.3	2.7		2	5	°C
Temperature Error with 25°C Calibration	$T_{MIN} \leq T_C \leq T_{MAX}$, $I_R = 1$ mA		0.3	1		0.5	1.5	°C
Calibrated Error at Extended Temperatures	$T_C = T_{MAX}$ (Intermittent)		2			2		°C
Non-Linearity	$I_R = 1$ mA		0.3	0.5		0.3	1	°C

Temperature Accuracy LM335, LM335A (Note 1)

Parameter	Conditions	LM335A			LM335			Units
		Min	Typ	Max	Min	Typ	Max	
Operating Output Voltage	$T_C = 25°C$, $I_R = 1$ mA	2.95	2.98	3.01	2.92	2.98	3.04	V
Uncalibrated Temperature Error	$T_C = 25°C$, $I_R = 1$ mA		1	3		2	6	°C
Uncalibrated Temperature Error	$T_{MIN} \leq T_C \leq T_{MAX}$, $I_R = 1$ mA		2	5		4	9	°C
Temperature Error with 25°C Calibration	$T_{MIN} \leq T_C \leq T_{MAX}$, $I_R = 1$ mA		0.5	1		1	2	°C
Calibrated Error at Extended Temperatures	$T_C = T_{MAX}$ (Intermittent)		2			2		°C
Non-Linearity	$I_R = 1$ mA		0.3	1.5		0.3	1.5	°C

Electrical Characteristics (Note 1)

Parameter	Conditions	LM135/LM235 LM135A/LM235A			LM335 LM335A			Units
		Min	Typ	Max	Min	Typ	Max	
Operating Output Voltage Change with Current	$400\ \mu A \leq I_R \leq 5$ mA At Constant Temperature		2.5	10		3	14	mV
Dynamic Impedance	$I_R = 1$ mA		0.5			0.6		Ω
Output Voltage Temperature Coefficient			$+10$			$+10$		mV/°C
Time Constant	Still Air		80			80		sec
	100 ft/Min Air		10			10		sec
	Stirred Oil		1			1		sec
Time Stability	$T_C = 125°C$		0.2			0.2		°C/khr

Note 1: Accuracy measurements are made in a well-stirred oil bath. For other conditions, self heating must be considered.

Note 2: Continuous operation at these temperatures for 10,000 hours for H package and 5,000 hours for Z package may decrease life expectancy of the device.

Note 3: Thermal Resistance

	TO-92	TO-46
θ_{JA} (junction to ambient)	202°C/W	400°C/W
θ_{JC} (junction to case)	170°C/W	N/A

Note 4: Refer to RETS135H for military specifications.

(B)

FIGURE H-5 (*Continued*) National Semiconductor data sheets for the LM135/235/335 series of precision temperature sensors.

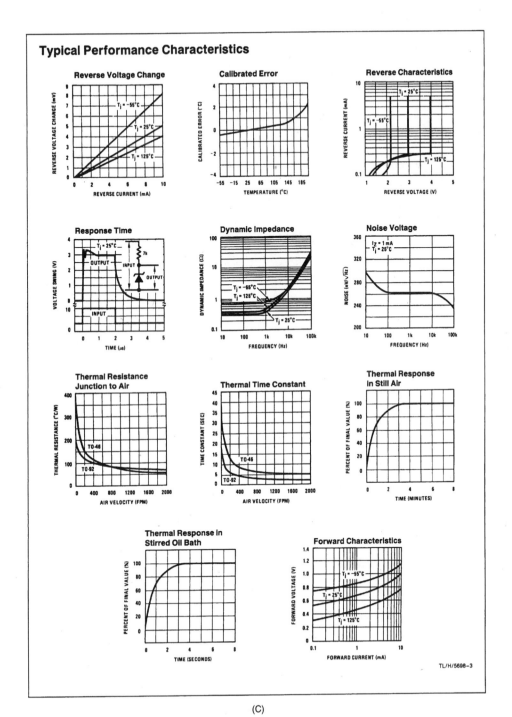

(C)

FIGURE H-5 (*Continued*) National Semiconductor data sheets for the LM135/235/335 series of precision temperature sensors.

Application Hints

CALIBRATING THE LM135

Included on the LM135 chip is an easy method of calibrating the device for higher accuracies. A pot connected across the LM135 with the arm tied to the adjustment terminal allows a 1-point calibration of the sensor that corrects for inaccuracy over the full temperature range.

This single point calibration works because the output of the LM135 is proportional to absolute temperature with the extrapolated output of sensor going to 0V output at 0°K (−273. 15°C). Errors in output voltage versus temperature are only slope (or scale factor) errors so a slope calibration at one temperature corrects at all temperatures.

The output of the device (calibrated or uncalibrated) can be expressed as:

$$V_{OUT_T} = V_{OUT_{T_o}} \times \frac{T}{T_o}$$

where T is the unknown temperature and T_o is a reference temperature, both expressed in degrees Kelvin. By calibrating the output to read correctly at one temperature the output at all temperatures is correct. Nominally the output is calibrated at 10 mV/°K.

To insure good sensing accuracy several precautions must be taken. Like any temperature sensing device, self heating can reduce accuracy. The LM135 should be operated at the lowest current suitable for the application. Sufficient current, of course, must be available to drive both the sensor and the calibration pot at the maximum operating temperature as well as any external loads.

If the sensor is used in an ambient where the thermal resistance is constant, self heating errors can be calibrated out. This is possible if the device is run with a temperature stable current. Heating will then be proportional to zener voltage and therefore temperature. This makes the self heating error proportional to absolute temperature the same as scale factor errors.

WATERPROOFING SENSORS

Meltable inner core heat shrinkable tubing such as manufactured by Raychem can be used to make low-cost waterproof sensors. The LM335 is inserted into the tubing about ½″ from the end and the tubing heated above the melting point of the core. The unfilled ½″ end melts and provides a seal over the device.

Typical Applications

Basic Temperature Sensor

TL/H/5698–2

Calibrated Sensor

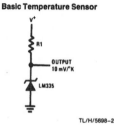

TL/H/5698–9

*Calibrate for 2.982V at 25°C

Wide Operating Supply

TL/H/5698–10

Minimum Temperature Sensing

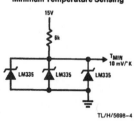

TL/H/5698–4

Average Temperature Sensing

TL/H/5698–18

Remote Temperature Sensing

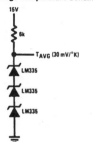

TL/H/5698–19

Wire length for 1°C error due to wire drop

AWG	$I_R = 1$ mA FEET	$I_R = 0.5$ mA* FEET
14	4000	8000
16	2500	5000
18	1600	3200
20	1000	2000
22	625	1250
24	400	800

*For I_R = 0.5 mA, the trim pot must be deleted.

(D)

FIGURE H-5 (*Continued*) National Semiconductor data sheets for the LM135/235/335 series of precision temperature sensors.

National
Semiconductor
Corporation

LM2931 Series Low Dropout Regulators

General Description

The LM2931 positive voltage regulator features a very low quiescent current of 1 mA or less when supplying 10 mA loads. This unique characteristic and the extremely low input-output differential required for proper regulation (0.2V for output currents of 10 mA) make the LM2931 the ideal regulator for standby power systems. Applications include memory standby circuits, CMOS and other low power processor power supplies as well as systems demanding as much as 100 mA of output current.

Designed originally for automotive applications, the LM2931 and all regulated circuitry are protected from reverse battery installations or 2 battery jumps. During line transients, such as a load dump (60V) when the input voltage to the regulator can momentarily exceed the specified maximum operating voltage, the regulator will automatically shut down to protect both internal circuits and the load. The LM2931 cannot be harmed by temporary mirror-image insertion. Familiar regulator features such as short circuit and thermal overload protection are also provided.

Fixed output of 5V is available in the plastic TO-220 power package or the popular TO-92 package. An adjustable output version, with on/off switch, is available in a 5-lead TO-220 package.

Features

■ Very low quiescent current
■ Output current in excess of 100 mA
■ Input-output differential less than 0.6V
■ Reverse battery protection
■ 60V load dump protection
■ −50V reverse transient protection
■ Short circuit protection
■ Internal thermal overload protection
■ Mirror-image insertion protection
■ Available in plastic TO-220 or TO-92
■ Available as adjustable with TTL compatible switch
■ 100% electrical burn-in in thermal limit

Output Voltage Options

LM2931T-5.0	5V	LM2931AT-5.0	5V
LM2931Z-5.0	5V	LM2931AZ-5.0	5V
LM2931CT	Adjustable		

Schematic and Connection Diagrams

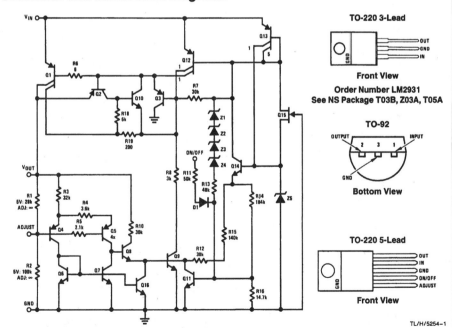

TL/H/5254–1

(A)

FIGURE H-6 National Semiconductor data sheets for the LM2931 series of low-dropout regulators.

Absolute Maximum Ratings

If Military/Aerospace specified devices are required, contact the National Semiconductor Sales Office/Distributors for availability and specifications.

Input Voltage	26V
Operating Range	
Overvoltage Protection	
LM2931A, LM2931CT Adjustable	60V
LM2931	50V

Internal Power Dissipation (Note 1)	Internally Limited
Operating Temperature Range	$-40°C$ to $+85°C$
Maximum Junction Temperature	125°C
Storage Temperature Range	$-65°C$ to $+150°C$
Lead Temp. (Soldering, 10 seconds)	230°C

Electrical Characteristics

$V_{IN} = 14V$, $I_O = 10$ mA, $T_J = 25°C$ (Note 1), C2 = 100 μF (unless otherwise specified)

Parameter	Conditions	LM2931A-5.0			LM2931-5.0			Units Limit
		Typ	Test Limit (Note 2)	Design Limit (Note 3)	Typ	Tested Limit (Note 2)	Design Limit (Note 3)	
Output Voltage		5	5.19			5.25		V_{MAX}
			4.81			4.75		V_{MIN}
	$6.0V \le V_{IN} \le 26V$, I_O 100 mA			5.25			5.5	V_{MAX}
	$-40°C \le T_j \le 125°C$			4.75			4.5	V_{MIN}
Line Regulation	$9V \le V_{IN} \le 16V$	2	10		2	10		mV_{MAX}
	$6V \le V_{IN} \le 26V$	4	30		4	30		mV_{MAX}
Load Regulation	5 mA $\le I_O \le 100$ mA	14	50		14	50		mV_{MAX}
Output Impedance	100 mA$_{DC}$ and 10 mA$_{rms}$, 100 Hz–10 kHz	200		600	200			$m\Omega_{MAX}$
Quiescent Current	$I_O \le 10$ mA, $6V \le V_{IN} \le 26V$	0.4	1.0	1.0	0.4	1.0	1.0	mA_{MAX}
	$-40°C \le T_j \le 125°C$							mA_{MIN}
	$I_O = 100$ mA, $V_{IN} = 14V$, $T_j = 25°C$	15		30	15			mA_{MAX}
				5				mA_{MIN}
Output Noise Voltage	10 Hz–100 kHz, $C_{OUT} = 100$ μF	500		1000	500			$\mu V_{rms\,MAX}$
Long Term Stability		20		50	20			mV/1000 hr
Ripple Rejection	$f_O = 120$ Hz	80		55	80			dB_{MIN}
Dropout Voltage	$I_O = 10$ mA	0.05	0.2		0.05	0.2		V_{MAX}
	$I_O = 100$ mA	0.3	0.6		0.3	0.6		V_{MAX}
Maximum Operational Input Voltage		33			33			V_{MAX}
			26			26		V_{MIN}
Maximum Line Transient	$R_L = 500\Omega$, $V_O \le 5.5V$, 100 ms	70	60		70	50		V_{MIN}
Reverse Polarity Input Voltage, DC	$V_O \ge -0.3V$, $R_L = 500\Omega$	-30	-15		-30	-15		V_{MIN}
Reverse Polarity Input Voltage, Transient	1% Duty Cycle, $\tau \le 100$ ms, $R_L = 500\Omega$	-80	-50		-80	-50		V_{MIN}

Note 1: To ensure constant junction temperature, low duty cycle pulse testing is used.

Note 2: Guaranteed and 100% production tested.

Note 3: Guaranteed (but not 100% production tested) over the operating temperature and input current ranges. These limits are not used to calculate outgoing quality levels.

Note 4: Thermal resistance junction-to-case (θ_{jC}) is 3°C/W; case-to-ambient is 50°C/W.

(B)

FIGURE H-6 (*Continued*) National Semiconductor data sheets for the LM2931 series of low-dropout regulators.

Electrical Characteristics for Adjustable LM2931CT

$V_{IN} = 14V$, $V_{OUT} = 3V$, $I_O = 10$ mA, $T_J = 25°C$ (Note 1), $R1 = 27k$, $C2 = 100$ μF (unless otherwise specified)

Parameter	Conditions	Typ	Tested Limit	Design Limit	Units Limit
Reference Voltage		1.20	1.26 1.14		V_{MAX} V_{MIN}
	$I_O \leq 100$ mA, $-40°C \leq T_j = \leq 125°C$, $R1 = 27k$ Measured from V_{OUT} to Adjust Pin			1.32 1.08	V_{MAX} V_{MIN}
Output Voltage Range			24 3		V_{MAX} V_{MIN}
Line Regulation	$V_{OUT} + 0.6V \leq V_{IN} \leq 26V$	0.2	1.5		mV/V_{MAX}
Load Regulation	5 mA $\leq I_O \leq 100$ mA	0.3	1		%$_{MAX}$
Output Impedance	100 mA$_{DC}$ and 10 mA$_{rms}$, 100 Hz-10 kHz	40			mΩ/V
Quiescent Current	$I_O = 10$ mA $I_O = 100$ mA During Shutdown $R_L = 500\Omega$	0.4 15 0.8	1 1		mA$_{MAX}$ mA mA$_{MAX}$
Output Noise Voltage	10 Hz-100 kHz	100			μV_{rms}/V
Long Term Stability		0.4			%/1000 hr
Ripple Rejection	$f_O = 120$ Hz	0.02			%/V
Dropout Voltage	$I_O \leq 10$ mA $I_O = 100$ mA	0.05 0.3	0.2 0.6		V_{MAX} V_{MAX}
Maximum Operational Input Voltage		33	26		V_{MIN}
Maximum Line Transient	$I_O = 10$ mA, Reference Voltage $\leq 1.5V$	70	60		V_{MIN}
Reverse Polarity Input Voltage, DC	$V_O \geq -0.3V$, $R_L = 500\Omega$	-30	-15		V_{MIN}
Reverse Polarity Input Voltage, Transient	1% Duty Cycle, $T \leq 100$ ms, $R_L = 500\Omega$	-80	-50		V_{MIN}
On/Off Threshold Voltage On Off	$V_O = 3V$	 2.0 2.2	 1.2 3.25		 V_{MAX} V_{MIN}
On/Off Threshold Current		20	50		μA_{MAX}

(C)

FIGURE H-6 (*Continued*) National Semiconductor data sheets for the LM2931 series of low drop-out regulators.

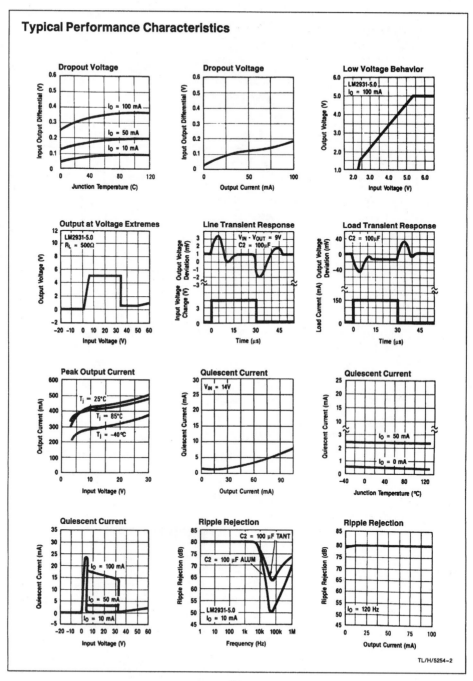

(D)

FIGURE H-6 (*Continued*) National Semiconductor data sheets for the LM2931 series of low drop-out regulators.

PRELIMINARY

LMC660AM /LMC660AI /LMC660C
CMOS Quad Operational Amplifier

General Description

The LMC660 CMOS Quad operational amplifier is ideal for operation from a single supply. It is fully specified for operation from +5V to +15V and features rail-to-rail output swing in addition to an input common-mode range that includes ground. Performance limitations that have plagued CMOS amplifiers in the past are not a problem with this design. Input V_{OS}, drift, and broadband noise as well as voltage gain into realistic loads (2 kΩ and 600Ω) are all equal to or better than widely accepted bipolar equivalents.

This chip is built with National's advanced Double-Poly Silicon-Gate CMOS process.

Features

- Rail-to-rail output swing
- Specified for 2 kΩ and 600Ω loads
- High voltage gain   126 dB
- Low input offset voltage   3 mV max
- Low offset voltage drift   1.3 μV/°C
- Ultra low input bias current   40 fA
- Input common-mode includes GND
- Operation guaranteed from +5V to +15V
- I_{SS} = 375 μA/amplifier; independent of V$^+$
- Low distortion   0.01% at 10 kHz
- Slew rate   1.1 V/μs
- Insensitive to latch-up

Connection Diagram

14 Pin DIP

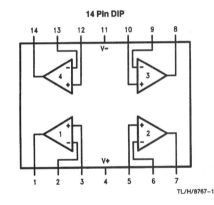

TL/H/8767–1

Order Number LMC660AMD or
LMC660AID
Order Number LMC660AIN or
LMC660CN
See NS Package Number D14E or N14A

Output Swing

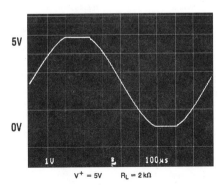

V$^+$ = 5V R_L = 2 kΩ

TL/H/8767–2

(A)

FIGURE H-7 National Semiconductor data sheets for the LMC660 series of CMOS quad operational amplifiers.

Absolute Maximum Ratings

If Military/Aerospace specified devices are required, contact the National Semiconductor Sales Office/Distributors for availability and specifications.

Differential Input Voltage	±Supply Voltage
Either Input beyond V^+ or V^-	0.7V
Supply Voltage	16V
Output Short Circuit to GND (Note 1)	Continuous
Lead Temperature (Soldering, 10 sec.)	260°C

Storage Temp. Range	−65°C to +150°C
Operating Temperature Range	
LMC660AM	−55°C to +125°C
LMC660AI	−40°C to +85°C
LMC660C	0°C to +70°C
Operating Supply Range	4.75V to 15.5V
Junction Temperature (Note 2)	150°C
ESD rating is to be determined.	

DC Electrical Characteristics (Note 3)

Parameter	Conditions	Typ	LMC660AM Tested Limit (Note 4)	LMC660AM Design Limit (Note 5)	LMC660AI Tested Limit (Note 4)	LMC660AI Design Limit (Note 5)	LMC660C Tested Limit (Note 4)	LMC660C Design Limit (Note 5)	Units
Input Offset Voltage		1	3		3	**3.3**	6	**6.3**	mV
			3.5						max
Input Offset Voltage Average Drift		1.3							µV/°C
Input Bias Current	(Note 9)	0.04	20		20	**4**		**2**	pA
			30						max
Input Offset Current	(Note 9)	0.01	20		20	**2**		**1**	pA
			30						max
Input Resistance		>1							TerraΩ
Common Mode Rejection Ratio	$0V \leq V_{CM} \leq 12.0V$ $V^+ = 15V$	83	70		72	**68**	63	**62**	dB
			68						min
Positive Power Supply Rejection Ratio	$5V \leq V^+ \leq 15V$ $V_O = 2.5V$	83	70		70	**68**	68	**62**	dB
			68						min
Negative Power Supply Rejection Ratio	$0V \leq V^- \leq -10V$	94	84		84	**83**	74	**73**	dB
			82						min
Input Common-Mode Voltage Range	$V^+ = 5V$ & $15V$ For CMRR $\geq$ 50 dB	−0.4	−0.1		−0.1	**0**	−0.1	**0**	V
			0						max
		$V^+ - 1.9$	$V^+ - 2.3$		$V^+ - 2.3$	$V^+ - 2.5$	$V^+ - 2.3$	$V^+ - 2.4$	V
			$V^+ - 2.6$						min
Large Signal Voltage Gain	$R_L = 2 k\Omega$ (Note 6)								
Sourcing		2000	400		400	**440**	200	**300**	V/mV
			300						min
Sinking		500	180		180	**120**	90	**80**	V/mV
			70						min
	$R_L = 600\Omega$ (Note 6)								
Sourcing		1000	200		200	**220**	100	**150**	V/mV
			150						min
Sinking		250	100		100	**60**	50	**40**	V/mV
			35						min

(B)

FIGURE H-7 (*Continued*) National Semiconductor data sheets for the LMC660 series of CMOS quad operational amplifiers.

DC Electrical Characteristics (Note 3) (Continued)

Parameter	Conditions	Typ	LMC660AM		LMC660AI		LMC660C		Units
			Tested Limit (Note 4)	Design Limit (Note 5)	Tested Limit (Note 4)	Design Limit (Note 5)	Tested Limit (Note 4)	Design Limit (Note 5)	
Output Swing	$V^+ = 5V$ $R_L = 2\ k\Omega$ to $V^+/2$	4.87	4.82	4.77	4.82	4.79	4.78	4.76	V min
		0.10	0.15	0.19	0.15	0.17	0.19	0.21	V max
	$V^+ = 5V$ $R_L = 600\Omega$ to $V^+/2$	4.61	4.41	4.24	4.41	4.31	4.27	4.21	V min
		0.30	0.50	0.63	0.50	0.56	0.63	0.69	V max
	$V^+ = 15V$ $R_L = 2\ k\Omega$ to $V^+/2$	14.63	14.50	14.40	14.50	14.44	14.37	14.32	V min
		0.26	0.35	0.43	0.35	0.40	0.44	0.48	V max
	$V^+ = 15V$ $R_L = 600\Omega$ to $V^+/2$	13.90	13.35	13.02	13.35	13.15	12.92	12.76	V min
		0.79	1.16	1.42	1.16	1.32	1.45	1.58	V max
Output Current $V^+ = 5V$	Sourcing, $V_O = 0V$	22	16	12	16	14	13	11	mA min
	Sinking, $V_O = 5V$	21	16	12	16	14	13	11	mA min
Output Current $V^+ = 15V$	Sourcing, $V_O = 0V$	40	19	19	28	25	23	21	mA min
	Sinking, $V_O = 13V$	39	19	19	28	24	23	20	mA min
Supply Current	All Four Amplifiers	1.5	2.2	2.9	2.2	2.6	2.7	2.9	mA min

(C)

FIGURE H-7 (*Continued*) National Semiconductor data sheets for the LMC660 series of CMOS quad operational amplifiers.

AC Electrical Characteristics (Note 3)

Parameter	Conditions	Typ	LMC660AM		LMC660AI		LMC660C		Units
			Tested Limit (Note 4)	Design Limit (Note 5)	Tested Limit (Note 4)	Design Limit (Note 5)	Tested Limit (Note 4)	Design Limit (Note 5)	
Slew Rate	(Note 7)	1.1	0.8		0.8	**0.6**		**0.7**	V/μs
			0.5						min
Gain-Bandwidth Product		1.4							MHz min
Phase Margin		50							Deg
Gain Margin		17							dB
Amp-to-Amp Isolation	(Note 8)	130							dB
Input Referred Voltage Noise	F = 1 kHz	22							nV/√Hz
Input Referred Current Noise	F = 1 kHz	0.0002							pA/√Hz
Total Harmonic Distortion	F = 10 kHz, $A_V = -10$ $R_L = 2$ kΩ, $V_O = 8$ V_{PP}	0.01							%

Note 1: Applies to both single supply and split supply operation. Continuous short circuit operation at elevated ambient temperature and/or multiple Op Amp shorts can result in exceeding the maximum allowed junction temperature of 150°C.

Note 2: The junction-to-ambient thermal resistance of the molded plastic DIP (N) is 75°C/W., the molded plastic SO (M) package is 105°C/W., and the cavity DIP (D) package is 92°C/W. All numbers apply for packages soldered directly into a PC board.

Note 3: Unless otherwise specified, all limits guaranteed for $T_A = T_J = 25$°C. **Boldface** limits apply at the temperature extremes. $V^+ = 5V$, $V^- = 0V$, $V_{CM} = 1.5V$, $V_O = V^+/2$, and $R_L > 1$ MΩ unless otherwise specified.

Note 4: These limits are guaranteed and are used in calculating outgoing AQL.

Note 5: These limits are guaranteed, but are not used in calculating outgoing AQL.

Note 6: $V^+ = 15V$, $V_{CM} = 7.5V$ and R_L connected to 7.5V. For Sourcing tests, $7.5V \leq V_O \leq 11.5V$. For Sinking tests, $2.5V \leq V_O \leq 7.5V$.

Note 7: $V^+ = 15V$. Connected as Voltage Follower with 10V step input. Number specified is the slower of the positive and negative slew rates.

Note 8: Input referred. $V^+ = 15V$ and $R_L = 10$ kΩ connected to $V^+/2$. Each amp excited in turn with 1 kHz to produce $V_O = 13$ V_{PP}.

Note 9: The specifications in the Design Limit column reflect the true performance of the part, while those in the Tested Limit column are degraded to allow for the unavoidable inaccuracies involved in cost-effective high-speed automatic testing.

(D)

FIGURE H-7 (*Continued*) National Semiconductor data sheets for the LMC660 series of CMOS quad operational amplifiers.

MOTOROLA
SEMICONDUCTOR TECHNICAL DATA

8-Bit Serial-Input/Serial or Parallel-Output Shift Register with Latched 3-State Outputs
High–Performance Silicon–Gate CMOS

The MC54/74HC595A is identical in pinout to the LS595. The device inputs are compatible with standard CMOS outputs; with pullup resistors, they are compatible with LSTTL outputs.

The HC595A consists of an 8–bit shift register and an 8–bit D–type latch with three–state parallel outputs. The shift register accepts serial data and provides a serial output. The shift register also provides parallel data to the 8–bit latch. The shift register and latch have independent clock inputs. This device also has an asynchronous reset for the shift register.

The HC595A directly interfaces with the Motorola SPI serial data port on CMOS MPUs and MCUs.

- Output Drive Capability: 15 LSTTL Loads
- Outputs Directly Interface to CMOS, NMOS, and TTL
- Operating Voltage Range: 2.0 to 6.0 V
- Low Input Current: 1.0 µA
- High Noise Immunity Characteristic of CMOS Devices
- In Compliance with the Requirements Defined by JEDEC Standard No. 7A
- Chip Complexity: 328 FETs or 82 Equivalent Gates
- Improvements over HC595
  - Improved Propagation Delays
  - 50% Lower Quiescent Power
  - Improved Input Noise and Latchup Immunity

MC54/74HC595A

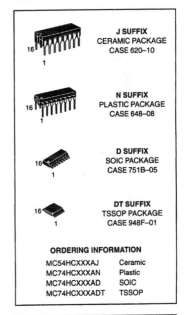

J SUFFIX
CERAMIC PACKAGE
CASE 620–10

N SUFFIX
PLASTIC PACKAGE
CASE 648–08

D SUFFIX
SOIC PACKAGE
CASE 751B–05

DT SUFFIX
TSSOP PACKAGE
CASE 948F–01

ORDERING INFORMATION

MC54HCXXXAJ	Ceramic
MC74HCXXXAN	Plastic
MC74HCXXXAD	SOIC
MC74HCXXXADT	TSSOP

PIN ASSIGNMENT

Q_B	1		16	V_{CC}
Q_C	2		15	Q_A
Q_D	3		14	A
Q_E	4		13	OUTPUT ENABLE
Q_F	5		12	LATCH CLOCK
Q_G	6		11	SHIFT CLOCK
Q_H	7		10	RESET
GND	8		9	SQ_H

LOGIC DIAGRAM

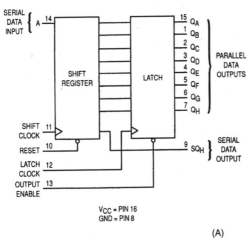

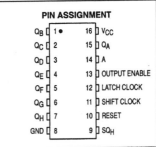

V_{CC} = PIN 16
GND = PIN 8

(A)

FIGURE H-8 Motorola data sheets for the MC54/74HC595A series of 8-bit serial-input/serial- or parallel-output shift register with latched three-state outputs.

MAXIMUM RATINGS*

Symbol	Parameter	Value	Unit
V_{CC}	DC Supply Voltage (Referenced to GND)	− 0.5 to + 7.0	V
V_{in}	DC Input Voltage (Referenced to GND)	− 1.5 to V_{CC} + 1.5	V
V_{out}	DC Output Voltage (Referenced to GND)	− 0.5 to V_{CC} + 0.5	V
I_{in}	DC Input Current, per Pin	± 20	mA
I_{out}	DC Output Current, per Pin	± 35	mA
I_{CC}	DC Supply Current, V_{CC} and GND Pins	± 75	mA
P_D	Power Dissipation in Still Air, Plastic or Ceramic DIP† SOIC Package† TSSOP Package†	750 500 450	mW
T_{stg}	Storage Temperature	− 65 to + 150	°C
T_L	Lead Temperature, 1 mm from Case for 10 Seconds (Plastic DIP, SOIC or TSSOP Package) (Ceramic DIP)	 260 300	°C

* Maximum Ratings are those values beyond which damage to the device may occur.
Functional operation should be restricted to the Recommended Operating Conditions.
†Derating — Plastic DIP: − 10 mW/°C from 65° to 125°C
Ceramic DIP: − 10 mW/°C from 100° to 125°C
SOIC Package: − 7 mW/°C from 65° to 125°C
TSSOP Package: − 6.1 mW/°C from 65° to 125°C
For high frequency or heavy load considerations, see Chapter 2.

This device contains protection circuitry to guard against damage due to high static voltages or electric fields. However, precautions must be taken to avoid applications of any voltage higher than maximum rated voltages to this high–impedance circuit. For proper operation, V_{in} and V_{out} should be constrained to the range GND ≤ (V_{in} or V_{out}) ≤ V_{CC}.
Unused inputs must always be tied to an appropriate logic voltage level (e.g., either GND or V_{CC}). Unused outputs must be left open.

RECOMMENDED OPERATING CONDITIONS

Symbol	Parameter	Min	Max	Unit
V_{CC}	DC Supply Voltage (Referenced to GND)	2.0	6.0	V
V_{in}, V_{out}	DC Input Voltage, Output Voltage (Referenced to GND)	0	V_{CC}	V
T_A	Operating Temperature, All Package Types	− 55	+ 125	°C
t_r, t_f	Input Rise and Fall Time (Figure 1) V_{CC} = 2.0 V V_{CC} = 4.5 V V_{CC} = 6.0 V	0 0 0	1000 500 400	ns

DC ELECTRICAL CHARACTERISTICS (Voltages Referenced to GND)

Symbol	Parameter	Test Conditions	V_{CC} V	Guaranteed Limit − 55 to 25°C	≤ 85°C	≤ 125°C	Unit
V_{IH}	Minimum High–Level Input Voltage	V_{out} = 0.1 V or V_{CC} − 0.1 V $|I_{out}|$ ≤ 20 µA	2.0 4.5 6.0	1.5 3.15 4.2	1.5 3.15 4.2	1.5 3.15 4.2	V
V_{IL}	Maximum Low–Level Input Voltage	V_{out} = 0.1 V or V_{CC} − 0.1 V $|I_{out}|$ ≤ 20 µA	2.0 4.5 6.0	0.5 1.35 1.8	0.5 1.35 1.8	0.5 1.35 1.8	V
V_{OH}	Minimum High–Level Output Voltage, Q_A − Q_H	V_{in} = V_{IH} or V_{IL} $|I_{out}|$ ≤ 20 µA	2.0 4.5 6.0	1.9 4.4 5.9	1.9 4.4 5.9	1.9 4.4 5.9	V
		V_{in} = V_{IH} or V_{IL} $|I_{out}|$ ≤ 6.0 mA $|I_{out}|$ ≤ 7.8 mA	4.5 6.0	3.98 5.48	3.84 5.34	3.7 5.2	
V_{OL}	Maximum Low–Level Output Voltage, Q_A − Q_H	V_{in} = V_{IH} or V_{IL} $|I_{out}|$ ≤ 20 µA	2.0 4.5 6.0	0.1 0.1 0.1	0.1 0.1 0.1	0.1 0.1 0.1	V
		V_{in} = V_{IH} or V_{IL} $|I_{out}|$ ≤ 6.0 mA $|I_{out}|$ ≤ 7.8 mA	4.5 6.0	0.26 0.26	0.33 0.33	0.4 0.4	

NOTE: Information on typical parametric values can be found in Chapter 2.

(B)

FIGURE H-8 (*Continued*) Motorola data sheets for the MC54/74HC595A series of 8-bit serial-input/serial- or parallel-output shift register with latched three-state outputs.

DC ELECTRICAL CHARACTERISTICS (Continued)

Symbol	Parameter	Test Conditions	V_{CC} V	Guaranteed Limit			Unit
				– 55 to 25°C	≤ 85°C	≤ 125°C	
V_{OH}	Minimum High–Level Output Voltage, SQ$_H$	$V_{in} = V_{IH}$ or V_{IL} $|I_{out}| \leq 20 \mu A$	2.0 4.5 6.0	1.9 4.4 5.9	1.9 4.4 5.9	1.9 4.4 5.9	V
		$V_{in} = V_{IH}$ or V_{IL} $|I_{out}| \leq 4.0$ mA $|I_{out}| \leq 5.2$ mA	4.5 6.0	3.98 5.48	3.84 5.34	3.7 5.2	
V_{OL}	Maximum Low–Level Output Voltage, SQ$_H$	$V_{in} = V_{IH}$ or V_{IL} $|I_{out}| \leq 20 \mu A$	2.0 4.5 6.0	0.1 0.1 0.1	0.1 0.1 0.1	0.1 0.1 0.1	V
		$V_{in} = V_{IH}$ or V_{IL} $|I_{out}| \leq 4.0$ mA $|I_{out}| \leq 5.2$ mA	4.5 6.0	0.26 0.26	0.33 0.33	0.4 0.4	
I_{in}	Maximum Input Leakage Current	$V_{in} = V_{CC}$ or GND	6.0	± 0.1	± 1.0	± 1.0	μA
I_{OZ}	Maximum Three–State Leakage Current, Q$_A$ – Q$_H$	Output in High–Impedance State $V_{in} = V_{IL}$ or V_{IH} $V_{out} = V_{CC}$ or GND	6.0	± 0.5	± 5.0	± 10	μA
I_{CC}	Maximum Quiescent Supply Current (per Package)	$V_{in} = V_{CC}$ or GND $I_{out} = 0 \mu A$	6.0	4.0	40	160	μA

AC ELECTRICAL CHARACTERISTICS (C_L = 50 pF, Input $t_r = t_f = 6.0$ ns)

Symbol	Parameter	V_{CC} V	Guaranteed Limit			Unit
			– 55 to 25°C	≤ 85°C	≤ 125°C	
f_{max}	Maximum Clock Frequency (50% Duty Cycle) (Figures 1 and 7)	2.0 4.5 6.0	6.0 30 35	4.8 24 28	4.0 20 24	MHz
$t_{PLH},$ t_{PHL}	Maximum Propagation Delay, Shift Clock to SQ$_H$ (Figures 1 and 7)	2.0 4.5 6.0	140 28 24	175 35 30	210 42 36	ns
t_{PHL}	Maximum Propagation Delay, Reset to SQ$_H$ (Figures 2 and 7)	2.0 4.5 6.0	145 29 25	180 36 31	220 44 38	ns
$t_{PLH},$ t_{PHL}	Maximum Propagation Delay, Latch Clock to Q$_A$ – Q$_H$ (Figures 3 and 7)	2.0 4.5 6.0	140 28 24	175 35 30	210 42 36	ns
$t_{PLZ},$ t_{PHZ}	Maximum Propagation Delay, Output Enable to Q$_A$ – Q$_H$ (Figures 4 and 8)	2.0 4.5 6.0	150 30 26	190 38 33	225 45 38	ns
$t_{PZL},$ t_{PZH}	Maximum Propagation Delay, Output Enable to Q$_A$ – Q$_H$ (Figures 4 and 8)	2.0 4.5 6.0	135 27 23	170 34 29	205 41 35	ns
$t_{TLH},$ t_{THL}	Maximum Output Transition Time, Q$_A$ – Q$_H$ (Figures 3 and 7)	2.0 4.5 6.0	60 12 10	75 15 13	90 18 15	ns
$t_{TLH},$ t_{THL}	Maximum Output Transition Time, SQ$_H$ (Figures 1 and 7)	2.0 4.5 6.0	75 15 13	95 19 16	110 22 19	ns
C_{in}	Maximum Input Capacitance	—	10	10	10	pF
C_{out}	Maximum Three–State Output Capacitance (Output in High–Impedance State), Q$_A$ – Q$_H$	—	15	15	15	pF

NOTE: For propagation delays with loads other than 50 pF, and information on typical parametric values, see Chapter 2.

		Typical @ 25°C, V_{CC} = 5.0 V	
C_{PD}	Power Dissipation Capacitance (Per Package)*	300	pF

* Used to determine the no–load dynamic power consumption: $P_D = C_{PD} V_{CC}{}^2 f + I_{CC} V_{CC}$. For load considerations, see Chapter 2.

(C)

FIGURE H-8 (*Continued*) **Motorola data sheets for the MC54/74HC595A series of 8-bit serial-input/serial- or parallel-output shift register with latched three-state outputs.**

MC54/74HC595A

TIMING REQUIREMENTS (Input $t_r = t_f = 6.0$ ns)

Symbol	Parameter	V_{CC} V	Guaranteed Limit			Unit
			25°C to −55°C	≤ 85°C	≤ 125°C	
t_{su}	Minimum Setup Time, Serial Data Input A to Shift Clock (Figure 5)	2.0 4.5 6.0	50 10 9.0	65 13 11	75 15 13	ns
t_{su}	Minimum Setup Time, Shift Clock to Latch Clock (Figure 6)	2.0 4.5 6.0	75 15 13	95 19 16	110 22 19	ns
t_h	Minimum Hold Time, Shift Clock to Serial Data Input A (Figure 5)	2.0 4.5 6.0	5.0 5.0 5.0	5.0 5.0 5.0	5.0 5.0 5.0	ns
t_{rec}	Minimum Recovery Time, Reset Inactive to Shift Clock (Figure 2)	2.0 4.5 6.0	50 10 9.0	65 13 11	75 15 13	ns
t_w	Minimum Pulse Width, Reset (Figure 2)	2.0 4.5 6.0	60 12 10	75 15 13	90 18 15	ns
t_w	Minimum Pulse Width, Shift Clock (Figure 1)	2.0 4.5 6.0	50 10 9.0	65 13 11	75 15 13	ns
t_w	Minimum Pulse Width, Latch Clock (Figure 6)	2.0 4.5 6.0	50 10 9.0	65 13 11	75 15 13	ns
t_r, t_f	Maximum Input Rise and Fall Times (Figure 1)	2.0 4.5 6.0	1000 500 400	1000 500 400	1000 500 400	ns

FUNCTION TABLE

Operation	Inputs					Resulting Function			
	Reset	Serial Input A	Shift Clock	Latch Clock	Output Enable	Shift Register Contents	Latch Register Contents	Serial Output SQ_H	Parallel Outputs Q_A – Q_H
Reset shift register	L	X	X	L, H, ⌐	L	L	U	L	U
Shift data into shift register	H	D	⌐	L, H, ⌐	L	$D \to SR_A$; $SR_N \to SR_{N+1}$	U	$SR_G \to SR_H$	U
Shift register remains unchanged	H	X	L, H, ⌐	L, H, ⌐	L	U	U	U	U
Transfer shift register contents to latch register	H	X	L, H, ⌐	⌐	L	U	$SR_N \to LR_N$	U	SR_N
Latch register remains unchanged	X	X	X	L, H, ⌐	L	*	U	*	U
Enable parallel outputs	X	X	X	X	L	*	**	*	Enabled
Force outputs into high impedance state	X	X	X	X	H	*	**	*	Z

SR = shift register contents   D = data (L, H) logic level   X = don't care   * = depends on Reset and Shift Clock inputs
LR = latch register contents   U = remains unchanged   Z = high impedance   ** = depends on Latch Clock input

(D)

FIGURE H-8 (*Continued*) Motorola data sheets for the MC54/74HC595A series of 8-bit serial-input/serial- or parallel-output shift register with latched three-state outputs.

MC54/74HC595A

PIN DESCRIPTIONS

INPUTS

A (Pin 14)

Serial Data Input. The data on this pin is shifted into the 8–bit serial shift register.

CONTROL INPUTS

Shift Clock (Pin 11)

Shift Register Clock Input. A low– to–high transition on this input causes the data at the Serial Input pin to be shifted into the 8–bit shift register.

Reset (Pin 10)

Active–low, Asynchronous, Shift Register Reset Input. A low on this pin resets the shift register portion of this device only. The 8–bit latch is not affected.

Latch Clock (Pin 12)

Storage Latch Clock Input. A low–to–high transition on this input latches the shift register data.

Output Enable (Pin 13)

Active–low Output Enable. A low on this input allows the data from the latches to be presented at the outputs. A high on this input forces the outputs (Q_A–Q_H) into the high–impedance state. The serial output is not affected by this control unit.

OUTPUTS

Q_A – Q_H (Pins 15, 1, 2, 3, 4, 5, 6, 7)

Noninverted, 3–state, latch outputs.

SQ_H (Pin 9)

Noninverted, Serial Data Output. This is the output of the eighth stage of the 8–bit shift register. This output does not have three–state capability.

(E)

FIGURE H-8 (*Continued*) **Motorola data sheets for the MC54/74HC595A series of 8-bit serial-input/serial- or parallel-output shift register with latched three-state outputs.**

SWITCHING WAVEFORMS

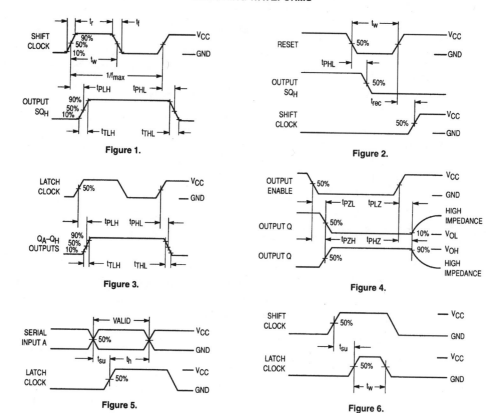

Figure 1.

Figure 2.

Figure 3.

Figure 4.

Figure 5.

Figure 6.

TEST CIRCUITS

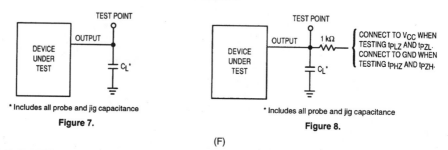

* Includes all probe and jig capacitance

Figure 7.

* Includes all probe and jig capacitance

Figure 8.

(F)

FIGURE H-8   (*Continued*) Motorola data sheets for the MC54/74HC595A series of 8-bit serial-input/serial- or parallel-output shift register with latched three-state outputs.

MC54/74HC595A

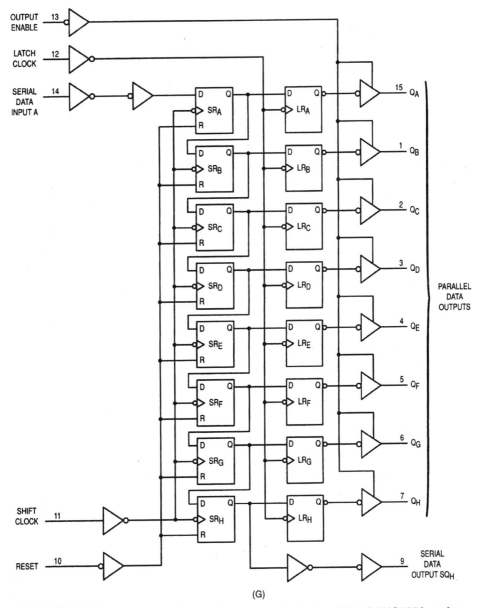

(G)

FIGURE H-8 (*Continued*) Motorola data sheets for the MC54/74HC595A series of 8-bit serial-input/serial- or parallel-output shift register with latched three-state outputs.

TIMING DIAGRAM

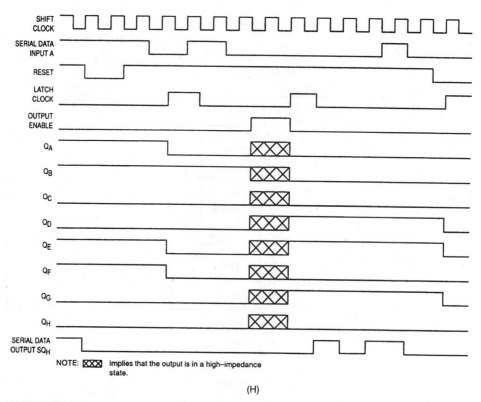

NOTE: ⊠⊠⊠ implies that the output is in a high–impedance state.

(H)

FIGURE H-8 (*Continued*) Motorola data sheets for the MC54/74HC595A series of 8-bit serial-input/serial- or parallel-output shift register with latched three-state outputs.

MOTOROLA
SEMICONDUCTOR TECHNICAL DATA

8-Bit Serial or Parallel-Input/ Serial-Output Shift Register with Input Latch
High–Performance Silicon–Gate CMOS

The MC54/74HC597 is identical in pinout to the LS597. The device inputs are compatible with standard CMOS outputs; with pullup resistors, they are compatible with LSTTL outputs.

This device consists of an 8–bit input latch which feeds parallel data to an 8–bit shift register. Data can also be loaded serially (see Function Table).

The HC597 is similar in function to the HC589, which is a 3–state device.

- Output Drive Capability: 10 LSTTL Loads
- Outputs Directly Interface to CMOS, NMOS, and TTL
- Operating Voltage Range: 2 to 6 V
- Low Input Current: 1 µA
- High Noise Immunity Characteristic of CMOS Devices
- In Compliance with the Requirements Defined by JEDEC Standard No. 7A
- Chip Complexity: 516 FETs or 129 Equivalent Gates

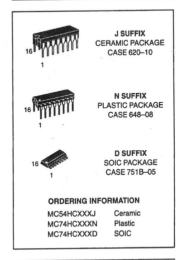

MC54/74HC597

J SUFFIX
CERAMIC PACKAGE
CASE 620–10

N SUFFIX
PLASTIC PACKAGE
CASE 648–08

D SUFFIX
SOIC PACKAGE
CASE 751B–05

ORDERING INFORMATION

MC54HCXXXJ	Ceramic
MC74HCXXXN	Plastic
MC74HCXXXD	SOIC

PIN ASSIGNMENT

B	1	16	V$_{CC}$
C	2	15	A
D	3	14	S$_A$
E	4	13	SERIAL SHIFT/ PARALLEL LOAD
F	5	12	LATCH CLOCK
G	6	11	SHIFT CLOCK
H	7	10	RESET
GND	8	9	Q$_H$

LOGIC DIAGRAM

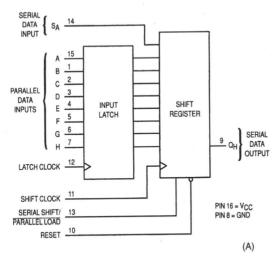

(A)

FIGURE H-9 Motorola data sheets for the MC54/74HC597 series of 8-bit serial- or parallel-input/serial-output shift register with input latch.

MAXIMUM RATINGS*

Symbol	Parameter	Value	Unit
V_{CC}	DC Supply Voltage (Referenced to GND)	− 0.5 to + 7.0	V
V_{in}	DC Input Voltage (Referenced to GND)	− 1.5 to V_{CC} + 1.5	V
V_{out}	DC Output Voltage (Referenced to GND)	− 0.5 to V_{CC} + 0.5	V
I_{in}	DC Input Current, per Pin	± 20	mA
I_{out}	DC Output Current, per Pin	± 25	mA
I_{CC}	DC Supply Current, V_{CC} and GND Pins	± 50	mA
P_D	Power Dissipation in Still Air, Plastic or Ceramic DIP† / SOIC Package†	750 / 500	mW
T_{stg}	Storage Temperature	− 65 to + 150	°C
T_L	Lead Temperature, 1 mm from Case for 10 Seconds (Plastic DIP or SOIC Package) (Ceramic DIP)	260 / 300	°C

This device contains protection circuitry to guard against damage due to high static voltages or electric fields. However, precautions must be taken to avoid applications of any voltage higher than maximum rated voltages to this high–impedance circuit. For proper operation, V_{in} and V_{out} should be constrained to the range GND ≤ (V_{in} or V_{out}) ≤ V_{CC}.

Unused inputs must always be tied to an appropriate logic voltage level (e.g., either GND or V_{CC}). Unused outputs must be left open.

* Maximum Ratings are those values beyond which damage to the device may occur.
Functional operation should be restricted to the Recommended Operating Conditions.
†Derating — Plastic DIP: − 10 mW/°C from 65° to 125°C
Ceramic DIP: − 10 mW/°C from 100° to 125°C
SOIC Package: − 7 mW/°C from 65° to 125°C
For high frequency or heavy load considerations, see Chapter 2.

RECOMMENDED OPERATING CONDITIONS

Symbol	Parameter		Min	Max	Unit
V_{CC}	DC Supply Voltage (Referenced to GND)		2.0	6.0	V
V_{in}, V_{out}	DC Input Voltage, Output Voltage (Referenced to GND)		0	V_{CC}	V
T_A	Operating Temperature, All Package Types		− 55	+ 125	°C
t_r, t_f	Input Rise and Fall Time (Figure 1)	V_{CC} = 2.0 V	0	1000	ns
		V_{CC} = 4.5 V	0	500	
		V_{CC} = 6.0 V	0	400	

DC ELECTRICAL CHARACTERISTICS (Voltages Referenced to GND)

Symbol	Parameter	Test Conditions	V_{CC} V	Guaranteed Limit			Unit
				− 55 to 25°C	≤ 85°C	≤ 125°C	
V_{IH}	Minimum High–Level Input Voltage	V_{out} = 0.1 V or V_{CC} − 0.1 V $|I_{out}| ≤ 20$ µA	2.0 / 4.5 / 6.0	1.5 / 3.15 / 4.2	1.5 / 3.15 / 4.2	1.5 / 3.15 / 4.2	V
V_{IL}	Maximum Low–Level Input Voltage	V_{out} = 0.1 V or V_{CC} − 0.1 V $|I_{out}| ≤ 20$ µA	2.0 / 4.5 / 6.0	0.3 / 0.9 / 1.2	0.3 / 0.9 / 1.2	0.3 / 0.9 / 1.2	V
V_{OH}	Minimum High–Level Output Voltage	V_{in} = V_{IH} or V_{IL} $|I_{out}| ≤ 20$ µA	2.0 / 4.5 / 6.0	1.9 / 4.4 / 5.9	1.9 / 4.4 / 5.9	1.9 / 4.4 / 5.9	V
		V_{in} = V_{IH} or V_{IL} $|I_{out}| ≤ 4.0$ mA / $|I_{out}| ≤ 5.2$ mA	4.5 / 6.0	3.98 / 5.48	3.84 / 5.34	3.70 / 5.20	
V_{OL}	Maximum Low–Level Output Voltage	V_{in} = V_{IH} or V_{IL} $|I_{out}| ≤ 20$ µA	2.0 / 4.5 / 6.0	0.1 / 0.1 / 0.1	0.1 / 0.1 / 0.1	0.1 / 0.1 / 0.1	V
		V_{in} = V_{IH} or V_{IL} $|I_{out}| ≤ 4.0$ mA / $|I_{out}| ≤ 5.2$ mA	4.5 / 6.0	0.26 / 0.26	0.33 / 0.33	0.40 / 0.40	
I_{in}	Maximum Input Leakage Current	V_{in} = V_{CC} or GND	6.0	± 0.1	± 1.0	± 1.0	µA
I_{CC}	Maximum Quiescent Supply Current (per Package)	V_{in} = V_{CC} or GND I_{out} = 0 µA	6.0	8	80	160	µA

NOTE: Information on typical parametric values can be found in Chapter 2.

(B)

FIGURE H-9 (*Continued*) Motorola data sheets for the MC54/74HC597 series of 8-bit serial- or parallel-input/serial-output shift register with input latch.

MC54/74HC597

AC ELECTRICAL CHARACTERISTICS (C_L = 50 pF, Input t_r = t_f = 6 ns)

Symbol	Parameter	V_{CC} V	Guaranteed Limit			Unit
			− 55 to 25°C	≤ 85°C	≤ 125°C	
f_{max}	Maximum Clock Frequency (50% Duty Cycle) (Figures 2 and 8)	2.0 4.5 6.0	6.0 30 35	4.8 24 28	4.0 20 24	MHz
t_{PLH}, t_{PHL}	Maximum Propagation Delay, Latch Clock to Q_H (Figures 1 and 8)	2.0 4.5 6.0	210 42 36	265 53 45	315 63 54	ns
t_{PLH}, t_{PHL}	Maximum Propagation Delay, Shift Clock to Q_H (Figures 2 and 8)	2.0 4.5 6.0	175 35 30	220 44 37	265 53 45	ns
t_{PHL}	Maximum Propagation Delay, Reset to Q_H (Figures 3 and 8)	2.0 4.5 6.0	175 35 30	220 44 37	265 53 45	ns
t_{PLH}, t_{PHL}	Maximum Propagation Delay, Serial Shift/Parallel Load to Q_H (Figures 4 and 8)	2.0 4.5 6.0	175 35 30	220 44 37	265 53 45	ns
t_{TLH}, t_{THL}	Maximum Output Transition Time, Any Output (Figures 1 and 8)	2.0 4.5 6.0	75 15 13	95 19 16	110 22 19	ns
C_{in}	Maximum Input Capacitance	—	10	10	10	pF

NOTES:
1. For propagation delays with loads other than 50 pF, see Chapter 2.
2. Information on typical parametric values can be found in Chapter 2.

		Typical @ 25°C, V_{CC} = 5.0 V	
C_{PD}	Power Dissipation Capacitance (Per Package)*	50	pF

* Used to determine the no–load dynamic power consumption: $P_D = C_{PD} V_{CC}^2 f + I_{CC} V_{CC}$. For load considerations, see Chapter 2.

PIN DESCRIPTIONS

DATA INPUTS

A, B, C, D, E, F, G, H (Pins 15, 1, 2, 3, 4, 5, 6, 7)

Parallel data inputs. Data on these inputs is stored in the input latch on the rising edge of the Latch Clock input.

S_A (Pin 14)

Serial data input. Data on this input is shifted into the shift register on the rising edge of the Shift Clock input it Serial Shift/Parallel Load is high. Data on this input is ignored when Serial Shift/Parallel Load is low.

CONTROL INPUTS

Serial Shift/Parallel Load (Pin 13)

Shift register mode control. When a high level is applied to this pin, the shift register is allowed to serially shift data. When a low level is applied to this pin, the shift register accepts parallel data from the input latch, and serial shifting is inhibited.

Reset (Pin 10)

Asynchronous, Active–low shift register reset. A low level applied to this input resets the shift register to a low level, but does not change the data in the input latch.

Shift Clock (Pin 11)

Serial shift register clock. A low–to–high transition on this input shifts data on the Serial Data Input into the shift register and data in stage H is shifted out Q_H, being replaced by the data previously stored in stage G.

Latch Clock (Pin 12)

Latch clock. A low–to–high transition on this input loads the parallel data on inputs A–H into the input latch.

OUTPUT

Q_H (Pin 9)

Serial data output. This pin is the output from the last stage of the shift register.

(C)

FIGURE H-9 *(Continued)* **Motorola data sheets for the MC54/74HC597 series of 8-bit serial- or parallel-input/serial-output shift register with input latch.**

TIMING REQUIREMENTS (Input $t_r = t_f = 6$ ns)

Symbol	Parameter	V_{CC} V	Guaranteed Limit − 55 to 25°C	≤ 85°C	≤ 125°C	Unit
t_{su}	Minimum Setup Time, Parallel Data inputs A–H to Latch Clock (Figure 5)	2.0 4.5 6.0	100 20 17	125 25 21	150 30 26	ns
t_{su}	Minimum Setup Time, Serial Data Input S_A to Shift Clock (Figure 6)	2.0 4.5 6.0	100 20 17	125 25 21	150 30 26	ns
t_{su}	Minimum Setup Time, Serial Shift/Parallel Load to Shift Clock (Figure 7)	2.0 4.5 6.0	100 20 17	125 25 21	150 30 26	ns
t_h	Minimum Hold Time, Latch Clock to Parallel Data Inputs A–H (Figure 5)	2.0 4.5 6.0	25 5 5	30 6 6	40 8 7	ns
t_h	Minimum Hold Time, Shift Clock to Serial Data Input S_A (Figure 6)	2.0 4.5 6.0	5 5 5	5 5 5	5 5 5	ns
t_{rec}	Minimum Recovery Time, Reset Inactive to Shift Clock (Figure 3)	2.0 4.5 6.0	100 20 17	125 25 21	150 30 26	ns
t_w	Minimum Pulse Width, Latch Clock and Shift Clock (Figures 1 and 2)	2.0 4.5 6.0	80 16 14	100 20 17	120 24 20	ns
t_w	Minimum Pulse Width, Reset (Figure 3)	2.0 4.5 6.0	80 16 14	100 20 17	120 24 20	ns
t_w	Minimum Pulse Width, Serial Shift/Parallel Load (Figure 4)	2.0 4.5 6.0	80 16 14	100 20 17	120 24 20	ns
t_r, t_f	Maximum Input Rise and Fall Times (Figure 1)	2.0 4.5 6.0	1000 500 400	1000 500 400	1000 500 400	ns

NOTE: Information on typical parametric values can be found in Chapter 2.

FUNCTION TABLE

Operation	Inputs Reset	Serial Shift/ Parallel Load	Latch Clock	Shift Clock	Serial Input S_A	Parallel Inputs A–H	Resulting Function Latch Contents	Shift Register Contents	Output Q_H
Reset shift register	L	X	L, H, ⤸	X	X	X	U	L	L
Reset shift register; load parallel data into data latch	L	X	⤒	X	X	a–h	a–h	L	L
Load parallel data into data latch	H	H	⤒	L,H, ⤸	X	a–h	a–h	U	U
Transfer latch contents to shift register	H	L	L, H, ⤸	X	X	X	U	$LR_N \rightarrow SR_N$	LR_H
Contents of data latch and shift register are unchanged	H	H	L, H, ⤸	L,H, ⤸	X	X	U	U	U
Load parallel data into data latch and shift register	H	L	⤒	X	X	a–h	a–h	a–h	h
Shift serial data into shift register	H	H	X	⤒	D	X	*	$SR_A = D$; $SR_N \rightarrow SR_{N+1}$	$SR_G \rightarrow SR_H$
Load parallel data into data latch and shift serial data into shift register	H	H	⤒	⤒	D	a–h	a–h	$SR_A = D$; $SR_N \rightarrow SR_{N+1}$	$SR_G \rightarrow SR_H$

LR = latch register contents
SR = shift register contents
* = depends on latch clock input

a–h = data at parallel data inputs A–H
D = data (L, H) at serial data input S_A

U = remains unchanged
X = don't care

(D)

FIGURE H-9 *(Continued)* Motorola data sheets for the MC54/74HC597 series of 8-bit serial- or parallel-input/serial-output shift register with input latch.

MC54/74HC597

SWITCHING WAVEFORMS

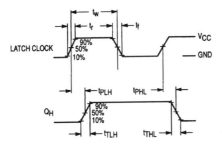

Figure 1. (Serial Shift/$\overline{\text{Parallel Load}}$ = L)

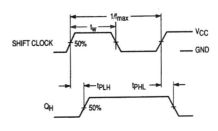

Figure 2. (Serial Shift/$\overline{\text{Parallel Load}}$ = H)

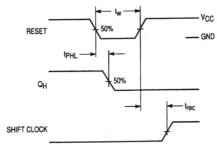

Figure 3.

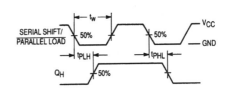

Figure 4.

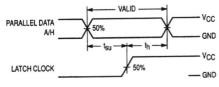

Figure 5.

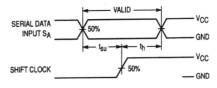

Figure 6.

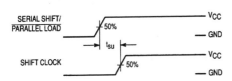

Figure 7.

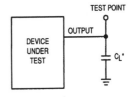

* Includes all probe and jig capacitance

Figure 8. Test Circuit

(E)

FIGURE H-9 (*Continued*) Motorola data sheets for the MC54/74HC597 series of 8-bit serial- or parallel-input/serial-output shift register with input latch.

MC54/74HC597

EXPANDED LOGIC DIAGRAM

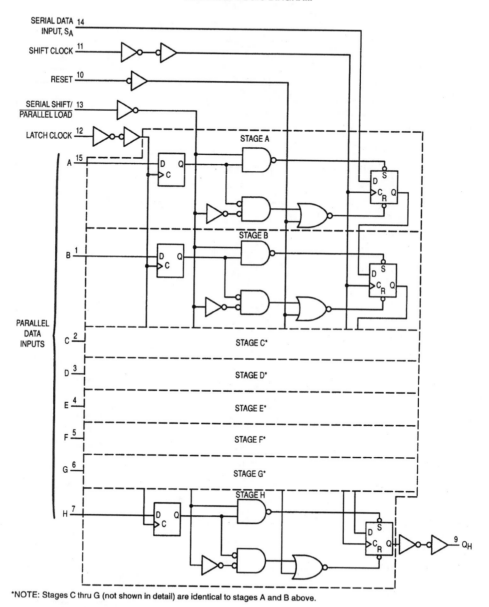

*NOTE: Stages C thru G (not shown in detail) are identical to stages A and B above.

(F)

FIGURE H-9   (*Continued*) Motorola data sheets for the MC54/74HC597 series of 8-bit serial- or parallel-input/serial-output shift register with input latch.

MC54/74HC597

TIMING DIAGRAM

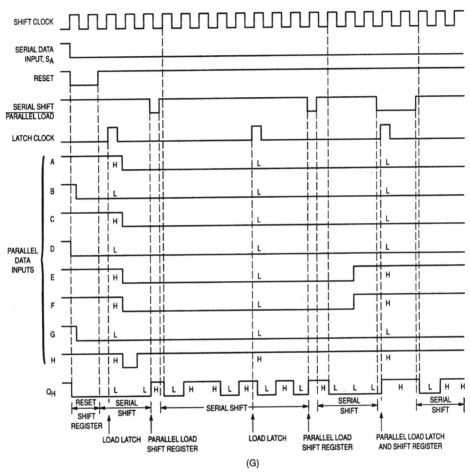

(G)

FIGURE H-9 (*Continued*) Motorola data sheets for the MC54/74HC597 series of 8-bit serial- or parallel-input/serial-output shift register with input latch.

INDEX

About the Author

Tom Fox has been researching, designing, and creating electronic circuits for over 25 years. The author of many articles on projects for magazines such as *Popular Electronics*, and *Computer Craft*, for which he recently wrote a five-part series on the HC11, he is also Workshop Editor for *Boys' Quest* magazine. Tom has incorporated many of his designs into products he sells through his company, Magicland. He can be reached at *tfox@ncats.net* or *tomfox@nccats.net*.

DISK WARRANTY

This software is protected by both United States copyright law and international copyright treaty provision. You must treat this software just like a book, except that you may copy it into a computer in order to be used and you may make archival copies of the software for the sole purpose of backing up our software and protecting your investment from loss.

By saying "just like a book," McGraw-Hill means, for example, that this software may be used by any number of people and may be freely moved from one computer location to another, so long as there is no possibility of its being used at one location or on one computer while it also is being used at another. Just as a book cannot be read by two different people in two different places at the same time, neither can the software be used by two different people in two different places at the same time (unless, of course, McGraw-Hill's copyright is being violated).

LIMITED WARRANTY

McGraw-Hill takes great care to provide you with top-quality software, thoroughly checked to prevent virus infections. McGraw-Hill warrants the physical diskette(s) contained herein to be free of defects in materials and workmanship for a period of sixty days from the purchase date. If McGraw-Hill receives written notification within the warranty period of defects in materials or workmanship, and such notification is determined by McGraw-Hill to be correct, McGraw-Hill will replace the defective diskette(s). Send requests to:

McGraw-Hill
Customer Services
P.O. Box 545
Blacklick, OH 43004-0545

The entire and exclusive liability and remedy for breach of this Limited Warranty shall be limited to replacement of defective diskette(s) and shall not include or extend to any claim for or right to cover any other damages, including but not limited to, loss of profit, data, or use of the software, or special, incidental, or consequential damages or other similar claims, even if McGraw-Hill has been specifically advised of the possibility of such damages. In no event will McGraw-Hill's liability for any damages to you or any other person ever exceed the lower of suggested list price or actual price paid for the license to use the software, regardless of any form of the claim.

McGRAW-HILL SPECIFICALLY DISCLAIMS ALL OTHER WARRANTIES, EXPRESS OR IMPLIED, INCLUDING, BUT NOT LIMITED TO, ANY IMPLIED WARRANTY OF MER-CHANTABILITY OR FITNESS FOR A PARTICULAR PURPOSE.

Specifically, McGraw-Hill makes no representation or warranty that the software is fit for any particular purpose and any implied warranty of merchantability is limited to the sixty-day duration of the Limited Warranty covering the physical diskette(s) only (and not the software) and is otherwise expressly and specifically disclaimed.

This limited warranty gives you specific legal rights; you may have others which may vary from state to state. Some states do not allow the exclusion of incidental or consequential damages, or the limitation on how long an implied warranty lasts, so some of the above may not apply to you.